ILLINOIS CENTRAL COLLEGE
PN4061.05
STACKS
Rhetoric, roman...   technology;

A129   282723

W9-BYW-847

WITHDRAWN

PN
4061   ONG   45399
.0 5   Rhetoric, romance, and
          technology

EGLI

Illinois Central College
Learning Resources Center

RHETORIC, ROMANCE, AND TECHNOLOGY

Studies in the Interaction
of Expression and Culture

Also by Walter J. Ong, S.J.

Frontiers in American Catholicism (1957)
Ramus, Method, and the Decay of Dialogue (1958)
Ramus and Talon Inventory (1958)
American Catholic Crossroads (1959)
Darwin's Vision and Christian Perspectives (1960)
  (Editor and Contributor)
The Barbarian Within (1962)
In the Human Grain (1967)
The Presence of the Word (1967)
Knowledge and the Future of Man (1968)
  (Editor and Contributor)
Petrus Ramus and Audomarus Talaeus, *Collectaneae praefationes, epistolae, orationes* (1969)
  (Editor)
Petrus Ramus, *Scholae in liberales artes* (1970)
  (Editor)

Rhetoric, Romance, and Technology

STUDIES IN THE INTERACTION
OF EXPRESSION AND CULTURE

by Walter J. Ong, S.J.

Cornell University Press
Ithaca and London

ILLINOIS CENTRAL COLLEGE
LEARNING RESOURCES CENTER

45399

PN
4061
.O.5

Copyright © 1971 by Cornell University

All rights reserved. Except for brief quotations in a review, this book, or parts thereof, must not be reproduced in any form without permission in writing from the publisher. For information address Cornell University Press, 124 Roberts Place, Ithaca, New York 14850.

First published 1971 by Cornell University Press.
Published in the United Kingdom by Cornell University Press Ltd.,
2-4 Brook Street, London W1Y 1AA.

International Standard Book Number 0-8014-0645-5
Library of Congress Catalog Card Number 74-153722

Printed in the United States of America
by Vail-Ballou Press, Inc.

Clarentio Harvey Miller
atque
Ricardo Standish Sylvester
filiis Universitatis Sancti Ludovici
fide sapientiaque Moreana iamdudum imbutis
ac studiorum rhetoricorum peritissimis
necnon uxoribus
Ioannae Zimmer Miller
eiusdem Universitatis filiae
et
Laetitiae Gurian Sylvester
auctor hoc libelli qualecumque
ex animo
dat donat dedicat

Preface

Cicero used to make the point that the orator needed to know everything that could be known. Hence rhetoric, the art of oratory or public speaking, ultimately took all knowledge as its province. Cicero was not voicing merely a private hope or theory. For most of classical antiquity rhetoric was the focus of learning and intelligence, the foundation and culmination of the humanities and of a liberal education.

Today rhetoric has become again a focus of learning in another way. As we have moved away from the rhetorical world of classical antiquity, a world which remained very much alive until the age of romanticism, we have found rhetoric a more and more fascinating subject of study—not so much as a skill to be used by ourselves (Cicero's way of studying it) but rather as a historical phenomenon. Seen through history in its full sociological and noetic context, rhetoric throws a great deal of light on much in the past which is otherwise obscure and mystifying.

The mid-twentieth century first turned to the history of rhetoric because the rhetorical tradition enfolded a life style so different from the one in which modern technological man operates, a curious and intriguing life style, rich and strange.

The United States has moved further from the old rhetorical tradition than any other country in the West—though all have moved away from it to a degree—and American scholars have been particularly active in bringing to light the rhetorical core of ancient culture and interpreting its meaning. This rhetorical core contrasts markedly with the technological culture so active in the United States.

As we have come to understand it more, the rhetorical tradition has thrown light not merely on the past but almost equally on the present, for it has enabled us better to see the significance of the shifts in media and in modes of knowledge storage and retrieval which have both resulted from and produced our present technological age. The ancient rhetoricians were the first media buffs. Study of the rhetorical tradition enables us to interpret the past on its own terms and thus to discover many of the real roots out of which the present grows. It is thus one way to keep history from degenerating into antiquarianism.

In its wide ambience and manifold significance, rhetoric as it operated at the center of Western life is now known to connect with a great variety of noetic and cultural phenomena —including phenomena at first blush so unlike it as the romantic movement and technology itself. For this reason rhetoric is a useful pivot for the studies brought together in this book. These studies relate to the rhetorical tradition in varying degrees of explicitness and are here arranged chronologically by subject. They focus individually on periods or persons from the Renaissance through the present age, but they also reach far back beyond the Renaissance into antiquity and even into prehistory, where rhetoric had its beginnings in the primal oral, preliterate culture of mankind. I hope that this volume will be of help to the growing number of scholars across the world who are directing their attention and research efforts to the rhetorical tradition as one key to understanding much that went on in the past and much that is going on in our own times, as well as much that may come about in the future.

Of the studies in the present volume the following were first printed in the publications here indicated, the editors or publishers of which have kindly granted permission for reprinting: "Oral Residue in Tudor Prose Style," *PMLA*, LXXX (1965), 145–154; "Tudor Writings on Rhetoric, Poetic, and Literary Theory" (exclusive of Section IV, "Criticism and Literary Theory," here printed for the first time), *Studies in the Renaissance*, XV (1968), 36–69; "Memory as Art," *Renaissance Quarterly*, XX (1967), 253–260; "Latin Language Study as a Renaissance Puberty Rite," *Studies in Philology*, LVI (1959), 103–124; "Ramist Classroom Procedure and the Nature of Reality," *Studies in English Literature*, I (1961), 31–47; "Ramist Method and the Commercial Mind," *Studies in the Renaissance*, VIII (1961), 155–172; "Swift on the Mind: Satire in a Closed Field," *Modern Language Quarterly*, XV (1954), 208–221; "Psyche and the Geometers: Associationist Critical Theory," *Modern Philology*, XLIX (1951, University of Chicago Press), 16–27; "J. S. Mill's Pariah Poet," *Philological Quarterly*, XXIX (1950), 333–344; "Crisis and Understanding in the Humanities," *Dædalus: Proceedings of the American Academy of Arts and Sciences*, XCVIII (1969), 617–640. Some of these have been revised for the present volume, but in general I have not made reference to scholarly works which appeared after the original printings of these studies except where such works have significantly modified my own original statements or positions.

Chapter 2 grew in part from work made possible by a travel grant from the Penrose Fund of the American Phil-

osophical Society, for which I am grateful. Part of this
chapter was presented as a paper at the 1963 Annual Meeting
of the Modern Language Association of America. Chapter 6
was presented as a paper, in French, at the International Conference on Renaissance Educators and Jurists held at the
Centre d'Etudes Supérieures de la Renaissance at Tours, France,
July 4 to 23, 1960. A French version of Chapter 7 was presented at the same conference. Chapter 11, here published for
the first time, was presented as a paper at the English Institute,
Columbia University, in September 1970.

W. J. O.

Saint Louis University

Contents

RHETORIC, ROMANCE, AND TECHNOLOGY

Studies in the Interaction
of Expression and Culture

I Rhetoric and the Origins of Consciousness

I

Until the modern technological age, which effectively began with the industrial revolution and romanticism, Western culture in its intellectual and academic manifestations can be meaningfully described as rhetorical culture. Any number of scholars have borne witness to the pervasiveness of rhetoric in the West, such as Ernst Robert Curtius, Leo Spitzer, Pedro Laín Entralgo, and the late C. S. Lewis. Near the beginning of his *English Literature in the Sixteenth Century excluding Drama*, Lewis states that "in rhetoric, more than in anything else, the continuity of the old European tradition was embodied," adding that rhetoric was "older than the Church, older than Roman Law, older than all Latin literature" and that it "penetrates far into the eighteenth century." [1] He also notes that rhetoric "is the greatest barrier between us and our ancestors" and thereupon surprisingly drops the subject forever, despite his avowal that the entire literary

---

[1] C. S. Lewis, *English Literature in the Sixteenth Century excluding Drama*, Vol. III of *The Oxford History of English Literature*, ed. by F. P. Wilson and Bonamy Dobrée (Oxford: Clarendon Press, 1954), p. 61.

history he is writing is "certainly . . . vitiated by our lack of sympathy on this point" and that "probably all our literary histories" are similarly vitiated. (Significantly, the rhetorical age did not engage in writing literary histories.)

There has been a good deal of water under the bridge since this avowal, which remains the more impressive because it is so reluctant and so stubborn. "Rhetoric" is the anglicized Greek word for public speaking, and thus refers primarily to oral verbalization, not to writing. It comes from the Greek term *rhēma*, a word or saying, which in turn derives from the Proto-Indo-European *wer*, the source of the Latin *verbum* and of our "word." All human culture was of course initially rhetorical in the sense that before the introduction of writing all culture was oral. This means not merely that all verbal communication—there are obviously other kinds of communication—was oral, effectively limited to sound, but also that the economy of thought was oral. For human thought structures are tied in with verbalization and must fit available media of communication: there is no way for persons with no experience of writing to put their minds through the continuous linear sequence of thought such as goes, for example, into an encyclopedia article. Lengthy verbal performances in oral cultures are never analytic but formulaic. Until writing, most of the kinds of thoughts we are used to thinking today simply could not be thought. Orality is a pervasive affair.

But when we say that Western culture until recently was rhetorical, we are saying something more specific than that it was oral. We mean also that Western culture, after the invention of writing and before the industrial revolution, made a science or "art" of its orality.

After the invention of script (around 3500 B.C.) the central verbal activity to which systematic attention was at first given was the art of public speaking, not the art of written composition. Scribes learned how to commit discourse to writing, but basically composition as such remained an oral matter. Early written prose is more or less like a transcribed oration, and early poetry is even more oral in its economy. This fixation on the oral diminished only slowly. From antiquity through the Renaissance and to the beginnings of romanticism, under all teaching about the art of verbal expression there lies the more or less dominant supposition that the paradigm of all expression is the oration. With the exception of the letter-writing taught in the medieval *ars dictaminis*, virtually the only genre of expression formally taught in schools was the oration, with its various parts, numbering from a minimal two to four or even seven or more. Even the art of letter-writing, maximized in the highly literate culture of the Middle Ages, was conceived of by analogy with an oration: as will be detailed in Chapter 3, the letter commonly began with the equivalent of the oration's exordium, next set down the *petitio* or statement of what was to be asked for (corresponding to the oration's *narratio*, or statement of what was to be proved), the reasons or "proofs" bearing on the *petitio*, the refutation of counter-reasons (in the oration, refutation of adversaries), and the conclusion.[2]

This focusing of attention on speech rather than writing we can now understand. In preliterate ancient Greek culture, as in probably every early culture, oral performance had been held in high esteem and cultivated with great skill. But it had not been possible to codify its procedures systematically, to produce a science or art of oratory. An oral culture can produce—that is, perform—lengthy oral epics, for these are made up of memorable thematic and formulary elements, but it has

[2] See Chapter 3, pp. 54, 73–74, below.

no way of putting together a linear analysis such as Aristotle's *Art of Rhetoric*. For an oral culture can produce lengthy verbalization only orally, and there is no way to compose an *Art of Rhetoric* on one's feet. No one could remember it even if it could be so composed. Even today, when oral performance has the advantage of being able to echo the analytic writing styles and accompanying thought patterns which saturate us, it is still impossible to extemporize a lengthy scientific treatise orally with no reliance at all on writing. At best an oral culture can produce a set of memorable sayings or aphorisms pertaining to a subject. But a collection of sayings such as "Feed a cold and starve a fever" is hardly a treatise on medicine. With writing, attempts to organize such sayings in meaningful sequence, such as the Hippocratic *Aphorisms*, can begin to move toward the treatise form. Once writing made feasible the codification of knowledge and of skills, oral performance, enjoying the high prestige that it did, was one of the first things scientifically codified. To use an expression still current in the sixteenth century, oral performance was "technologized" (made into a *technē* or art), earlier by the Sophists, later by Aristotle and others.

Others cultures, too, once they had writing, at least in many cases gave systematic attention to oral performance. But there was a difference. What was distinctive about the organization of rhetoric among the Greeks was the close alliance of the art with another subject which, if we except a considerably later and much less developed discipline in India,[3] was an exclusively Western invention: formal logic. Logic and rhetoric have always been uneasy bedfellows, but in the West they

[3] See I. M. Bochenski, *A History of Formal Logic*, tr. from the German by Ivo Thomas (Notre Dame, Ind.: University of Notre Dame Press, 1961), pp. 415–447.

have been bedfellows nevertheless pretty well from the beginning. Coming out of an oral background, logic was regarded from antiquity on through the Middle Ages not as concerned simply with private thinking but rather as allied to dialectic. It was typically defined the way Cicero defines it, as *ars disserendi* or the "art of discourse," an art of communication, not of solipsistic (and by implication wordless) thought such as is implied by "the art of thinking"—a favorite definition after the invention of print. There was a technical distinction between formal logic as "necessary" or scientific logic and dialectic as the logic of the more probable, governing debate or discussion, but this distinction was most often of little operative value.[4] In effect, "dialectic" and "logic" came to the same thing, and dialectic or logic were commonly grouped together as the *artes sermocinales*, the speech arts.

Aristotle associates rhetoric closely with dialectic (he never uses as a noun the term "logic"—*logike*, in Greek—nor even the term *logike techne*, "logical art").[5] "Rhetoric is the counterpart of dialectic," for both have to do with "things

[4] See Walter J. Ong, *Ramus, Method, and the Decay of Dialogue* (Cambridge, Mass.: Harvard University Press, 1958), pp. 59–63, 101–103, 175–176, 217–218. For a fresh and thorough treatment of the pervasiveness of rhetoric in all thought and an exposure of the inadequacy of the view, dominant since Descartes, that human reason is truly at home only in the domain of the self-evident, see Ch. Perelman and L. Olbrechts-Tyteca, *The New Rhetoric: A Treatise on Argumentation*, tr. from the French by John Wilkinson and Purcell Weaver (Notre Dame, Ind.: University of Notre Dame Press, 1969); see also the discussion of this and related works in Vasile Florescu, "Rhetoric and Its Rehabilitation in Contemporary Philosophy," tr. from the French by Barbara Johnstone, *Philosophy and Rhetoric*, III (1970), 193–224.

[5] "Logic" comes into English via Cicero's *logica* (*ars*), which follows late Greek usage. For what is formally logical in the strict sense, Aristotle uses the term "analytic" (*analytikos*).

that do not belong to any one science."[6] Rhetoric covers any subject matter, for it is the faculty of "discovering in the particular case what are the available means of persuasion."[7] The two means of proof in rhetoric, enthymeme and example, Aristotle defines by analogy with the two means of proof in formal logic, syllogism and induction. Topics or commonplaces he finds both in rhetoric and in dialectic.[8]

Rhetoric, moreover, for all practical purposes, from antiquity until the eighteenth century, included poetic. This it did often, though not always, in theory,[9] and even more often in practice, for poetry enjoyed no particular status as an independent academic discipline whereas rhetoric enjoyed enormous academic prestige, so that any use of words for effects other than the strictly logical was thought of as formally governed by rhetoric.[10] Because, as we have seen, rhetoric was closely allied with logic, the association of poetry with rhetoric tended to assimilate poetry, too, to logic all through the Western tradition until romanticism,[11] when this association was rather abruptly discredited.

Rhetoric and formal logic could not, of course, simply merge. As the art of persuasion, moving men to action, rhetoric is ordered to decision making. And often decisions must be made when the grounds for decision are not under full logical control. Rhetoric has to deal often with probabilities. Still, it

---

[6] *Art of Rhetoric* i. 1. 1354a.   [7] *Ibid.* i. 2. 1355b.

[8] *Ibid.* i. 2. 1356b, 1358a, ii. 22. 1395b–1397a.

[9] See Chapter 3 below.

[10] See the still highly informative monograph by Edmond Faral, *Les Arts poétiques du XIIᵉ et du XIIIᵉ siècle* (Paris: E. Champion, 1924).

[11] See Walter J. Ong, "The Province of Rhetoric and Poetic," in *The Province of Rhetoric*, ed. by Joseph Schwartz and John A. Rycenga (New York: Ronald Press, 1965), pp. 48–56.

drives toward the certainty of formal logic as far as possible. The computor cannot tell us everything we need to know to make a decision, but it is good to have it tell us all that it feasibly can. Given the existence of formal logic, rhetoric in fact availed itself of everything logic had to offer, as the history of both disciplines in the West abundantly shows. In cultures which did not have formal logic, rhetoric or its equivalent had to take other turns.

The relationship between rhetoric and logic over the ages has been partly reinforcing and partly competitive. Rhetoric overshadowed logic in the patristic age, yielded to it more or less in the Middle Ages (though rather less than even scholarly mythology today commonly assumes), and overshadowed it again in a different way in the Renaissance. The interaction of these two of the *artes sermocinales* is complex, sensitive to a great many forces in the culture—intellectual, pedagogical, social, religious, political, economic. Certain features of this interaction are the core subject of the studies here.

But there is far more to say than this book can attend to. Indeed the interaction of logic and rhetoric is far too large a subject ever to be exhausted. Since logic and rhetoric correspond to the basic polarity in life represented in other ways by contemplation (theory) and action, or intellect and will, and since logic and rhetoric have come into being not in the hollows of men's minds but in the density of history, it is quite possible to analyze almost anything in Western culture (and perhaps in all cultures) in terms of its relationship to the logical and rhetorical poles. Needless to say, there is no total theoretical statement of the nature of either rhetoric or logic, much less of their interrelation. Conceivably such a statement might finally be achieved at the end of history, when rhetoric and logic would be outmoded.

With the advent of the age which from one point of view we call the technological age and from the other point of view the romantic age, rhetoric was not wiped out or supplanted, but rather disrupted, displaced, and rearranged. It became a bad word—as did many of the formerly good words associated with it, such as art, artificial, commonplace, and so on. Rhetoric was a bad word for those given to technology because it represented "soft" thinking, thinking attuned to unpredictable human actuality and decisions, whereas technology, based on science, was devoted to "hard" thinking, that is, formally logical thinking, attunable to unvarying physical laws (which, however, are no more real than variable human free acts). Rhetoric was a bad word also for those given to romanticism because it seemed to hint that the controlling element in life was a contrivance rather than freedom in the sense of purely "spontaneous" or unmotivated action, sprung up unsolicited from the interior wells of being. (In support of rhetoric, it might be noted that no such choice is possible: psychology can identify real motives always underlying the seemingly random or whimsical choice—which is precisely unfree because its motives are not under conscious control. Free action is not unmotivated action but action from motives consciously known.)

The displacement and rearrangement of rhetoric is, from one point of view, the story of the modern world. By now this great depository of culture has exfoliated into a variety of seemingly disparate things. In the academic world, for example, in one or another guise or avatar rhetoric is now taught in elementary and high school English courses, in freshman college English courses, in courses in marketing and advertising and creative writing (which is never entirely creative, but always to some degree persuasive, as Wayne Booth has so well shown ¹²). As persuasion, the operation of rhetoric has become in some ways more indirect (in marketing and advertising— though we must never underestimate the indirect methods of classical orators such as Cicero). Modern rhetoric has become more visualist than the older verbal rhetoric, not merely through the use of pictures for persuasion but also through the presentation of words as objects, with "display type" in "display advertising." Until the early eighteenth century, it was at best extremely uncommon to find any sign display of lettered words as such: a tradesman's name or business was not advertised in words on the outside of his shop; instead, the old iconography of the pretypographic world was used, an ivy bush for a tavern, a barber's pole, a pawnbroker's three balls, and to distinguish individual shops, easily represented and easily remembered designs such as a Cheshire cheese, a Turk's head, three casks (the Triple Tun of Ben Jonson's festive gatherings). In the new rhetoric words themselves are treated as designs and even as physical objects. Movies and television even set words in motion across a visual field, commanding their alphabetic components, like Ariel, "to fly, / To swim, to dive into the fire, to ride / On the curl'd clouds." But at the same time with the help of the electronic media words are also made more active aurally as a means of rhetorical persuasion in advertising jingles, mottoes, and slogans. In the present world, the relationship of persuasion to the totality of human existence thus differs radically from what it was in the past. But only those who have no knowledge of the changes in culture itself will think that the changes in the role of rhetoric have been chaotic. The history of rhetoric simply mirrors the evolution of society.

¹² Wayne C. Booth, *The Rhetoric of Fiction* (Chicago and London: University of Chicago Press, 1961).

As might be expected if rhetoric is as central to the human lifeworld as it is made out to be, the historical development of rhetoric ties in with the stages through which the human consciousness evolves out of the unconscious as these stages are revealed in the myths successively asserting themselves through the history of art and literature and ritual and cultural institutions. These stages, which run both phylogenetically through human history and ontogenetically through the history of the individual, are described by various cultural historians in similar, if competing, fashions. Here I can do no more than indicate *grosso modo* some of the relationships of rhetoric to this psychic evolution, particularly in its phylogenetic or social phase, without attempting to arbitrate differences between the various psychologists and psychiatrists and cultural historians. I shall take as a point of reference the psychic states as described in Erich Neumann's rich and insightful work, *The Origins and History of Consciousness*.[13] Neumann treats both phylogenetic and ontogenetic psychic evolution.

The stages of psychic development as treated by Neumann are successively (1) the infantile undifferentiated self-contained whole symbolized by the uroboros (tail-eater), the serpent with its tail in its mouth, as well as by other circular or global mythological figures, (2) the Great Mother (the impersonal womb from which each human infant, male or female, comes, the impersonal femininity which may swallow him up again), (3) the separation of the world parents (the principle of opposites, differentiation, possibility of change), (4) the birth of the hero (rise of masculinity and of the per-

[13] With a Foreword by C. G. Jung, tr. from the German by R. F. C. Hull, Bollingen Series, XLII (New York: Pantheon Books, 1954).

sonalized ego) with its sequels in (5) the slaying of the mother (fight with the dragon: victory over primal creative but consuming femininity, chthonic forces), and (6) the slaying of the father (symbol of thwarting obstruction of individual achievement, to what is new), (7) the freeing of the captive (liberation of the ego from endogamous kinship libido and emergence of the higher femininity, with woman now as person, anima-sister, related positively to ego consciousness), and finally (8) the transformation (new unity in self-conscious individualization, higher masculinity, expressed primordially in the Osiris myth but today entering into new phases with the heightened individualism—or, more properly, personalism—of modern man).

These stages can be discerned in phylogenetic psychic evolution. Earlier culture finds itself bound to the earlier stages. Thus, the myths of earlier cultures are concerned more with the uroboros or its equivalent, later ones successively more with the Great Mother and on to the hero and finally the Osiris figure. Corresponding stages are to be found in ontogenetic or individual psychic history: the uroboros is the world of the infant, the birth of the hero that of adolescence, the transformation stage that of the adult (and correspondingly for the intervening stages). The stages are traversed by both males and females, but in different fashions. Neumann's structures constitute a revision of earlier suggestions of Freud and Jung (Jung wrote the Foreword to Neumann's book) and of others. They are drastically oversimplified here. But they are representative of a vast amount of psychoanalytic thought and will serve provisionally here to indicate briefly some relationships of the history of rhetoric to the phylogenetic and ontogenetic history of consciousness.

Rhetoric clearly occupies an intermediary stage between

the unconscious and the conscious. This is more than suggested in Aristotle's statement that in rhetoric the equivalent of the formal syllogism of strictly scientific logical reasoning is the enthymeme.[14] Since Boethius (c. 470–525), an enthymeme has commonly meant a scientific syllogism with one of the premises unexpressed, but for Aristotle it always means rather a subscientific syllogistic or parasyllogistic argument from probable premises to a probable conclusion—the kind of argument which ordinarily governs decisions regarding human actions. It is thought of as concluding because of something unexpressed, unarticulated: *enthymēma* primarily signifies something within one's soul, mind, heart, feelings, hence something not uttered or "outered" and to this extent not a fully conscious argument, legitimate though it may be. Aristotle's term here thus clearly acknowledges the operation of something at least very like what we today would call a subconscious element.

From within our present intellectual milieu, Gilbert Durand, in his brilliant psychocultural study, *Les Structures anthropologiques de l'imaginaire*, points out that it is rhetoric which effects the transit from the polysemous sign language of symbols to the formalism of logic, where signs have a "proper meaning." With rhetoric, man sets himself against fate: rhetoric is a movement of hope; an upward movement—a "euphemism," Durand calls it, "which will color in the large all the activity of formalization of thought."[15] Rhetoric does not grow directly out of the preconceptual but is antithetic to it; it is not so fragmenting, however, as the logic toward which

[14] *Art of Rhetoric* i. 1 1355a, i. 2 1356b.
[15] Gilbert Durand, *Les Structures anthropologiques de l'imaginaire* (Paris: Presses Universitaires de France, 1960), p. 451; cf. pp. 453–459.

it moves will be. Rhetoric is less preoccupied with distinctions, rather more unifying. It works through the imagination, which euphemizes actuality through hyperbole and antithesis.[16] Rhetoric also schematizes what would otherwise be too fantastic into identifiable figures of style that can be made out to be simple embellishments on formal signification [17] (this of course is never so formal as extreme analytic philosophers would make it out to be). Being thus intermediary between stages of the noetic world, rhetoric is full of ambiguities and thus difficult to study in depth, which is why, Durand regretfully suggests, it is given so little philosophical and anthropological attention.[18]

In various ways which can only be touched on here, rhetoric as described by Durand and as seen historically through the studies in this book is intermediate in the psychic evolutionary processes discussed by Neumann. Rhetoric as a formal discipline arises, we have seen, out of the primary oral world, fixing attention initially on oral performance rather than writing even though it comes into being as a formal discipline only through the use of writing. To this extent, in its oral grounding, rhetoric stands on a psychologically primitive base. The three stages in the development of verbal communications media (oral, chirographic-typographic, and electronic) bear uneven resemblance to the Freudian psychosexual stages (oral, anal, and genital), as I have attempted to show in *The Presence of the Word.*[19] Correspondence is least between the last stages, electronic and genital; in the first or oral stages, however, the resemblance is striking, though not so overwhelming as in the second (chirographic-typographic and

---

[16] *Ibid.*, pp. 453, 455.   [17] *Ibid.*, p. 457.   [18] *Ibid.*, p. 459.
[19] New Haven and London: Yale University Press, 1967, pp. 92–110.

anal). The world of primary oral (preliterate) verbalization in which rhetoric is rooted thus has some affinity with infantile and juvenile orality, with the world of the self-contained uroboros and of the overwhelming Great Mother. Rhetoric is typically an overwhelming phenomenon, implemented by what the classical world and the Renaissance called *copia*, abundance, plenty, unstinted flow. Our fears of it resemble our fears of the Great Mother, who swallows her children. One can drown in rhetoric. Its world is commonly and aptly described in the water symbolism associated with primitive impersonalist femininity.

But rhetoric has affinity also with the succeeding stages. It introduces the principle of opposites, differentiation, as had the myth of the separation of the world parents. It is part and parcel of the heroic age and its myths. The hero disappears as rhetoric falls somewhat into discredit. Romanticism, as we have suggested, marks the end of rhetorical culture. There are romantic heroes, but the romantic hero is always slightly preposterous. His rhetoric lacks the credibilty of that of Achilles or Ulysses or Aeneas or Roland or Milton's Adam. Byron had it right: tongue-in-cheek is the best approach to both his romantic heroes, Don Juan and Childe Harold. Today the hero lives in his antitheses, typified in Samuel Beckett's characters, whose verbalization is ostentatiously logical and whose rhetoric is secret.

Rhetoric at its most impressive peak was heroic and masculinizing through its association with puberty rites. In the West, as several of the studies in this book will detail,[20] the study of Latin had the characteristics of a male puberty ritual. Until romanticism matured, rhetoric as a formal discipline was

[20] See Chapters 2, 3, and especially 5.

studied as part of the study of Latin, for English as such was not a curriculum subject. When Latin gradually disappeared and concomitantly schools began to admit girls, formal rhetoric also disappeared. Today's "freshman rhetoric," even when it is called this, ordinarily bears only weak resemblance to the severe discipline of the past, which had been geared to ceremonial and taxing male polemic and the formation of diplomatic jousters. Renaissance educators, such as Sir Thomas Elyot, were particularly explicit in connecting the study of rhetoric to the development of masculine courage.

The ego dominance fostered by rhetoric is evident particularly in the Renaissance, which in many ways represents the phylogenetic high point of rhetoric development. The extreme egoism and egotism of the highly proficient, often baroque Renaissance rhetorician is typified in personages such as the Scaligers, whose genealogy itself, very likely, was the product of rhetoric rather than of ordinary procreation. Other comparable characters abound at this time, for example Justus Lipsius (first a Catholic, then a Protestant, then a Catholic again, and yet author of a famous work, *De Constantia*), Paracelsus (Theophrastus Bombastus von Hohenheim, 1493–1541), even to a degree Erasmus himself, who contrasts on this point quite markedly with his friend St. Thomas More: More had far greater confidence in his own masculinity, though, with Erasmus, he also scorned formal logic with a passion and to this extent remained in the rhetorician's camp. To these and a host of other real characters could be added Gargantua and Pantagruel, Don Quixote, Falstaff and Polonius and their congeners in Shakespeare, and countless others in the literature that represents and often spoofs this characteristic personality type of the age.

II

The studies in this collection, some first published here, others produced earlier, are not restricted to rhetoric in any narrow sense but concern its general ambience as well as some of its quite specific manifestations. For reasons indicated above, they include some reference to logic or dialectic and to poetry, the concomitants of rhetoric through the ages. Grammar, the third of the *artes sermocinales*, is treated only in passing, partly because I have covered some relevant aspects of grammar elsewhere,[21] but chiefly to keep discussion within feasible limits. Grammar is very closely connected with writing (the term comes from the Greek *technē grammatikē*, "the art of letters of the alphabet") and its relationship to culture is a vast story all by itself.

The studies here move chronologically from the Renaissance, which fossilized much of the ancient oral past in certain stylistic phenomena, up to the present time. As the studies progress, they become less restrictively focused on rhetoric, in keeping with the suggestions above that rhetoric today has diffused itself in many forms through our culture and no longer has the neater contours of earlier, less exfoliated civilizations.

Chapter 2 shows the continuity of verbal expression during the English Renaissance with earlier speech and thought patterns before the invention of writing (and by implication similar connections in Continental European milieus). The third chapter gives a detailed report on the entire production

---

[21] See Walter J. Ong, *The Barbarian Within* (New York: Macmillan, 1962), chaps. ix, xii, and *The Presence of the Word*, pp. 78–79, 209–210, 293; cf. Walter J. Ong, *In the Human Grain* (New York: Macmillan, 1967), pp. 52–59.

of English-language books on rhetoric and poetic and literary criticism or theory during the Tudor age, from the late fifteenth through the beginning of the seventeenth century. The brief fourth chapter indicates the curiously central significance of the art of memory, which, with the help of writing and later of print, had been elaborated to provide special devices for the knowledge storage and retrieval demanded by oral cultures although these very devices had in fact been rendered obsolescent by the simple existence of writing and print.

The next chapters treat more directly the interrelationships between social institutions and modes of thought and expression. An institution as widespread as Learned Latin was more than merely a linguistic phenomenon. Learned Latin, the old classical Latin which remained in the schools after ancient Latin had fragmented in the home and nonacademic world into hundreds of vernaculars, was soon structured into a whole series of social institutions, some treated in Chapter 5 and others suggested in later chapters as occasion offers. Learned Latin was permanently aligned with the primary oral or rhetorical tradition and helped keep this tradition alive long past the development not only of writing but even of print. A language spoken by millions but only by those who could write it, Learned Latin paradoxically also built up an extreme deference for the written word which verged on superstition and was to affect the aims of lexicography down to our present day. Used only by males and under the sway of the old oral dialectical-rhetorical tradition, Learned Latin was a ceremonial polemic instrument which from classical antiquity until the beginnings of romanticism helped keep the entire academic curriculum programmed as a form of ritual male combat centered on disputation.

These effects of Latin suggest still further connections with cultural institutions. The use of Learned Latin and the self-image and style of life it automatically fostered tended to strengthen the wide-spread opinion that war was not only inevitable in human society but in many ways was even good. (It kept society from softness and effeteness—vices which can lead only to war!) If boys went to school to war ceremonially with each other (and with the teacher), combat was a necessary and admirable condition of existence. It is evident that this view of life helped keep the ideal of the hero, especially the martial hero, alive in men's impossible dreams long after the social conditions of oral and residually oral culture which originally generated the fictional hero had disappeared.

In the intimate connection it sustains with the heroic age, Learned Latin—and the cult of dialectic and rhetoric which for historical reasons the use of Latin supported—is built into the social and psychological structures earlier mentioned here as studied by Neumann, Carl Jung, and others. That is to say, the use of Learned Latin for scientific and scholarly thinking over nearly a millennium and a half had a great deal to do with the development of the collective and individual psyche in the Western European world. Indeed, since there are more or less contemporaneous parallels in other cultures which have used learned languages discrete from the vernacular, perhaps such languages belong to a certain stage in cultural and psychological development.

The sixth chapter undertakes to show one of the ways in which the academic tradition, dominated by the dialectic or logic which was a medieval legacy, projected itself into views of the cosmos in the sixteenth century: the actual tended to be defined as the logical and the readily teachable. In the Ramist movement, which came into being as an attractive simplification of the scholastic philosophy grown up with the universities, academic interpretation of actuality took on a form particularly appealing to the commercial mind, the following chapter suggests. By reducing metaphysics to something very like bookkeeping, Ramism in effect put all of actuality into ledger books and thereby developed its distinctive appeal for the upward-mobile commercial classes who were its chief proponents.

Moving into the eighteenth century, the chapter on Swift and the one on the critical theory of the associationist philosophers advance evidence of the strong propensity of this age to interpret the mind and the world itself in quantitative terms. This propensity connects with the forces which had been at work in producing the arts of memory and Ramist philosophy, and fosters a further move away from the orality of rhetorical culture to visualist scientism. The next chapter examines a significant turn in John Stuart Mill's thought which connects with the same line of development. Poetry, which in an oral culture had been identified with actual oral performance, typically before a sizable and highly responsive audience, for Mill has become soliloquy. "Eloquence is *heard;* poetry is overheard." The psyche has withdrawn to this extent for the old social, exteriorized oral world into its own isolated spaces.

The isolated individualism of Mill's poet suggests the romantic movement, which has persisted through his day and into ours. Romanticism is not ordinarily examined in its relationship to the development of communications media, as it is in Chapter 11 here. Seen in terms of the build-up of knowledge which the media of writing and print had effected by the time of the Enlightenment, certain features of romanticism become more understandable. In the earlier oral or residually oral ages of mankind, when knowledge was in short supply and in constant danger of being lost, the romantic celebration of the mysterious and the unknown was at best a luxury and could prove a serious psychological threat. Hence earlier ages show only limited traces of romanticism, which can break out as a large movement only when vast supplies of knowledge have been stored in the readily retrievable form made possible by print and typified in the polymath productions, such as Diderot's *Encyclopédie*, which appeared in quantity in the century and a half immediately preceding the romantic movement. Examining romanticism in terms of knowledge storage and retrieval problems, one can suggest some new and plausible reasons as to why the romantic movement arose when it did and not earlier or later. Romanticism appears as a result of man's noetic control over nature, a counterpart of technology, which matures in the same regions in the West as romanticism and at about the same time, and which likewise derives from control over nature made possible by writing and even more by print as means of knowledge storage and retrieval.

We live still in the romantic era, which probably will persist through the entire foreseeable future. We live in an age which also in its forms of expression is evidently more oral than the age of print which immediately preceded it. Astonishing, and to some persons alarming, similarities have been pointed out between our secondarily oral culture and primary (preliterate) oral cultures: our sense of togetherness, for example, matches in surprising detail that of early man before the development of individualism fostered by writing and print. Without attempting to canvass the whole assemblage of likenesses between primary and secondary orality—to use the terms here suggested—the next to last chapter compares the use of formulas in works of popular art with the use of formulas in the oral culture which, as several earlier chapters of this book explain, had persisted as part of the old oral heritage not only through the sixteenth but also into the nineteenth century. Our present popular literature is highly formulary and thus is notably assimilable to the productions of primary oral culture. But its attitude toward formulas betrays its permanently literary state of mind: when popular literature today consciously adverts to the formulas it relies on, it tends to take the formulas ironically. The depreciation of the cliché which marked the once New Criticism thus appears to have a much wider base than even this widespread critical movement: strong disapproval for the cliché is a regular concomitant of the romantic state of mind, which is a literate state of mind, subconsciously convinced that what is already known does not require repetition because what is known is stored in books, whereas art is necessarily a venture into the unknown. The present age is thus in a dilemma, on the one hand driven to the use of clichés by its new orality and on the other driven to mock them because of its relentless literacy.

This paradoxical state of affairs suggests that the humanities themselves, which are centered upon the arts of expression, verbal and other, are undergoing drastic reorganization in our day. The final chapter here looks back over the history of the humanities to point out that they have not been the same in all ages. The humanities in modern times, as Ernst Robert Curtius suggests, have developed a close alliance with writing and print. Writing and print are here to stay and will certainly amount to even more in the future than in the past, but since they are now interwoven with newer media—radio, television, the cinema, and other devices—with which they interact and which transform them, both in content and form the humanities grow and reshape themselves with the evolution of culture. The humanities are certainly in an unsatisfactory state, but are also probably better off than at any time before in history. For, if they are alive, the humanities are always in a state of crisis. Man's life is one of crises, and the humanities do not stand outside man's life but live within it. The unchallenged life is not worth living.

2 Oral Residue in Tudor Prose Style

I

We have recently been growing more aware of the differences between oral cultures and literate cultures.[1] The effects on modes of thought inherent in the successive media of expression—oral speech, analphabetic writing, alphabetic writing, letterpress printing, the electronic media, wired and wireless—have been studied in some detail, and we know something of the effect of (alphabetic) writing on the ability

[1] For the emergence of literature out of preliterate oral performance, see the massive work of H. Munro Chadwick and Nora Kershaw Chadwick, *The Growth of Literature*, 3 vols. (Cambridge: The University Press, 1932-40). In *The Singer of Tales*, Harvard Studies in Comparative Literature, Vol. 24 (Cambridge, Mass.: Harvard University Press, 1960), Albert B. Lord follows up the work of Milman Parry on Yugoslavian singers of epic tales and brings the results to bear on Homer, *Beowulf*, and other epics, showing conclusively that not only is the oral epic independent of literacy but that it is also dependent on illiteracy—if an individual knows how to read, at least in twentieth-century fashion, he cannot render the truly oral form. In his brilliant *Preface to Plato* (Cambridge, Mass.: Belknap Press of Harvard University Press, 1963), Eric A. Havelock traces the emergence of the Greek philosophical mind to the shift from an oral (Homeric) culture to alphabetization.

to perform abstract analysis and to exercise individually controlled thinking as against communally controlled thinking.[2]

The full effects of our new sensitivity to the shift in media, however, have hardly begun to be felt in literary history and criticism. For example, in his magnificent and germinal work *European Literature and the Latin Middle Ages*, Ernst Robert Curtius pays almost no attention to the media as roots of culture and/or style. Yet, with out being aware of it, Curtius is deeply involved in shifts in the media. If by nothing else, this is shown by his casual statement that "every true humanism delights simultaneously in the world and in the book." [3] Such a statement links the very existence of humanism to the medium of writing and simultaneously disallows any claim which most of mankind in the past might have to be humanistic. Thus the biases generated in chirographic and typographic culture will assert themselves until they are quelled by reflection. Today, devotion to books, if it is unreflective, does not liberate but enslaves our minds. A liberal education today must include reflection on the significance of writing and print, situating these media and all their works in their historically patterned sequences.

When we speak of a sequence of media, we do not mean that new media of communications annihilate their antecedents. When men learned to write, they continued to talk. When

[2] See J. C. Carothers, "Culture, Psychiatry, and the Written Word," *Psychiatry*, XXII (1959), 307–320, and the many studies cited there; Marvin K. Opler, *Culture, Psychiatry, and Human Values* (Springfield, Ill.: Charles C. Thomas, 1956); Gilbert Durand, *Les Structures anthropologiques de l'imaginaire* (Paris: Presses Universitaires de France, 1960).

[3] *European Literature and the Latin Middle Ages*, tr. from the German by Willard Trask, Bollingen Series, XXXVI (New York: Pantheon Books, 1953), p. 315.

they learned letterpress printing, they continued both to talk and write. Since they have invented radio and television, they have continued to talk and write and print. But the advent of newer media alters the meaning and relevance of the older. Media overlap, or, as Marshall McLuhan has put it, move through one another as do galaxies of stars, each maintaining its own basic integrity but also bearing the marks of the encounter ever after. Manuscript and even typographic cultures thus sustain traces of oral culture, but they do so to varying degrees. Generally speaking, literature becomes itself slowly, and the closer in time a literature is to an antecedent oral culture, the less literary or "lettered" and the more oral-aural it will be. Thus, it is to be expected that the oral residue in Tudor literature is, by contrast with most writing in comparable genres today, heavy in the extreme. Indeed, the interaction between oral tradition, writing, and the relatively new mode of alphabetic typography can be used to define some of the most salient characteristics of Tudor style, both as this was consciously cultivated in the highly rhetorical Tudor milieu and as it was actually achieved.

This chapter is concerned with oral phenomena in Tudor literature, but its main interest is in oral residue rather than in consciously cultivated oral effects. Oral effects, such as conversational elements in literary style, approximations of informal speech patterns, the use of dialogue, can be imported into written composition at will. They occur in Tudor writing, too, although, with the exception of dialogue, rather less frequently than in comparable writing today. Oral residue is another thing. By oral residue I mean habits of thought and expression tracing back to preliterate situations or practice, or deriving from the dominance of the oral as a medium in a given culture, or indicating a reluctance or inability to dissociate the written medium from the spoken. Such residue is not especially contrived and seldom conscious at all. Habits of thought and expression inseparable from the older, more familiar medium are simply assumed to belong equally to the new until this is sufficiently "interiorized" [4] for its own techniques to emerge from the chrysalis and for those more distinctive of the older medium to atrophy. Thus, for example, early type designers laboriously cut punches for the myriad ligatures which had been a godsend to scribes but were only encumbrances in typography.

In assessing the forces at work to preserve oral states of mind and the techniques which went with them, we must not forget the extreme novelty of alphabetic writing in the total duration of human cultures. Oral composition, more or less formalized, apparently belongs to the entire time span of the human race, going back perhaps well over 500,000 years. By contrast, in Tudor times alphabetic writing was around three thousand, or perhaps three thousand five hundred, years old. The "interiorization" of alphabetic writing under such conditions could quite understandably be far from complete, and could remain far from complete even with the extraordinary intensification of chirographic effects being brought about by printing. Print was closer to chirography than to speech and reinforced some of the psychological effects of chirography at the very time it was succeeding only poorly in interiorizing itself as a genuinely new medium (poets felt it a profanation of their written lines to have them printed, just as Plato had once felt it a profanation of the spoken word to have it written).

[4] I borrow this term from Marshall McLuhan, *The Gutenberg Galaxy* (Toronto: University of Toronto Press, 1962), pp. 24, 54, 58, etc.

The study of oral residue in Tudor prose could be carried much further than it will be here. Here we shall concentrate on the Latin problem faced by the humanists in its relation to the oral set of mind and on certain features of Tudor style, some macroscopic and some microscopic, connected with the cult of *copia* and of the commonplaces, a cult itself involving highly oral elements connected with the humanist Latin tradition. This approach to style differs from that taken by Croll, Williamson, and others concerned with the Ciceronian, Senecan, baroque, and other parallel classifications.[5] It does not dispense with Croll's type of analysis but simply looks at the situation in other perspectives.

II

The rhetorical tradition which in the academic world had so largely controlled the concept and practice of expression from antiquity strongly supported the oral set of mind in Renaissance culture. Rhetoric, which is at root simply the Greek word for oratory, governed far more than oratory as such. Poetry, highly "rhetoricized" in the Middle Ages, was commonly thought of as more or less ruled by the orator's art,[6] and letter-writing manuals, which proliferated in the sixteenth century under the influence of Erasmus, Vives, Melanchthon, Junius, Lipsius, Macropedius, and Hegendorff commonly prescribed, in accord with medieval *artes dictaminis*, that letters themselves be organized in the same fashion as orations,

[5] Morris W. Croll's principal studies have recently appeared in collected form as *Style, Rhetoric, and Rhythm*, ed. by J. Max Patrick and Robert O. Evans, with John M. Wallace and R. J. Schoeck (Princeton, N.J.: Princeton University Press, 1966).

[6] See Richard McKeon, "Rhetoric in the Middle Ages," *Speculum*, XVII (1942), 28–29; cf. Wilbur Samuel Howell, *Logic and Rhetoric in England, 1500–1700* (Princeton, N.J.: Princeton University Press, 1956), pp. 4–6, 40, 274–276, etc.

proceeding from exordium, through statement of proposition to be proved, proof, and refutation of adversaries, to a peroration or conclusion.[7] Scholarly treatises of all sorts had an oratorical coloring. For example. Rudolph Agricola's work on the art of dialectic or logic, *De inventione dialectica libri tres*, like other comparable humanist treatises, in the 1521 Strasbourg edition and others as well, dutifully featured a *peroratio*, labeled as such, at its conclusion.

In accord with the rhetorical outlook, students were almost never taught objective description or reportorial narration: the object of education was to get them to take a stand, as an orator might, and defend it, or to attack the stand of another. Everyone is now aware that partisanship was encouraged by dialectic, the art of formal debate, but even scholars often fail to observe that it was encouraged even more by addiction, real or fictional, to oratory. In either case, the partisanship was thought of as functioning in an oral setting: debate or persuasion was felt as an oral-aural undertaking. Over all the teaching of expression, even though writing was much employed, there hung a feeling that what was being taught was an oral rather than a written mode.

Rhetoric, despite its deep involvement in the written medium, retained its earlier, expressly oral contours intact: normally it included as one of its five parts *pronuntiatio* or delivery—which meant oral delivery—as well as memory. The Ramists managed finally to discard this latter (memory), but not the former (oral delivery), although, like the others, they actually devoted little enough space to delivery in their rhetoric textbooks. This avowed commitment to the oral flew

[7] See Sister Mary Humiliata, "Standards of Taste Advocated for Feminine Letter Writing, 1640–1797," *Huntington Library Quarterly*, XIII (1950), 261–277, and references there cited, pp. 261–262.

in the face of the fact that Latin, the language of the schools (and *a fortiori* its satellite Greek and far dimmer satellite Hebrew), had been completely controlled by chirography for almost a thousand years—Latin here meaning that part of the ancient linguistic tradition which had been artificially preserved from the normal oral processes of development resulting in the romance vernaculars.

Apart from the association of Latin with rhetoric as an art, this last mentioned fact, that Latin was totally controlled by writing no matter how much it was used for speech, produced other special kinds of drives toward the oral within the academic world. This can be seen in the attention given to promoting *copia*. *Copia* does not translate readily into English. It means abundance, rich flow, as well as ability, power, resources, or means of doing things. The quest for *copia* could express the whole humanist dream of "an eloquence which would cover the whole range of the human mind." [8] But there is something curiously relevant to the oral-aural outlook suggested by the humanists' affection for *copia* as expressive of their educational and intellectual ideals. Fear of failure in *copia* is close to the oral performer's fear of hesitancy—for abundant flow is more critical to the orator, who cannot call time out, than it is to the writer, who can and does. Renaissance works proposing expressly to develop *copia* are often curiously elementary. The *copia* which they assure is often that which would come in great part from normal nonacademic oral activity in the case of the vernaculars.

This is evident in the works of Erasmus, the largest and most important collection of material undertaking to supply *copia*. Significantly, Erasmus' more than four thousand *Adages*

[8] R. R. Bolgar, *The Classical Heritage and Its Beneficiaries* (Cambridge: The University Press, 1958), p. 273.

and his eight books of *Apophthegms* are collections not of writings but specifically and professedly of sayings (which have been written down, to be sure, but are thought of still as things spoken). The adages are proverbs or anonymous popular sayings; the apothegms are sayings attributed to individual persons, more carefully tooled, frequently—but by no means always—gnomic, sententious, and witty. Often, however, the so-called adages prove not to be proverbial at all: Erasmus puts into this collection merely ways of saying things, *modus loquendi,* the Latin equivalents, for example, of "whiter than snow," "blacker than pitch," "sweeter than honey," "softer than the ear lobe." He is providing the small change picked up in the vernaculars largely through oral chit-chat but hard to come by when one learns a language so chirographically and typographically committed as Latin was.

Erasmus' most basic workbook for Latin learners, his *De duplici copia verborum et rerum commentarii duo* (in some editions, . . . *libri duo*) is less ostensibly but quite as really meant to serve in place of the informal oral patter from which basic fluency in a language normally derives. It is a collection of phrases and of recipes for accumulating and exploiting expressions, examples, and indeed any sort of material which might later be worked into what one might want to say. Made up chiefly of informal jottings, it hurls at the teacher or pupil in cloying abundance thousands of ways of varying Latin words or phrases and of varying thought. For example, over one hundred and fifty different ways of varying in Latin the term *delectarunt* ("has delighted") in the statement, "Your letter has delighted me very much" are shot at the reader in one uninterrupted blast, followed by some two hundred ways of saying in Latin, "I shall remember you as long as I live." [9]

[9] Under the heading "Delectarunt," in Erasmus' *De copia verbo-*

Erasmus' *Colloquies* or sample dialogues have a similar oral orientation. The humanists were desperate to get youngsters to speak Latin, as is evident from school statutes [10] and from individual boyhood careers, such as that of the young Thomas More, whom William Roper writes of not as being "taught Latin" but as "brought upp in the Latyne tongue" at St. Anthony's in London.[11] The *Colloquies* provide models of what little boys might talk about—or so it was hoped—once they were sent away from their families to be free of the company of women, who normally knew no Latin.

Erasmus' other works extend and buttress this make-up program for fluency. Significantly, most of the classical works he edited were collections of one sort or another: Aesop's *Fables*, Epiphanius' *Lives of the Prophets*, Sophronius' *Lives of the Evangelists*, St. Jerome's *Lives of the Evangelists*, Lucian of Samosata's *Dialogues* and other works, Pliny's *Natural History*, Plutarch's *Works*, and Suetonius' *Lives of the Twelve Caesars*. These—among other purposes they obviously serve—help supply the anecdotal substructure always deficient in a language only derivatively oral.

Erasmus undertook to make antiquity current by moving it as close to the oral world as he could. Although he himself was a textual critic, he was irresistibly driven to establish a

---

rum ac rerum commentarius primus . . . secundus [table of contents, *De copia verborum ac rerum libri duo*] Lib. 1, cap. xxxiii, in his *Opera omnia* (Leyden, 1703–06), 1, cols. 23–30.

[10] See the sources in T. W. Baldwin, *William Shakspere's Small Latine and Lesse Greeke* (Urbana: University of Illinois Press, 1944), 1, 75.

[11] William Roper, *The Lyfe of Sir Thomas More, Knighte*, ed. by Elsie Vaughan Hitchcock, Early English Text Society, Original Series, No. 197 (London: Humphrey Milford, Oxford University Press, 1935), p. 5.

community of conversation and of personal interchange—in Latin, of course, for this was his medium of communication not only in letters and learned treatises but in face-to-face contact with his friends in the More circle and elsewhere. Despite his residence in England, we must remember that Erasmus never troubled to learn English.

Erasmus' educational purposes were of course complex, and it was certainly his intention that the springs of invention, which he had pooled for his age largely single-handed, should give life not merely to speech but to writing and print as well. Indeed, the humanists' programs stress written rather than oral performance. Our point here is not that the humanists were as intent on oratory as the classical rhetoricians and writers had been but rather that the humanist program gives evidence of concern—very often unacknowledged—with building up the oral presence of Latin, partly because the humanists were influenced by the oratorical orientation of the ancients whom they read, partly because the literary tradition was in their day still continuous to some extent with the ancient oratorical tradition, and partly because one of the problems with Latin was the atrophy of its oral roots over a millennium. Erasmus' quest of *copia* was tied in with all these considerations—most often not at all expressly.

Erasmus' quest of *copia* was that of the age. From his collections, the waters of invention flowed off in all directions and seemingly without stop. On the basis of the admittedly tentative listings in the *Bibliotheca Erasmiana*, presented by its editors in 1893 (reprinted in facsimile by B. De Graaf, Nieuwkoop, 1961) with its thousands upon thousands of entries as a "simple questionnaire," one can estimate that well over six thousand *editions* of one or another of Erasmus' works have been put into print. Erasmus was not merely an individual writer: he was an institution, and as such a relict of an oral culture which despite the humanists, and partly because of them, was no longer viable.

III

In his extraordinarily well documented work on epic song, Albert Lord has contrasted the oral with the literary in terms of text, grammatical structure, and ideas or themes. He notes that in the oral "text" expression is chiefly by means of formulas, with the bulk of the remainder formula-like or "formulaic" and very little nonformulaic expression. In a literary text, on the contrary, we find few formulas and only a bit of the formulaic, leaving us with nonformulaic composition. Oral composition or grammatical structure is typically nonperiodic, proceeding in the "adding" style; literary composition tends more to the periodic. Oral composition uses ideas and themes which are well established and can be rapidly maneuvered in standard patterns by the bard; literary composition uses typically newer themes or combines older themes in ways more novel than are usually found in oral composition.[12] The differences between oral and literary are, of course, not absolute, in that any single feature of either kind of composition can be matched from time to time in the other, but the differences are clear in the sense that in sufficiently extended passages the incidence of distinctive features is notably different in the two kinds of composition.

Although Lord is writing primarily about the epic, what he has to say here is based on the known mechanics and psychology of oral and literary composition taken more or less generally. In the light of his observations and of what has just been noted about the state of Latin as a medium and focus of

[12] Lord, pp. 130–131.

training in expression, it becomes possible to reinterpret much in Tudor English prose style as residual oralism, in part endemic to the native tongue because of the still largely non-literary, oral grounding of this tongue, and in part due to positive encouragement from the oral residue and emphases in Latin.

Here we can resuscitate a concept long moribund and unexploited by Lord but highly serviceable to sum up and focus Lord's points for our present purpose. Oral composition is essentially "rhapsody" (Greek *rhapsōidia*), that is, a stitching together, in the original meaning of this term as applied by the Greeks to their epic song, and by Englishmen well through the seventeenth century to loose collections of material of all sorts.[13] The epic singer is not a memorizer in our post-Gutenberg sense of the word, but a skilled collector. He works unavoidably with a deep sense of tradition, which preserves the essential meaning of stories. But he has no fixed text to reproduce, such as we take for granted in a typographic culture. Instead, he possesses an armory consisting of formulas or metrically malleable phrases (together with near-formulas or "formulaic" expressions), and of themes or situations—the banquet, the messenger, the demand for surrender, the challenge, the invitation, the boast, the departure, the arrival, the

[13] "For Metaphysicks, I say that Aristotles Metaphysicks is the most impertinent Booke (sit venia) in all his works; indeed, a rapsodie of Logicall scraps."—[Thomas Barlow?], '*A Library for Younger Schollers' Compiled by an English Scholar-Priest about 1655*, ed. by Alma DeJordy and Harris Francis Fletcher, Illinois Studies in Language and Literature, Vol. XLVIII (Urbana: University of Illinois Press, 1961), 4. This assertion clearly echoes Ramus' accusation that Aristotle's metaphysics was only theology sieved through dialectic (or logic—the two terms were synonymous for Ramus). See Walter J. Ong, *Ramus, Method, and the Decay of Dialogue* (Cambridge, Mass.: Harvard University Press, 1958), p. 190.

recognition, and so on (most of these still occur today in the Western, clearly a regressive art form). Formula and theme are the stuff which the epic singer rhapsodizes or "stitches" into his oral epic fabric, never worded exactly the same on any two occasions.

But rhapsody was not restricted to poetry. The classic orator proceeded in much the same rhapsodic way as the ancient bard, and so, in theory, did his Renaissance followers. Doctrines of invention, rhetorical and dialectical or logical, had encouraged the view that composition was largely, if not esentially, an assembling of previously readied material. The humanists had reinforced this view with their doctrine of imitation and their insistence—not new in actuality but only in conscious emphasis—that antiquity was the storehouse of knowledge and eloquence. Bolgar has shown how humanist educational procedures enforced the assumption that the classics were writings which could be dismembered into bite-size pieces for reassemblage into new configurations. Indeed, he has made the point that the humanist achievement consisted largely in transferring into the modern consciousness the best of classical antiquity largely by just such a process of decomposition and recomposition.[14]

Operationally, this rhapsodic view of literary composition found its center in the doctrine of the commonplaces to which one can relate so many of the tactics advocated and practiced by Tudor prose writers. The doctrine of the places was applied to poetry, too (generally taken as a form of rhetoric), but was developed mostly for prose use. It was a complicated and at certain points inconsistent set of assumptions, views, and prescriptions,[15] but the part of it relevant here is its ad-

[14] Bolgar, pp. 271–275, 295–301.
[15] Sister Joan Marie Lechner, O.S.U., *Renaissance Concepts of the*

vocacy of the advance preparation of material which could be inserted as occasion offered into whatever one was composing. The Greeks, and, following them, Cicero and Quintilian, advocated that the orator get up in advance a repertory of *loci communes* or commonplaces—little purple patches on loyalty or treachery or friendship or decadence ("O tempora! O mores!") or other themes "common" to any number of cases or occasions for insertion into an oration as opportunity offered. As has been seen, the classical oration was the product of a situation, typically an oral performance even in the case of so literate an orator as Cicero, who wrote out his orations only after he had delivered them—sometimes, it appears, years afterwards.[16] For such a performance a stock of commonplaces was the equivalent of the epic singer's stock of formulas and themes, although a literate orator such as Cicero might make incidental use of writing in ways in which illiterate bards and orators did not. Quintilian reports on the writing out of commonplaces in order to decry it, but Cicero appears to have known this practice and also to have taken notes during trials at which he was to speak (although of course he did not use them in the course of the speech itself).[17] Commonplaces, however, even when written, belonged to the oral tradition in that they stocked the imagination with material accessible on the spur of the moment. Moreover, in their use,

---

Commonplaces (New York: Pageant Press, 1962), pp. 65-131, and *passim*.

[16] Torsten Petersson, *Cicero: A Biography* (Berkeley: University of California Press, 1920), pp. 92-94.

[17] Quintilian, *The Institutio Oratoria* . . . with an English translation by H. E. Butler, ii. 4. 30-32, Loeb Classical Library (London: William Heinemann, 1920), I, 239-241; Petersson, pp. 92-93.

as in all oral performance, the question of originality as a virtue does not even arise. The oral traffics in the already known. Only in the early nineteenth century, when the residue of oralism had greatly diminished as its major depository, Latin, lost effectiveness, does "commonplace" become a generally derogatory term. Commonplaces could of course be composed by the orator himself or garnered from other writers—the germ of the commonplace book, so much in use in Tudor times, lies here.

Following Aristotle and other Greeks, Cicero and Quintilian also wrote of *loci communes* or commonplaces, in another, deviously related, sense, as headings, sources, or "seats" (*sedes argumenti*) "common" to all sorts of subjects, to which one could betake oneself to discover arguments to prove one's point: headings such as genus, species, cause, effect, related things, opposed things, and so on. Special or "private" places or headings had been devised to supply arguments specifically for law, physics, medicine, and other subjects. In practice, common and special places or topics were often intertwined in the many competing lists of places which filled the dialectic or logic and rhetorical textbooks. The use of the places was far more attended to than the often flimsy theory got up to support their existence.[18]

In either this latter meaning (headings) or that earlier mentioned (prefabricated purple passages), the commonplaces were an answer to the need for fluency which the orator, like the epic singer, felt much more acutely than the writer, since, as has been seen, oral performance, once begun, cannot be interrupted with impunity as written composition can. Thus the Tudor retention and intensification of interest in the com-

[18] Ong, *Ramus, Method, and the Decay of Dialogue*, p. 116.

monplaces suggests a need, felt if not articulated, to attend to expression in an oral rather than chirographic frame of reference.

IV

It is well known that one of the major features of Tudor prose—and poetry, although this is not our direct concern here—was a tendency to a loosely strung-out, episodic style. Episodic, loosely serial organization is observable in a great spectrum of forms from controversial literature (Fish, More, Tyndall) through the chroniclers who for the time serve as historians, in prose fiction (Nashe's narrative in *The Unfortunate Traveller*, the romances of John Lyly, Robert Greene, and Thomas Lodge), and on through "characters" and that most distinctively Tudor product, the essay. This tendency to feature strung-outness, the collection as such, under various guises, is not taken into account in the ordinary analysis of Tudor prose in terms of Ciceronian, Attic, Senecan, Euphuistic, or baroque style. The tendency suggests the somewhat disjointed, nonperiodic Senecan style, but chiefly in the latter's meandering phrases rather than in its moments of gnomic intensity.

The perspectives earlier sketched here invite us to see the looseness of Tudor style as residual oralism, even when it may also be in part an imitation of written models. The Tudor writer was by literary and cultural heredity in great part rhapsodic. For him written expression had never been so fully detached from the oral as it was to be shortly when the new invention of letterpress typography had its fuller effect. His was thus a marginal position, and it deserves more study as such. Here we shall undertake only a survey of some typical manifestations of loose, "adding," organization.

One of the most prominently oral phenomena of this prose
is its reliance on the "formulary" structures discussed by
Howell.[19] It should be noted that we are referring here not to
a formula in the sense of Lord and Parry, as "a group of words
which is regularly employed under the same metrical conditions to express a given essential idea," [20] but rather to formulas for a sequence of elements in a passage of some length, to
organizational formulas. Such organizational formulas provided, among other things, for an ordered progression through
a list of topics in praising or blaming a character: for example,
Richard Rainolde's *The Foundacion of Rhetorike* (1563), an
adaptation of the *Progymnasmata* of the fourth-century
Greek rhetorician Aphthonius, a favorite textbook in Tudor
schools, specifies that for the praise of an individual person
one should eulogize in succession his country, ancestors, education, "actes" (*res gestae* or achievements, that is, his use of
gifts of the soul, of the body, of fortune), concluding with
some comparison which would show to the advantage of the
subject of eulogy.[21] The relationship between this approach to
composition and the oral procedures discussed by Lord as
mentioned above is evident enough. Aphthonius has codified
in writing an oral institution.

The *de casibus virorum illustrium* (the fall of great men)
pattern which persists in works such as St. Thomas More's
Richard III can be regarded as a special type of formulary
writing: a foreordained sequence of rise, triumph, and fall.
The persistence of *de casibus* formulas in the Middle Ages
(alongside other equally schematized patterns for lives of the

---

[19] Howell. pp. 138–146.   [20] Lord, p. 30.
[21] Richard Rainolde, *The Foundacion of Rhetorike* (1563), with
an introduction by Francis R. Johnson (New York: Scholars' Facsimiles and Reprints, 1945), fols. xl^r–xliii^v.

saints) [22] and its gradual demise through the Renaissance can be regarded from one point of view as marking the transit to a less orally controlled culture.

The formula, in a larger sense, goes even further than this, controlling the pattern of the oration proper, the master prose genre. For the oration, approved rhetorics prescribed from two parts (Aristotle's minimum in the *Rhetoric*) to seven parts (Thomas Wilson's prescription in *The Arte of Rhetorique*), the minimal two being always statement of the case plus proof. To these two can be added an introduction, a summary of the case, division of the proof, refutation of adversaries, and conclusion. Within this set formulary order other subformulas tend to grow. These consist of special groups of commonplaces (in the sense of headings or classes of arguments). For, although the places as sources of "arguments" properly belong only to the proof, such is the force of the commonplace frame of mind that the commonplaces serve to structure other parts of the oration as well. Thus Curtius discusses the special topics or places used for the introduction or exordium (the first part of the oration when four or more parts are taken as normative) and the peroration or conclusion (the last part).[23] Curtius is concerned directly with the Middle Ages, but there is no significant difference here between medieval and Tudor practice.

The parts of the oration imposed formulary structure on much other prose writing besides the oration itself, including not only letters (as already mentioned) but also learned treatises and narrative. Sidney's *Apology for Poetry*, for which Sidney also used the title *The Defense of Poesie*,

---

[22] See Donald A. Stauffer, *English Biography before 1700* (Cambridge, Mass.: Harvard University Press, 1930), pp. 3–32.

[23] Curtius, pp. 85–91.

and which consists largely of literary theory, is organized as an oration. This organization is neither a happenstance nor a tour de force. Sidney really thought the work out in oratorical form, as its title, *Apology* or *Defense*, a technical term, makes clear. Here Tudor criticism of literature, despite the fact that it is itself literature, written composition, and concerned essentially with literature as such (for Sidney views poetry as basically, although not exclusively, literature, something normally composed in writing) is nevertheless cast in a preliterate, oral form. For examples of oratorical organization in narrative, there are Nashe, Lyly, and many others.

Allied to the use of formulas for over-all structural organization is the use of epithets and of material from commonplace collections. This use suggests medieval techniques of amplification and in varying degrees marks virtually all styles of the Tudor period, poetry as well as prose. From More through Hooker, the reader is showered with adages, apothegms, standard incidents of all sorts to illustrate a moral or prove by example—the fall of Icarus as an example of the overreacher, Pyramus and Thisbe, Brutus the betrayer, and so on, ad infinitum. Every possibility in life was covered by hosts of standard sayings or incidents, which obtrude more in some styles than in others and are satirized in some, as in Nashe's, but are almost always before the reader's eyes. The shower of epithets and examples is pleasant and even beautiful, rainbow-hued at times, but it is curiously unlike what is advocated in postromantic writing. Our knowledge of the sources of examples and sententious sayings is improving with works such as that of Professors DeWitt T. Starnes and Ernest William Talbert on Renaissance dictionaries. But we have hardly scratched the surface of the vast collections of quotations available in the sixteenth century—the greatest number of them, I should estimate, deriving in one way or another through publishers at Basel, the long-standing headquarters for Erasmian forces. We are becoming more aware of the reliance of many Tudor writers on handbooks for their classicad echoes and allusions. It would appear that the glut of collections from classical authors which flooded the Renaissance—and gave it further continuity with the Middle Ages and their passion for florilegia—fed an appetite which was still largely created by a residual oralism in expression. The oral performer favors use of a well-known phraseology. The humanists' insistence on imitation and the typographers' ability to multiply lists of cullings effectively and cheaply combined for the moment to give the orally oriented mind a new lease on life, although ultimately typography was to spell its doom.

Another type of oralism marks the highly figured styles of the Tudor period, from More and Lyly through Nashe. These are clearly devised for their effect on the ear and thus are oral in a real sense, but one not exactly our direct concern in treating of rhapsodic oralism. Titillation of the ear is not necessarily residual oralism: is can be a new and conscious sophistication. More akin to rhapsodic tradition are certain special features of the figured styles, most notably their use of epithets, found so abundantly in Lyly, as in many of his contemporaries: "this new kind of kindness, such *sweet* meat, such *sour* sauce, such *fair* words, such *faint* promises, such *hot* love, such *cold* desire, such *certain* hope, such *sudden* change," or "the *soft* drops of rain pierce the *hard* marble." [24] Such usage is very close to the oral formulas of which Lord has made such careful analyses.

[24] John Lyly, *Euphues: The Anatomy of Wit, Euphues and His England*, ed. by Morris William Croll and Henry Clemons (London: George Routledge and Sons, 1916), pp. 65–66.

In part, of course, Lyly's style connects directly with oral performance, deriving not only from figured oratorical style generally but in particular, very probably, from the style in the orally presented lectures of John Rainolds at Oxford.[25] But Lyly's use of epithets has also direct sanction in the Tudor collections of epithets got up to assure *copia*. In one of the most influential collections, the *Epitheta* of the Parisian rhetorician Ioannes Ravisius Textor (1480?–1524), Elizabethan schoolboys, for whom Textor's *Epitheta* and his related collection *Officina* were prescribed,[26] could find epithets by the thousands for all things under heaven and many above: Apollo is accorded 113 epithets, love 163, death 95, and so on through entries which include Africa, the river Alpheus, the arts themselves, to touch only on a few of the hundreds under the letter "A."[27] This approach (designed directly for Latin, but strongly influencing English) strikes us as weird. But it is, among other things, designed to sate the oral appetite for adequate and proper formulas.

Many other instances of oral residue could be discussed here. One could note the formulary aspects of decorum, or the very insistence on decorum itself as evidence of a formulary frame of mind. Deloney uses Lyly's euphuistic formulas to

[25] William A. Ringler, Jr., "The Immediate Source of Euphuism," *PMLA*, LIII (1938), 678–686.

[26] Baldwin, pp. 349, 394–395, 519, etc., cites school statutes and British editions of these two works. Of course, the far more numerous Continental editions were also known to schoolmasters in the British Isles, as can be seen from inscriptions in surviving copies today.

[27] See Ioannes Ravisius Textor, *Specimen epithetorum* (Paris: Henricus Stephanus, 1518; the second edition (Paris: Reginaldus Chauldiere, 1524), published posthumously under the editorship of Ioannes' brother Iacobus, and the many subsequent editions were entitled *Epitheta*.

cue in the speech of his rare characters from the nobility, such as the Earl of Shrewsbury's daughter, Margaret, in *Thomas of Reading*. Or one could also note the large-scale use of parallelisms which are formulaic in a highly oral way, often incremental, certainly "adding" in form. Such are many of Sir Thomas Elyot's expressions in *The Boke Named the Governour:* God shows "wisedome, bounte, and magnificence," man's estate is "distributed into sondry uses, faculties, and offices," man is exposed to "the fraude and deceitfull imaginations of sondry and subtile and crafty wittes," one is urged "to enserche and perceive the maners and conditions." [28] This multiplication of synonyms is of course due to nervousness about the adequacy of English words—it is a mannerism in translators—but it appears to be also in part residual oralism. One could note also the use of dialogue, which is not inserted into the narrative to give color and "interest," as it might be today, but which serves to cast up sociological or economic or psychological issues—for example, in More's *Richard III* or, in Deloney, the speeches of characters summing up the plight of the country or their own problems, social, economic, or psychological. In the absence of appropriate sciences such issues could hardly be caught in any other way than in direct, oratorical address stating more than merely personal issues in the guise of personal concerns. But multiplication of instances carries us too far afield for a brief survey. The examples we have adduced, however, may provide sufficient background for some possible conclusion about oral residue in Tudor style.

[28] Sir Thomas Elyot, *The Boke Named the Governour*, ed. by Henry Herbert Stephen Croft (London: Kegan, Paul, Trench, and Co., 1883), 1, 4, 6 (Book 1, chap. i), 60 (Book 1, chap. x).

v

This summary discussion of what we have styled oral residue in Tudor prose has not been concerned with what might
ordinarily be thought of as typically oral elements in prose today. Oral elements today are typically those introduced more
or less openly into a medium of expression felt primarily as
written. They are, in a way, extraneous, although they are by
the same token more obviously oral than was the case with the
oral residue we have here touched on. Today's prose is often
shot through with idioms calculatingly echoing oral performance at any and all levels—for example, in Joyce or Heming-
way or Faulkner. It has a special kind of oral richness which
decorum inhibited in the most permissive Tudor prose, where
slang and colloquialisms are mostly avoided insofar as they can
be identified in a language still little subject to written control.
(One must make some exception for Thomas Nashe and for
the playwrights.) At the same time, however, these same
twentieth-century writers are definitely using a written or
typographical medium rather than the oral medium as their
over-all matrix: although *Finnegans Wake* can be read aloud
and listened to with great enjoyment, and *The Sound and the
Fury* echoes voice within voice like an aeolian Chinese puzzle,
neither could have been conceived or executed as oral performances. By contrast, the *Iliad* and the *Odyssey* not only
were so conceived, but had to be in order to be what they are.
    The oral elements we have been considering in Tudor prose
are akin to those in the *Iliad* and the *Odyssey*. They are not
among the oral elements which Tudor writers may have consciously introduced into their prose as oral. Rather, they are
elements which Tudor writers automatically favored largely

because these writers were close to a far more primitive—and by this I do not mean unskilled—preliterate mode of expression. These oral elements survived centuries of manuscript culture because in the teaching and practice of expression this manuscript culture retained very live connections with the old preliterate oral-aural world. In particular, the persistence of oral residue was favored by the study of Latin which governed so largely the writing of English prose. Latin sustained the rhetorical—which is basically oratorical and thus oral—cast of mind. The stress on *copia* which marked Latin teaching and which was closely associated with rhetorical invention and the oral performer's need for an uninterrupted supply of material favored exploitation of commonplaces. The use of these in turn made for the "adding" or "rhapsodic" style which survives in so much Tudor writing as it had in medieval "amplification," and which is not adequately accounted for by the ordinary discussion of Tudor stylistics.

It should be subjoined that certain institutions surviving from oral-aural culture other than those concerned with formal language study also favored the oral cast of mind and the survival of oral elements. The most important of such institutions was the orientation of all academic instruction toward oral performance: written exercises were in use (and that a restricted use) only for the learning of Latin (and Greek and Hebrew), after which the student, whether in arts (philosophy), medicine, law, or theology, was tested only by his oral performance in disputation or oral examination. Outside the academic world, the oral cast of mind was also sustained by a special vernacular practices, such as the singing of ballads.

Much of what we have called oral residue in Tudor prose does not, as we have treated it, differentiate Tudor prose from that of earlier ages, although closer study could very likely show that there were indeed many points of difference. But what we have treated as oral residue here does differentiate Tudor prose from what was soon to follow it. The features here noted become more and more attenuated through the seventeenth and eighteenth centuries until their virtual disappearance with romanticism. The loose episodic structure, in particular, which is one of the central characteristics of oral style, was on its way out as narrative worked its way toward the modern novel and short story, and as devices such as Ramist "method" charmed the mind with prospects of organization which were basically visual, chirographic and, even more, typographic.

Despite the fact that it can be discerned readily in preceding ages, oral residue is of special importance on the Tudor scene, for, in one genre at least, the world of the Tudor writer shows a more massive concentration of oral residue than that of earlier ages. This is in its commonplace collections. The sequel of medieval collections of exempla and other such material, these collections reach their apogee in the two centuries following the invention of printing. The new typographic medium offered previously unheard of opportunities to the impulse of the orally oriented performer to have as much as possible on hand so that he would be prepared to extemporize in absolutely any eventuality. Sixteenth-century and seventeenth-century printed collections of commonplace material are so utterly countless that no one has ever attempted even a preliminary survey. They were the last flash of activity from the orally oriented mind, which had proved viable in manuscript culture but was soon to be reduced to insignificance when the visualization induced by writing was both supplemented and eventually transformed by print.

3 | Tudor Writings on Rhetoric, Poetic, and Literary Theory

I. RHETORICAL TRAINING AND LITERATURE

The literature of the Tudor age, like that of earlier and immediately subsequent ages, has some of its deepest roots in the rhetorical tradition. Though originally concerned with oratory, as has been seen, rhetoric was also intimately associated with what we today would call works of creative imagination, and never more effectively than through the sixteenth century. Its relationship to such works was sometimes straightforward, sometimes devious, but always pervasive. Writings about rhetoric itself flooding Tudor book stalls are often, it is true, too businesslike, too practical in tone, too free of the touch of play necessary for aesthetic performance to qualify in themselves as belles-lettres. But this does not keep them from exerting massive influence on all genres of writing. And many of the rhetoric works have true literary merit of their own. As Louis B. Wright has shown in *Middle-Class Culture in Elizabethan England*,[1] the collections of emblems, apothegms, and related material fostered by the rhetorical tradition formed the staple of the ordinary man's

[1] Chapel Hill: University of North Carolina Press, 1935.

[48]

reading. Books built around a rhetorical concern for "invention" such as Tottel's *Songs and Sonnets* (1557) or *England's Helicon* (1600), preserve the best texts we have for many Tudor poems, and indeed sometimes the only texts. Other collections, such as Francis Meres' *Palladis Tamia: Wit's Treasury* (1598), provide invaluable biographical detail on contemporary literary figures. And at least one new genre comes into existence as a metamorphosis of a rhetoric workbook: the much-touted essay, as we find Ben Jonson grumbling in his *Timber*, is at root simply a presentation of material garnered under one or another heading in a commonplace collection such as rhetoric encouraged writers to accumulate.

Tudor works on rhetoric and allied subjects, such as poetics and literary theory, of course cannot be understood apart from the classical heritage. More than at any other time in English literary history, in the Tudor age, the golden age of the great grammar schools such as St. Paul's, the classical rhetorical heritage took possession of literature and society itself. This heritage appears simultaneously as theory, as pedagogical practice, and as a determinant of the whole culture. From typographical usuage to court manners, from drama to Bacon's reform of science, the influence of rhetoric is clearly discernible not merely in style of expression but also deeply ingrained in ways of thought and world outlook.

The original Greek *rhētorikē* refers directly not to writing but to oral performance, public speaking, skill in which had constituted the major objective of intellectual training for the elite of ancient Greece. Rhetoric is thus the "art" developed by a literate culture to formalize the oral communication skills which had helped determine the structures of thought and society before literacy. Quite early, however, the term was generalized to include other than oral expression, but the fact that a term specific to oral verbalization came to be the ordinary one referring to the management of other forms of expression suggests that rhetoric may well have preserved early oral-aural cultural attitudes, as it did indeed through the Renaissance and beyond. As a teachable body of knowledge, rhetoric is defined by Aristotle in his *Rhetoric* (i. 1. 14. 1355b) as the art of discovering the available means of persuasion for any subject matter whatsoever. Largely through Cicero's great example and his treatises on the orator's profession, the formal study of rhetoric became established as the focus of academic education also in imperial Rome. Cicero's short treatise, *De inventione*, and the longer *Rhetorica ad Herennium* long ascribed to him were the backbone of rhetorical training until well along in the sixteenth century, abetted after the advent of humanism by Quintilian's *Institutio oratoria*.

In the Middle Ages, rhetoric was the second art of the trivium, that is, of the sequence of grammar, rhetoric, and dialectic (strictly, an art of disputation tolerating argument from probability, but in fact more or less equated with logic, the art of strictly scientific argumentation), which together constituted the lower academic curriculum. These subjects were followed by "philosophy" (natural philosophy chiefly, with a touch of metaphysics and some moral philosophy) to lead in English as in other universities to the degree of master of arts. In point of fact, dialectic or logic tended to be detached from the trivium and annexed to the higher curriculum of philosophy, and thus in a way to outclass rhetoric. Yet, as Richard McKeon has shown in his classic *Speculum* article on "Rhetoric in the Middle Ages," despite the ascendancy of logic in medieval times, both the theory and practice of rhetoric contributed massively from the fourth through the fourteenth centuries not only to expression but also to vast areas of medieval intellectual achievement, such as to the development of the scholastic method and of scientific inquiry as well as to psychology and medicine.

In Tudor England, despite humanist endeavors to pull it into the higher ranges of the curriculum, rhetoric in general retained its medieval position in the curriculum at a level lower than logic or dialectic.[2] Grammar and rhetoric, with only

[2] Medieval manuals of rhetoric and their use in the curriculum are well accounted for by Charles Sears Baldwin, *Medieval Rhetoric and Poetic* (New York: Macmillan, 1928), which is supplemented by the brief but extraordinarily comprehensive article by Richard McKeon, "Rhetoric in the Middle Ages," *Speculum*, xvii (1942), 1–32. The Tudor grammar schools and their studies are lavishly described by Thomas W. Baldwin, *William Shakspere's Small Latine and Lesse Greeke*, 2 vols. (Urbana: University of Illinois Press, 1944). The university curricula have been less well studied, but a new beginning has been made by Mark H. Curtis, *Oxford and Cambridge in Transition, 1558–1642* (Oxford: Clarendon Press, 1959). The books known and used in England have been studied in full detail by Wilbur Samuel Howell, *Logic and Rhetoric in England, 1500–1700* (Princeton: Princeton University Press, 1956) to which the present treatment is much indebted. Other useful works include: Charles Sears Baldwin, *Renaissance Literary Theory and Practice* (New York: Columbia University Press, 1939); William Garrett Crane, *Wit and Rhetoric in the Renaissance* (New York: Columbia University Press, 1937), still very informative; Sister Miriam Joseph [Rauh], C.S.C., *Shakespeare's Use of the Arts of Language* (New York: Columbia University Press, 1947); Donald Lemen Clark, "Ancient Rhetoric and English Renaissance Literature," *Shakespeare Quarterly*, ii (1951), 195–204; *The Province of Rhetoric and Poetic*, ed. by Joseph Schwartz and John A. Rycenga (New York: Ronald Press, 1965), an invaluable collection of historical and theoretical studies; Morris W. Croll (1872–1943), *Style, Rhetoric, and Rhythm*, ed. by J. Max Patrick and Robert O. Evans, with John M. Wallace and R. J. Schoeck (Princeton: Princeton University Press, 1966), a posthumous collection of pioneering studies. Helpful listings, with definitions, of the hundreds of rhetorical terms familiar in the schools will be

such elementary or "petty" logic as was needed for rhetoric, were studied in elementary or "middle" schools, leaving most of logic and all of philosophy to the universities. Grammar involved the study of language and literature, all but exclusively Latin of course, with gestures toward Greek, and took the form of parsing and translating and Latin prose composition, including Latin letter-writing; it also included *poetria*, or the study of metrics and versification, which was often considered simply a specialized part of rhetoric. Rhetoric was no longer focused so dominantly as it had been in antiquity on oral performance but had become more or less continuous with advanced instruction in grammar, leading to what is still called "theme" writing as well as to declamations or orations. Some study of rhetoric was continued into the university, but it seems to have been limited chiefly to lectures on theory and to the analysis of classical orations; for the disputations stressed

---

found in Lee Ann Sonnino, *A Handbook to Sixteenth-Century Rhetoric* (London: Routledge and Kegan Paul, 1968), and Richard A. Lanham, *A Handlist of Rhetorical Terms* (Berkeley and Los Angeles: University of California Press, 1969). On the relationship of English to Latin, see Richard Foster Jones, *The Triumph of the English Language* (Stanford: Stanford University Press, 1953). A group of more recent works concerned with early oral performance, including Eric Havelock, *Preface to Plato* (Cambridge, Mass.: Belknap Press of Harvard University Press, 1963), Robert Scholes and Robert Kellogg, *The Nature of Narrative* (New York: Oxford University Press, 1966), and the present author's *The Presence of the Word* (New Haven and London: Yale University Press, 1967), give rhetoric still fuller meaning by relating it to larger cultural and psychological developments concerned with the evolution of the media of communication. In general, references readily traceable through the foregoing works or other basic works to which the foregoing refer are not given in detail here. I should like to thank Professor Franklin B. Williams, Jr., of Georgetown University for calling my attention by letter to several details about works treated here below.

in university work were logical rather than rhetorical exercises. In fact, however, rhetoric still functioned in university work, for the disputant or commentator on a text on many occasions digressed rhetorically from his straight and narrow logical path.

Merely to list these various modes of language studies does not give a full idea of their method. A glance at the texts in use, whether classical or medieval or contemporary Tudor, for all coexisted, reveals an extraordinarily strict discipline in composition. It reveals also the degree to which the oration as such tyrannized over ideas of what expression as such— literary or other—was. The usual theory acknowledged three kinds of orations: the judicial (or courtroom), the deliberative, and the occasional or epideictic or demonstrative (encomium, consolatory, etc.). Orations of any kind were composed in a sequence of parts, which varied in the manuals from a minimal two to seven. In the *Rhetoric* (iii. 13. 1414b) Aristotle had listed four: the exordium, the narration or proposition (statement of what one is to prove), the proof, and the conclusion, indicating that the two essential parts were narration and proof. Cicero lists the parts differently in different places, and in the *De inventione* (i. 14–56) and *De oratore* (ii. 19) increases Aristotle's four to six: exordium, narration, division (of the subject matter), proof, refutation of adversaries, and conclusion. To these, between narration and division, Thomas Wilson in *The Arte of Rhetorique* (1553) adds a seventh part, the "proposition," which is "a pithie sentence [sententious saying] comprehendyng in a small roume, the some of the whole matter." [3] If we add a digression just before

[3] Wilson, *The Arte of Rhetorique*, ed. in facsimile by Robert Hood Bowers (Gainesville, Fla.: Scholars' Facsimiles and Reprints, 1962), p. 20.

the peroration, as Cicero (*De oratore* ii. 19) states some authors do, we can even have eight parts.

The art of letter writing, part of the *ars dictaminis* developed in the medieval schools for notaries and officials, had picked up this oratorical structure and applied it to letters. These were to have, after the proper *salutatio,* in succession an *exordium* or *benevolentiae captatio* (the winning of good will), a *narratio* or statement of the fact, a *petitio* or request (corresponding to the proof in the oration), and a *conclusio.*[4] Moreover, even the classification of kinds of letters most often echoed the kinds of the oration: in Erasmus' *De ratione conscribendi epistolas,* a common schoolbook after 1521, we find letters divided into persuasive (deliberative), laudatory (demonstrative), and judicial, plus a fourth type, which was nonoratorical, the familiar.[5] But there were other more elaborate classifications, as will be seen.

The writing of school themes was governed by as strict a discipline as the writing of letters and was likewise thought of—partly by oversight—in oratorical terms. Set formulas for various thematic orations were to be found in the *progymnasmata* or school exercises of the Greek rhetoricians Hermogenes (fl. A.D. 161–180) and Aphthonius (fl. A.D. 315). Aphthonius' *Progymnasmata* was current in a Latin version by Rudolph Agricola and Ioannes Maria Cataneus with scholia by Reinhard Lorich (despite humanists' campaigning, few schoolmasters could really read Greek with facility), and an English paraphrase with English examples was published in 1563 by an Oxford fellow, Richard Rainolde, as *The Found-*

---

[4] C. S. Baldwin gives an abstract of the *Candelabrum,* a thirteenth-century manual of *dictamen, Medieval Rhetoric and Poetic,* pp. 216–227.

[5] T. W. Baldwin, II, 251.

acion of Rhetorike. Rainolde lists Aphthonius' fourteen ways of "making" an oration as: fable (in the Aesopian sense), narration or tale, chria (praise or blame of a word or deed), sentence or gnomic saying, confutation or refutation, confirmation or proof, commonplace or amplification of a virtue or vice, praise or encomium, dispraise or vituperation, comparison, ethopeia or character portrayal, visual description, thesis or generalization, and *legislatio* or a plea for or against a law.[6] Schoolboys writing themes cast them in one or another of these molds or types. Each type had its subtypes and special formulary requirements. Thus:

This parte of Rhetorike called praise is either a particular praise of one, as of kyng Henry the fifte, Plato, Tullie, Demosthenes, Cyrus, Darius, Alexander the greate; or a generalle and universalle praise, as the praise of all the Britaines or of all the citezeins of London.

The order to make this Oracion is thus declared. First, for the enteryng of the matter, you shall place a [*sic*] *exordium*, or beginnyng. The seconde place, you shall bryng to his praise *Genus eius,* that is to saie, of what kinde he came of, which dooeth consiste in fower poinctes: of what nacion, of what countrie, of what auncetours, of what parentes. After that you shall declare his educacion. The educacion is conteined in three poinctes: in institucion, arte, lawes. Then put there to that, whiche is the chief grounde of al praise: his actes doen, which doe procede out of the giftes and excellencies of the minde, as the fortitude of the mynde, wisedome, and magnanimitee; of the

---

[6] Richard Rainolde, *The Foundacion of Rhetorike,* with an introduction by Francis R. Johnson (New York: Scholars' Facsimiles and Reprints, 1945), fol. iiij ff. For other recipes, see A. L. Bennett, "The Principal Rhetorical Conventions in the Renaissance Personal Elegy," *Studies in Philology,* LI (1954), 107–126.

bodie, as a beautifull face, amiable countenaunce, swiftnesse, the might and strength of the same; the excellencies of fortune, as his dignitée, power, aucthoritee, riches, substaunce, frendes. In the fifte place use a comparison, wherein that whiche you praise maie be advaunced to the uttermoste. Laste of all, use the *Epilogus* or conclusion.[7]

The other thirteen kinds of thematic orations demanded procedures of comparable complexity. Of these themes, those of praise (*encomium*) and dispraise (*vituperatio*) were certainly the most important, since ancient, medieval, and Renaissance literary performance in practice and even more in theory hinged on these two activities to a degree quite incredible today.

The formulas in Rainolde's Aphthonius give some idea of what went on in actual schoolroom practice, but the Renaissance vision of rhetoric extended far beyond such schoolroom exercises. Most manuals in use present Cicero's vision of rhetoric as consisting of five "parts": *inventio* or discovery of "arguments" to prove a point, *dispositio* or arrangement of the arguments found, *elocutio* or style, *memoria* or the use of memory, and *pronuntiatio* or delivery. These "parts," as a matter of fact, were not taught in strict sequence nor with equal emphasis by Tudor rhetoricians any more than they had been by ancient rhetoricians.[8] Cicero, whose work *De inventione* includes a great deal of material evidently belonging to the other parts, which he never got around to treating, suggests in his *Brutus* (vi.) that the five parts may really be five separate arts rather than divisions of a single art, coming close to the historical fact that they had originally been not

---

[7] Rainolde, fol. xxxvij (*sic* for xxxix)^v–xl^r (paragraphing and punctuation adjusted to modern usage).

[8] See Howell, pp. 66 ff.

"parts" of an "art" but more or less successive activities involved in ancient Greek liberal education.[9]

From the beginning in antiquity, *inventio* had received the lion's share of attention. It was particularly important insofar as rhetoric affected the writing of literature as such, for *inventio* corresponded roughly to what our postromantic world would call "use of the creative imagination," although it was implemented chiefly by exploitation of the highly conventional "places" or commonplaces (*loci* or *loci communes*). These have been mentioned in Chapter 2 and are explained at greater length below in this present chapter. In one sense of the term commonplaces were headings suggesting thoughts for any and all subjects and available in various competing lists. The formulas for "praise" and the other thirteen kinds of composition discussed above can be accurately viewed as lists of suitable commonplace headings ranged in effective order for fourteen particular purposes.

Except for such formulary arrangement of headings and some remarks on the parts of the oration, *dispositio* was given less attention in the manuals and the classroom. *Elocutio* or style was commonly interpreted in terms of ornament: the writer or speaker was thought of as "decorating" his otherwise plain thought with tropes or figures or schemes (the terminology varied) which, like the commonplaces, were classified in numerous competing lists in the various rhetorical manuals, partly overlapping and partly contradicting one another. England's earliest significant contribution to such catalogues of rhetorical ornaments had been Bede's *Liber de schematibus et tropis in Scriptura Sacra*. Medieval writers had

---

[9] H. I. Marrou, *A History of Education in Antiquity*, tr. from the French by George Lamb (New York: Sheed and Ward, 1956), pp. 194–205.

also developed concern with style in a special sense related to Cicero's concern with decorum. They wrote of three styles, which in the sixteenth century Sherry, Wilson, and Puttenham call "characters" of style: the high style was to be used in treating of noble or epic characters, the middle for middle-class characters (such as the landed gentry), the low for the lowest orders, the three styles being exemplified respectively by Virgil's *Aeneid*, *Georgics*, and *Eclogues*. Sixteenth-century rhetoricians commonly concern themselves with the high style only, although writers self-consciously did use all three.[10] Thomas Deloney's *Thomas of Reading* (c. 1599), as noted in Chapter 2, mixes tradesmen's middle style with the high-style Euphuism always employed by Margaret, daughter of the Earl of Shrewsbury, by Prince William, Prince Richard, and Duke Robert. The plain style favored by Ramism, as will be seen, is related to the middle and low styles, but is not the same as either. It consists of unadorned, flat statement. Memory, and even more *pronuntiatio*, were somewhat half-heartedly retained; they were in fact chiefly relics of the more truly rhetorical age of antiquity, when expression had been more typically an oral performance and less concerned with writing than in post-Gutenberg Tudor England.

We are today struck with amazement at the variety and rigidity of Tudor training in rhetoric, the more remarkable because it was imposed in a second language, Latin, with a sprinkling of a third language, Greek, upon schoolboys of ten to fourteen years of age. Rainolde's English version of Aphthonius, cited above because it is a contemporary translation, was in fact not a typical textbook; for these in principle, and

---

[10] Walter F. Staton, Jr., "The Characters of Style in Elizabethan Prose," *Journal of English and Germanic Philology*, LVII (1958), 197–207.

generally in actuality, were themselves in Latin. School statutes, although of course not always observed, typically imposed the speaking of Latin by boys and masters at all times on the school premises, aiming at creating the total Latin environment in which Cicero had lived when Latin was the vernacular. English appeared only indirectly and incidentally in the program: it was used, as occasion offered, simply to better the boys' Latin and Greek, as in the procedure advocated by Ascham in *The Scholemaster*, whereby the student translated a Latin passage into English so that he could translate the English back into his own Latin, thus perfecting his control of the ancient tongue. That such a rhetorical system could have helped produce the great writers of Tudor England appears bizarre today, but the fact that it did so is incontestable. The indelible marks of the system on Shakespeare, for example, often observable in his most effective and moving and seemingly most unaffectedly "natural" writing, have been conclusively spelt out by T. W. Baldwin in his *William Shakspere's Small Latine and Lesse Greeke*. Since Latin, with a dash of Greek, was virtually the only school subject, studied daily all day long for a period of seven to ten years, it is little wonder that skill in the language occasioned skill in the vernaculars. Perhaps never before or after was training in language skill so vigorous in England as in Tudor times. No apt student so relentlessly drilled in any language could fail to acquire some effectiveness in his own related vernacular.

Since rhetoric was studied at what today would be the elementary school, or at best the junior high school, level, it appears puzzling how the young boys subject to this training could have had anything at all to say worthy of the intricate amplifications provided by the system. Steps were taken, however, to provide them with something to say. These steps were two: stocking the mind with abundance of material or "copie" (*copia*) which could be drawn on by *inventio*, and simultaneously implementing *inventio* by training in the use of the "places," already mentioned. The humanist doctrine of imitation, which encouraged careful echoing of expressions or whole passages out of the best writing of antiquity, helped stock the mind with both ideas and words. Often the ideas and words came directly from readings in the classics themselves: Aesop, Terence, Ovid, Virgil, Horace, Plautus, Cicero, and the historians, together with a very few Neo-Latin writers such as the Carmelite priest and pastoral poet Ioannes Baptista Mantuanus, mentioned with warmth by Shakespeare and others.

Out of such authors expressions as well as ideas could be culled and written into one's own commonplace book or "copie" book (copybook). Countless such handwritten commonplace books survive, sometimes curiously abetted by the new medium of print. John Foxe's folio volume, *Pandectae locorum communium* (1572), provides 604 leaves (1208 pages), blank save for printed Latin headings under which commonplace material could be entered for retrieval by means of the printed alphabetic index of the headings at the end of the book: *Absolutio, Absurditas, Adolescentia, Adulterium, Aer*, etc. This publication, basically no more than an index to what is not yet on hand but may be (like the Dewey or Library of Congress library classifications), shows some of the visualist or spatialized control which print could have on the mind and imagination, but it apparently elicited little interest: the *Short-Title Catalogue* lists only the Cambridge University Library copy, which has almost no handwritten entries at all. Most users of commonplaces—and everyone used them—were more content with the printed books which provided not only indexes but thousands of pages of accompanying excerpts already printed out under the appropriate headings. Of such printed "copie" books, the most massive and representative and influential were those of Erasmus, in their original or enlarged forms, treated in Chapter 2 above. But there are unnumbered other collectors. General collections retailing all kinds of material were supplemented by specialized collections restricted to such things as proverbs, apothegms, anecdotes, examples, and similes. One of the best known of these last in English (most were in Latin) is Francis Meres' *Palladis Tamia: Wit's Treasury* (1598), which presents 666 pages of similes following a preface in which every sentence is itself entirely composed of triple similes—a tour de force not too difficult for one trained in this rhetorical tradition.

Means of exploiting the store of material accumulated in these manuals and, it was to be hoped, in the boy's mind, were found in the lists of commonplace headings elaborated from ancient and medieval writers. In order to "find" something to say on a "theme" (an idea, relatively abstract, such as bravery; or concrete, such as a palace) or on a "thesis" (a statement, such as "Emperors should be brave"), one betook oneself to an assortment of "places" or *loci* or "topics" (Greek *topoi*, places; adjective *topikos*). We would think of these today as "headings," but the Tudor mind, like the ancient and medieval, thought of them as somehow locales or compartments or, very often, hunting grounds in which all possible things one could say about a subject were considered to be lodged. Cicero (*Topics* ii) had defined a *locus* in terms of locale as the "seat of an argument" (*sedes argumenti*).

As has already been stated, lists of these "places" or *loci* varied greatly in make-up and length. Generally, since antiquity, *loci* for rhetoric had been considered to differ in principle from those for dialectic, although the two had been more or less confounded in practice. Here the Renaissance was to witness a revolution. In his *De inventione dialectica libri tres* (completed c. 1479, but printed only posthumously in 1515), the Dutch humanist Rudolph Agricola (Roelof Huusman), whom Thomas Wilson followed in *The Arte of Rhetorique*, had grouped all the places of invention without exception under dialectic. Using his list to develop, for example, the theme of bravery, the writer would consider in order the definition of bravery, its genus, species, properties, whole, parts, conjugates or closely linked matters, adjacents or loosely associated matters, acts of bravery, subjects of bravery, the efficient cause, end, consequences, and intended effects of bravery, the places and times when bravery was to be practiced, its connections, contingents, names, pronunciation, things comparable to it, things like it, opposites, and differences. These *loci* or topics provided "artificial" arguments, that is, arguments intrinsic to the subject and thus available through "art." In addition, there was the "inartificial" or extrinsic *locus* of "testimony" or "witness," less esteemed because it provided as an argument only what someone had said, that is, something lying outside the analysis intrinsic to the subject itself. By running through these "places," in whole or in part, the writer or speaker could bring to mind relevant material in the "copie" stored up from earlier reading, either in his own mind or in his notes or in the printed collections of excerpts from ancient (and a few contemporary) authors. Following the far from clear discussion of the places in Aristotle's *Topics* and *Rhetoric*, there existed through the Renaissance a tendency to distinguish "common" places, which provided arguments for any and all subjects (as do those just listed above), from "special" or "private" places, headings for arguments peculiar to a special subject such as law or politics or ethics or physics. But in fact "commonplace" (*locus communis*) was often used generically for both kinds of places. The *loci* were different historically and conceptually from the Aristotelian categories or predicaments, with which they were, however, occasionally confused—as by Ralph Lever in *The Art of Reason* (1573).[11]

If we ask what effects this Tudor training in composition had upon the prospective writer, we should note first of all the obvious emphasis upon both play of the mind and word-play. The grammar-school boy should never have been at a loss to play with any word or idea or—what was much the same—to develop any word or idea systematically. Tudor exuberance of language and expression was not accidental, but programmed. Since the student had read and imitated almost exclusively Latin authors, the style of his expression was necessarily Latinate, complex in form and vocabulary if not completely Ciceronian. (The slavish use of only Ciceronian words and expressions which Erasmus vigorously contested as pedantic in his *Ciceronianus* [1528] was rare indeed in England.) Since the student had been trained in one rhetorical (and logical) pattern after another, we should expect his speech

[11] Ralph Lever, *The Arte of Reason, rightly termed, Witcraft* (1573), p. 7. The Greek word *katēgoria* (after which the Latin equivalent, *praedicamentum*, is modeled) means an accusation or charge, not a class or storehouse; in Aristotle it refers to the predicate in a proposition or assertion, thought of as a "charge" brought against a subject. Hence Lever refers to the ten "demaunders." The *loci* or topics, on the other hand, are classes, subject to logical quantification. Lever, after referring to the "demaunders," proceeds to consider them erroneously as *loci*. See Sister Joan Marie Lechner, O.S.U. *Renaissance Concepts of the Commonplaces* (New York: Pageant Press, 1962), pp. 90–91.

or writing to be mannered jargon. It often is, and Shakespeare, Nashe, and others frequently poke fun at it for being so. But since the Latin models for imitation were good, the results were, at their optimum, utterly convincing and natural, and we find ourselves surprised to discover for example that Othello's "round unvarnish'd tale" is set in a strictly patterned *exordium* or introduction which comes straight out of the textbook.[12]

Furthermore, the study of rhetoric gave the most diverse literary genres a more or less oratorical cast, largely because the dominance of oratory in ancient culture had never been effectively challenged. We are not surprised that Tudor monuments to oral expression include obviously oral exhibits such as secular orations (most of the carefully wrought ones in Latin) or the great sermons of the age headed by John Colets' 1512 *Sermon . . . Made to the Convocation at Paul's* (delivered in Latin but printed in English in 1530) and by Hugh Latimer's 1548 English sermon commonly known as "The Sermon of the Plow." It is somewhat surprising, however, to note how far oratory infiltrated genres which we consider nonoratorical. Fiction writers made their characters speak to one another in orations or quasi orations even in private conversation, as we see in John Lyly's *Euphues: The Anatomy of Wit* (1578) and *Euphues and His England* (1580), Nashe's *The Unfortunate Traveller* (1594), or Sidney's *Arcadia* begun 1580 (published posthumously 1590 and 1593). Plays such as Shakespeare's *Henry V* feature lengthy stretches of highly oratorical declamation. Treatises such as Sidney's *Defense of Poesie*, together with essays, letters, the prefaces, and dedicatory pieces with which the age abounds, as well as epic and lyric poetry are all organized in oratorical form probably

---

[12] See T. W. Baldwin, II, 198–200, and *passim*.

more often than not. Indeed, praise and blame, the objectives of the epideictic orator, were often identified as equally the objectives of literature generally.[13]

The deeper effect of rhetorical teaching on the literary sensibilty is connected with this omnipresence of the oratorical frame of mind. It was an effect as real and sweeping as it was doubtless unpremeditated. A rhetorically dominated education gave a boy no training whatsoever in uncommitted, "objective," neutral exposition or narrative. It was not dialectic alone which gave the Tudor age its argumentative cast. Rhetoric is the art of persuasion, and the orator who exemplifies its training is a committed man, one who speaks for a side. The forensic orator prosecutes or defends, the deliberative orator pleads for or against the passage of the law or measure he discusses, and even the epideictic or demonstrative orator, the speaker who merely displays his mastery of a subject (but always, Renaissance writers resolutely maintained, to incite his hearers to virtue), does so in Tudor as in earlier theory and practice by judicious distribution of praise and blame. Rhetoric produced individuals predisposed to approach any subject by taking a side, because they were not formally trained to do anything else: any side, perhaps, but some side certainly. The polemic outlook was further intensified by the fact that the schools and the very language of the schools, Latin, were only for the boys and men. Academic aims are often formulated in the jargon of the aristocratic fighter-hero, as in Sir Thomas Elyot's *The Boke Named the Governor*.[14] The "lettered" women who knew Latin, as Sir Thomas More's

[13] See O. B. Hardison, Jr., *The Enduring Monument: A Study of the Idea of Praise in Renaissance Literary Theory and Practice* (Chapel Hill: University of North Carolina Press, 1962).

[14] See Chapter 5 below.

daughter Margaret and Lady Jane Grey and Queen Elizabeth did, had no discernible mollifying influence on the contentious academic climate: such women were very few, and they studied with tutors, away from the halls of disputation, at home, where other girls who learned some reading and writing did so almost always by working with the more peaceable vernacular.

The polemic rhetorical setting may have been bad for science, but it was good for many kinds of literature. The life of the mind was exciting because it was framed in conflict. Characters with words in their mouths put there by writers trained in rhetoric were sure to generate dramatic friction when they met together on the stage, and intellectuals engaged in any controversy were spurred on to making the most of their cause and indeed often to regrettable virulence. The combative basis of rhetorical (and dialectical) training is certainly one of the reasons for the effectiveness of the late Tudor and Jacobean drama, as well as of the great lyric poetry of the age. The polemic cast of expression continued far past Tudor times. Milton's essays on public affairs are structured in controversy, and his *Paradise Lost* was conceived as an oration to "justifie the wayes of God to men."

II. ENGLISH WORKS ON RHETORIC AND THEIR SOURCES

The revival of rhetoric in Tudor England was a part of the general Renaissance revival of the art. Like most Renaissance phenomena, this revival appeared in England much later than on the Continent. As on the Continent, when it did appear, it took the form of an antischolastic movement. During the scholastic Middle Ages, in Northern Europe particularly, the ancient focus on rhetoric had yielded to a focus on logic or dialectic, largely under the influence of the scientizing proclivities which developed with the universities and their scholastic philosophy from the twelfth century on. Since antiquity, the West had known an art of discourse (*ars disserendi*) which Cicero and others called *dialectica* and which the Middle Ages generally tended to identify more or less with logic (*logica*), although this latter was sometimes restricted to strictly "necessary" or scientific reasoning (such as in mathematics) as against the merely more probable reasoning which might win for one side in a dialectical disputation or debate.

Cicero's dialectic, and, following it, medieval dialectic, was thought of as made up of *inventio* or invention and *iudicium* or judgment (also called *dispositio* or arrangement), parts which were strikingly similar to the first two parts of Ciceronian rhetoric. Although more astute thinkers tried to differentiate logical or dialectical invention and disposition from rhetorical, the Middle Ages often tended to reduce the province of rhetoric by assigning invention and disposition, in effect, to logic or dialetic, leaving rhetoric with only *elocutio* or style as its province—memory and delivery being, as has been seen, minor matters. The *Metalogicon* of John of Salisbury states that rhetoric provides luster and resplendence to the arguments of logic, and elsewhere assigns to dialectic succinct expression, generally in syllogisms, and to rhetoric induction and amplification.[15]

Although rhetoric was thus often narrowed in scope in the Middle Ages, it was by no means completely forgotten, even in the North. We remember Chaucer's praise of Petrarch,

[15] Daniel D. McGarry, *The Metalogicon of John of Salisbury* (Berkeley and Los Angeles: University of California Press, 1955), pp. 67, 102 (Book I, chap. xxiv; Book II, chap. xii); the *Metologicon* is summarized in C. S. Baldwin, *Medieval Rhetoric and Poetic*, pp. 158–172.

"the laureate poete . . . whos rethorike sweete / Enlumyned al Ytaille of poetrie," and his other laudatory reference to the "rethor" who could "faire endite." [16] The rhetorical fires banked through the Middle Ages flared up with the same Petrarch's intensified passion for Cicero and with the influx into Italy of humanist educators from Greece, led by Manuel Chrysoloras, who came to Florence as a municipally paid lecturer in 1396 and had as his most influential pupil Guarino of Verona. The method of education perfected by Guarino was to become that of humanism generally: reading and composition to assure detailed assimilation of content from classical writers and meticulous imitation of their form.[17] Works from Greek antiquity were made increasingly available to the West in Latin. Aristotle's *Rhetoric* was translated into Latin during the fifteenth century by George of Trebizond (probably between 1447 and 1455) and Ermolao Barbaro (after 1480).[18]

Englishmen visiting Italy in Guarino's day and in contact with humanists there, such as William Grey, later Bishop of Ely, or John Free, later Bishop of Bath, brought back many of the new works, but these Englishmen were drawn commonly into public life before they were able themselves to add significantly to the new learning. Continental humanists in England were responisble for the early publications there concerned with rhetoric. Lorenzo Guglielmo Traversagni, a

[16] *The Canterbury Tales*, "Prologue of the Clerkes Tale," 31–33; "The Nun's Priest's Tale," 3207.

[17] See the careful and perceptive account in R. R. Bolgar, *The Classical Heritage and Its Beneficiaries* (Cambridge: The University Press, 1958), pp. 266–272, etc. Cf. Roberto Weiss, *Humanism in England during the Fifteenth Century*, 2d ed. (Oxford: Blackwell, 1957).

[18] Bolgar, p. 434.

Franciscan friar of Savona, lecturer in theology at various places in Europe, finished his *Nova rhetorica* at Cambridge University in 1478, as he tells us at the end of the work itself, which was published by Caxton around 1479 and by the St. Albans printer in 1480. This work is a large and substantial treatise (362 pages), based still on the pseudo-Ciceronian medieval favorite, the *Ad Herennium*, and oriented to preaching. Other Continental visitors doing some of their work on rhetoric in England include Erasmus and the Spanish lecturer in rhetoric at Oxford, Juan Luis Vives. But the influx of works published on the Continent, particularly those of the German humanist educators, was greater than the local British production, even abetted by immigrés. The *Epitome troporum ac schematum* of Iohannes Susenbrotus (Zurich, 1541) became one of the standard English school texts, together with rhetoric works by Philip Melanchthon, Petrus Mosellanus, Ioannes Caesarius, and others.[19] These were, of course, Latin compositions.

School statutes uniformly mention textbooks in Latin for classroom use, but by the 1530's, and increasingly in the latter half of the century,[20] some of these found their way into English translation, probably for a variety of reasons. Some teachers might ignore school statutes and do at least part of their teaching in English. Others might translate a work from Latin to guarantee their mastery of it. Some works translated or written in English might be designed for those who, like the upper-class youth prescribed for in Elyot's *Governour* (1531), did not go to the university, or for law students in

[19] T. W. Baldwin, II, 1–28.

[20] I have found Bolgar's lists in his Appendix II very useful: "The Translations of the Greek and Roman Classical Authors before 1600," Bolgar, pp. 506–541.

London.[21] The first rhetoric in English was that of the school-master Leonard Cox, *The Art or Crafte of Rhetoryke* (London, c. 1530; second ed., 1532), derived from Cicero's elementary treatises and Melanchthon's *Institutiones rhetoricae* (1521) and treating only *inventio*. The next English treatise was by Thomas Wilson, who received his master of arts degree at Cambridge in 1549 and was to go on to the law and important public service. In *The Rule of Reason* (1551) Wilson had produced the first book on logic in English, which he complemented with *The Arte of Rhetorique* (1553), a full, five-part Ciceronian work which emphasized *inventio* and in its treatment of *elecutio* cautioned against the use of "straunge inkehorne termes" derived from foreign languages. Learned or academic works in English seldom went beyond one more or less experimental edition, and the extraordinary demand which produced eight editions of *The Arte of Rhetorique* by 1585 together with the fact that its numerous illustrative examples relate to the law, the pulpit, and public affairs, lends substance to the conjecture that Wilson intended it for the young gentlemen and noblemen studying law at the Inns of Court. With his logic, Wilson's rhetoric stands out as lively and intelligent, "the only English-language rhetoric of the sixteenth century which goes beyond translation or close paraphrase." [22] Other works on rhetoric after Wilson con-

[21] Richard J. Schoeck, "Rhetoric and Law in Sixteenth-Century England," *Studies in Philology*, L (1953), 110–127. In "Rhetoric and the Law Student in Sixteenth-Century England," *Studies in Philology*, LIV (1957), 498–508, D. S. Bland concludes that little rhetoric was taught in the Inns of Court, that law students learned by doing. The use of an English-language textbook would seem to fit Bland's conclusion, for it would appear to indicate that the subject had little formal academic standing: the English textbooks were for informal, private use.

[22] Crane, pp. 100–101.

tinue to be derivative, but their increasing number indicates a growing ease in expressing in English academic thought previously cast chiefly in Latin. Some are what Howell calls "stylistic rhetorics," treating only *elocutio* and limiting it to tropes and figures of speech. Richard Sherry's *A Treatise of Schemes and Tropes* (1550) by the headmaster of the Magdalen College School at Oxford, was a large compilation, the first in English, drawing chiefly on Erasmus; a new 1555 edition introduced Latin alternating with the English to make the work usable as a school textbook. A later English compilation was Henry Peacham's *The Garden of Eloquence Conteyning the Figures of Grammer and Rhetorick, from whence maye bee gathered all manner of Flowers, Colours, Ornaments, Exornations, Formes and Fashions of Speech* (1577).[23] Peacham was a clergyman, and his book for "studious youth" who had not the benefit of Latin was revised in a second edition to draw especially upon the Bible for illustration.

Brief treatises on tropes, figures, and schemes were also to be found in letter-writing manuals in English, for example in the 1592 revision of Angel Day's *The English Secretorie* (1586) and in John Hoskins' *Directions for Speech and Style*, written about 1600 and much plagiarized around that time, though not published until 1935.[24] Hoskins' work is competently and forcefully written, making skillful use of homely expressions to put into English rhetorical concepts ordinarily managed in Latin. Thus he describes metaphor as "the friendly and neighborly borrowing of one word to express a thing with more

[23] Ed. (1593 revision) by William G. Crane (Gainesville, Fla.: Scholars' Facsimiles and Reprints, 1954.)

[24] Ed. by Hoyt H. Hudson (Princeton: Princeton University Press, 1935); ed. by Louise Brown Osborn in *The Life, Letters, and Writings of John Hoskins, 1566–1638* (New Haven: Yale University Press, 1937), pp. 115–166.

light and better note, though not so directly and properly as the natural name of the thing meant would signify." [25]

Sensitivity to tropes and figures and schemes—grammatical, logical, or rhetorical—was cultivated not only in one's own writing but also for purposes of literary analysis. As we know from marginal annotations in books of the time, texts were worked over to discover and identify *hyperbaton, metonymy, aphaeresis* (the dropping of an initial letter or syllable), *concessio* (granting to an opponent a point which hurts him), and the countless other "ornaments" treated in works on grammar and rhetoric—Peacham provided fifty-six grammatical "schemates" or patterns to work with, fifty patterns of word and sentence, and sixty of amplification, and he by no means exhausted the possibilities. This is the setting which helps generate the style of John Lyly's *Euphues* (1578).

The intensity of interest in tropes, figures, and schemes shows itself even in typography, with the development of gnomic printing as an adjunct of rhetoric. Especially between 1550 and 1660, printed texts in considerable quantity use special pointing—asterisks, daggers, variations in type face, and other devices—to indicate in sidenotes the occurrence of rhetorical figures, especially sententiae, or even label figures verbally (*similitudo, exemplum*, etc.), as schoolmasters and schoolboys did in analyzing texts. The practice of gnomic printing shades into other kinds of emphatic printing (italics, upper case, etc.),[26] and our present-day use of italics and exclamation and quotation marks can be seen as the fag end of this once more complicated rhetorical tradition.

Other types of rhetoric textbooks contained model exam-

[25] *Directions for Speech and Style*, ed. Hudson, p. 8.
[26] G. K. Hunter, "The Marking of Sententiae in Elizabethan Printed Plays, Poems, and Romances," *Library*, VI (1952), 171–188.

ples. Rainolde's adaptation of Aphthonius has been mentioned earlier; it provided short model orations, carefully analyzed. Some of these formulary textbooks become, in effect, anthologies of very effective pieces of writing. Model speeches come close to fiction in *The Orator: Handling a Hundred Severall Discourses* (1596), Lazarus Piot's translation from the French of Alexandre van den Busche or Le Sylvain. The hundred exercises here include each an accusation and a reply, one of them discussing the pound-of-flesh contract: the translator hopes that lawyers, preachers, and others might profit by his models. In *The Defence of Contraries* (1593, reprinted 1616), a translation of Ortensio Landi's *Paradossi* (1541), the veteran hack writer Anthony Munday presents twelve of Landi's original thirty essays. These defend poverty, ignorance, foolishness, and so on, continuing the school tradition in which Erasmus' *Praise of Folly* both belongs and excels.

Models for imitation are likewise to be found in the letter-writers. The first of these was William Fulwood's *The Enimie of Idlenesse* (1568), a book of precepts and sample letters which is almost entirely a translation of *Le Stile et maniere de composer, dicter, et escrire toute sorte d'epistre* (Lyons, 1566),[27] one of several French letter-writers available at the time. The sample letters strike occasional notes of real pathos, but Fulwood perpetuates the rigid Latin formularies, with their division of letters into deliberative, demonstrative, and judicial. Abraham Fleming's *A Panoplie of Epistles* (1576), under the influence of the work of the ancient Greek Sophist Libanius, is even more formulaic, with twenty-one types of

[27] The letter-writers are treated in detail in Katherine Gee Hornbeak, *The Complete Letter Writer in English, 1568–1800*, Smith College Studies in Modern Languages, Vol. xv, Nos. 3–4 (Northampton, Mass.: [Smith College,] 1934).

letters. Fleming's section on precepts is a translation from the Latin of Christoff Hegendorff (Hegendorphinus), much used in England. Angel Day's *The English Secretorie*, mentioned above, is more original as well as erudite and useful. Day believes that something more than formulas should go into "Epistles Amatorie" but otherwise multiplies formulas, dividing Erasmus' four epistolary types into thirty-two subdivisions. Day refers to the printer's copy for a later pioneer commercial letter-writer, *The Merchants Aviso* (1607 is the earliest edition I have seen or found recorded, but the manuscript was completed by 1587) by "that hartie well-willer in Christ" I[ohn] B[rown], a Bristol merchant.[28]

The samples in these letter-writers mostly lack the charm and gusto of those provided in Nicholas Breton's *A Poste with a Madde Packet of Letters* (1602), which waives explicit concern with formularies, *exordium, narratio,* tropes, figures, and the rest of the ancient heritage while actually using consummate rhetorical skill to present letters in a "mad" style often reminiscent of Tom Nashe at his yeasty best. The only letter-writers of comparable literary importance appear a century and a half later, when Samuel Richardson's manuals for the nonrhetorical because non-Latinate feminine set appear and burgeon into novels.

The art of preaching had been given special attention by rhetoricians from the days of St. Augustine through the Middle Ages and on to Erasmus' *Ecclesiastes, sive concionator evangelicus* (1535), a five-step Ciceronian treatment known in England through Continental editions. Here, too, England relied on foreign sources. By the late sixteenth century sev-

---

[28] For the identity of the author, see T. S. Wilson, *Studies in Elizabethan Foreign Trade* (Manchester: Manchester University Press, 1959), p. 18.

eral other Continental treatises were being rendered into English: *The Preacher: or Methode of Preaching* was a translation (1574) by J[ohn] H[orsfall] of a work by a Danish classicist Niel Hemmingsen; *The Practis of Preaching, Otherwise Called the Pathway to the Pulpit* was John Ludham's translation (1577) of the *De formandis concionibus sacris* (1553) by the Marburg theologian Andreas Hyperius (or Gerard); and *The Art or Skill Well and Fruitfullie to Heare the Holy Sermons of the Church* a translation (1600?) of the *Ars habendi et audiendi conciones sacras* (Siegen, 1598) of the German Protestant theologian Wilhelm Zepper.[29] The one original work by an Englishman on pulpit oratory was composed in Latin by the famous preacher of Cambridge William Perkins as *Prophetica, sive de sacra et unica ratione concionandi* (1592), later translated by Thomas Tuke as *The Art of Prophecying* (1607—copy in Harvard University Library). Perkins' treatise shows the influence of Ramist rhetoric common in the late sixteenth-century Puritan milieu, but it was more concerned with content than with form.

Also related to works on rhetoric were the treatises on mnemonics, which trace their origins often to Cicero.[30] *The Art of Memory, That Otherwise is Called the Phenix* was translated by Robert Copland (1548?) from the French version of the Latin of Pietro Tommai of Ravenna, *Foenix Domini Petri Ravennatis memoriae magistri* (Venice, 1491), and Guglielmo Gratarolo's *De memoria* was translated by

[29] These treatises here mentioned are discussed by Lee S. Hultzen, "Aristotle's Rhetoric in England to 1600" (Ph.D. dissertation, Cornell University, 1932), pp. 162–163.

[30] See Frances A. Yates, *The Art of Memory* (Chicago: University of Chicago Press, 1966), for a brilliant historical account of the importance of the mnemonic tradition; also Chapter 4 below.

William Fulwood as *The Castel of Memorie* (1562, republished 1563 and 1573). Formulas used in memory schemes are often related to similar formulas used for the sequences of *loci* which were standard for writing on various subjects, as in Rainolde's thematic orations, mentioned above. A work on memory by one G. P. of Cambridge (*G. P. Cantabrigiensis*), *Libellus de memoria verissimaque bene recordandi scientia* (1584), associated memory explicitly with the second part of rhetoric, *dispositio*.

Perhaps the most interesting by-product of rhetoric is the steady stream of collected sayings and excerpts useful for writing and for general education. These collections, or printed commonplace books, often of vast size, result from two drives in Tudor times: the humanist desire to expedite *inventio* by having at hand massive stores of material for "imitation," both in content and style, and the habit of collecting commonplace material inherited from the Middle Ages, when florilegia and conflated commentaries multiplied beyond anything dreamed of in antiquity. Letterpress printing gave a new outlet to the collecting drive by facilitating not only multiplication of texts but also—what was more important—relatively thorough and exact indexing. The back-breaking work of indexing became worthwhile once typography provided the same pagination in any number of copies. The resulting collections are often—but not always—identifiable by their titles, which exploit the gathering or hunting imagery associated with rhetorical *inventio*. Although there is no guide to these collections as such, they are legion in the Tudor literary landscape, where title after title, such as Peacham's cited above, features terms such as flowers (*flores* in Latin), blossoms, posies, garlands, nosegays, gardens, anthologies (Greek for "gathering of flowers"); or, in another series, springs, sources, fonts, wellsprings, Helicons, Parnassuses; in another, *silvae*, woods, forest, underwoods. Some of the collections are specialized by rhetorical structure: one will consist entirely of epigrams or similes, another of aphorisms, or "sentences" (sententious sayings), adages, paradoxes,[31] apothegms, jests, and so on. Others are specialized by subject matter and thus, as has been seen above, are technically collections not of "common" places but of "private" places for medicine, law, theology, or other subjects.

Associated with printed commonplace collections are dictionaries of various sorts, heavily relied on by the writers of the best Renaissance literature,[32] as well as grammars, which both ransack and propagate commonplace sources in providing favorite passages from ancient authors as examples of syntax. Also allied to printed commonplace collections are the courtesy books or books on education and manners, collections of "colloquies" or sample conversations in Latin, such as Erasmus' *Colloquia familiaria* (1516) or the *Colloquia scholastica* (1563, etc.) of the French Genevan teacher Mathurin Cordier, and other Latin aids such as the later *gradus ad Parnassum* or poetic phrase-book. Students of Montaigne and of Bacon have noted that many of the essays of these authors consist simply of gnomic sayings strategically assembled on a given topic, the fruit of academically encouraged note-taking: indeed, each essay is little more than a commonplace col-

[31] See A. E. Malloch, "The Techniques and Functions of Renaissance Paradox," *Studies in Philology*, LIII (1956), 191–203. Names of a selection of printed commonplace collections can be found in Lechner, pp. 239–259.

[32] See DeWitt Starnes and Ernest William Talbert, *Classical Myth and Legend in Renaissance Dictionaries* (Chapel Hill: University of North Carolina Press, 1955); James Sledd, "A Note on the Use of Renaissance Dictionaries," *Modern Philology*, XLIX (1951–1952), 10–15.

lection. Ben Jonson—as has been noted—wryly makes this point about essay writers in his own commonplace collection, *Timber*.

In the course of the century, the printed collections develop from helps for students to something like small encyclopedias and proliferate in countless forms. Erasmus' *Adagia* and *Apophthegmata*, the nub of collecting activity through Western Europe, grow larger and larger in successive editions through his lifetime. These Erasmian collections were made partly available in English through the translation work of Nicholas Udall and Richard Taverner. Udall's Englishing of Erasmus' *Apophthegmes* appeared in 1542 shortly after Taverner's *Proverbes or Adagies with Newe Addicions Gathered Out of the Chiliades of Erasmus* (1539). Taverner provides even more mixed fare in *The Garden of Wysdom* (1539) and *The Second Booke of the Garden of Wysedome* (1539), which import further matter from mixed Greek and Latin sources into English. William Baldwin's misleadingly titled work, *A Treatise of Morall Philosophie* (1547), is a large collection of sayings and other multifarious commonplace material which established a publishing record in Renaissance England, with twenty-three editions (one now available in facsimile, edited by Robert Hood Bowers, 1967); this is more than double the editions of Lyly's popular *Euphues*. *A Schole of Wise Conceyts Set Forth in Common Places by Order of the Alphabet* (1569) was collected by Thomas Blage from the classics. John Parinchef drew from contemporary continental collections *An Extract of Examples, Apothegmes, and Histories* (London: H. Bynneman, n.d.). Compilers often liked to advertise the great range of their wares, as in *Beautiful Blossoms Gathered by John Bishop from the Best Trees of All Kyndes, Divine, Philosophicall, Astronomicall, Cosmograph-*

icall, Historical, and Humane (1577). Simon Robson presented a commonplace book in triplets (conceivably echoed in Francis Meres' Preface to his *Palladis Tamia: Wit's Treasury*, mentioned above). Robson breaks down each heading into three divisions, as "The body containeth 3. things: Good cheare, Sleepe, Mery talke," and gives his book a wonderful consumer-oriented title: *The Choice of Change containing the Triplicitie of Divinity, Philosophie, and Poetrie, Shorte for Memorie, Profitable for Knowledge, and Necessarie for Maners: Whereby the Learned May Be Confirmed, the Ignorant Instructed, and All Men Generally Recreated* (1585). The year before, William Fiston had published his translation from the Italian, *The Welspring of Wittie Conceites* (1583), and in 1590 Robert Hitchcock Englished the "conceites, maximies, and politicke devices selected and gathered together by Francisco Sansovino" under the title of *The Quintessense of Wit*.

The commonplace collections perhaps most important for English literature are those promoted and sponsored at the turn of the century by John Bodenham, several of which have since been edited by various scholarly hands and otherwise carefully studied.[33] *Politeuphuia: Wit's Commonwealth* (1597) provides a collection of some four or five thousand "sentences" or aphoristic citations; *Palladis Tamia: Wit's Treasury* (1598),[34] by Francis Meres, 666 pages of similitudes or comparisons; *Wit's Theater of the Little World* (1599) supplies

---

[33] See the Introduction to *England's Helicon, 1600, 1614,* ed. by Hyder Edward Rollins, 2 vols. (Cambridge: Harvard University Press, 1935), where, however, the commonplace-book pattern of the Bodenham series is somewhat overlooked.

[34] The section on literature has been separately edited by Don Cameron Allen, *Francis Meres' Treatise, "Poetrie,"* University of Illinois Studies in Language and Literature, Vol. xvi, Nos. 3–4 (Urbana: University of Illinois Press, 1933).

examples; and *Belvedere, or, The Garden of the Muses* (1600) all three, sentences, similitudes, and examples, in verse. *England's Helicon* (1600), in the same Bodenham series, is not quite of a piece with the other four, being a collection of more complete "inventions," that is, of some 150 English poems, including many of our best sixteenth-century lyrics with the names of the often otherwise anonymous authors appended by the compiler, apparently Nicholas Ling. But its title, and its preface as well, assimilates it to the commonplace collections: its contents are not only drawn from the proper "springs" of invention (Helicon), but are also presented as themselves sources for other writers. They have proved invaluable sources for literary historians as well.

Bodenham's collections appear to have inspired other more or less related compilations, especially that of William Wrednot entitled *Palladis Palatium: Wisdom's Pallace, or, The Fourth Part of Wit's Commonwealth* (1604). Robert Cawdrey's *A Treasury or Storehouse of Similes* suggests Meres' work. Cawdrey's entries are either taken from the Scriptures or refer pretty directly to a religious theme.

These collections and the countless others in Latin provided building blocks for writers throughout the century to an extent which recent scholarship is only beginning to make clear. From More to Shakespeare, adult Tudor authors turned to the collections for ideas, phrases, illustrations, and even plots, just as they had done when they were schoolboys. The most resounding and most quoted passages of Shakespeare are generally his reworked versions of what anyone could find here. Like Alexander Pope a century later, Shakespeare was less an originator than a consummately expert retooler of thought and expression. The commonplace tradition would undergo no serious deterioration until romanticism.

Related to commonplace collections and the rhetoric of invention is a special genre combining literature and the visual arts: the emblem books, which present tableaulike pictures often of gnomic or commonplace character, accompanied by appropriate mottoes, verses, and elaborate prose analyses. The *Emblemata* (Augsburg, 1531) of the eminent Italian lawyer Andrea Alciati began the vogue for such works, which reached England in the translation by Samuel Daniel (1585) of Paolo Giovio's *Imprese* (1555),[35] Geffrey Whitney's *A Choice of Emblems and Other Devices* was published at Leyden in 1586, and "P.S.'s" translation of *The Heroicall Devices* of the French writer Claude Paradin appeared in London in 1591. The genre was particularly influential in Spenser's circle, but the best known English emblem books, those of George Wither and Francis Quarles, belong to the Stuart period. The emblematist's concern with iconography and all sorts of symbolism is intimately related to rhetorical and dialectical word play and to rhetorical "ornament."

III. RAMISM: LOGIC-RHETORIC INTERACTION

The major revolution affecting rhetoric in Tudor England turned on the relationship of rhetoric to logic or dialectic. It was begun in Paris by the French professor of philosophy and rhetoric Petrus Ramus (Pierre de la Ramée, in English generally known as Peter Ramus). His new program for these arts began in his *Dialecticae partitiones* of 1543, which led to his later *Dialectique* (1555, with hundreds of subsequent editions and adaptations), and in the complementary works of his literary lieutenant Audomarus Talaeus (Omer Talon) in which Ramus himself had some hand, the *Institutiones ora-*

[35] For a history of the genre in England, see Rosemary Freeman, *English Emblem Books* (London: Chatto and Windus, 1948).

toriae (1545) and the *Rhetorica* (1548).[36] For reasons basically pedagogical rather than philosophical, Ramus was particularly annoyed by the confusion arising from the fact that from antiquity *inventio* and *iudicium* or *dispositio* had belonged to both logic and rhetoric. His efforts at reform were to be a continuation of those by Rudolph Agricola, who, as has been seen above, had by 1479 proposed a dialectic or logic cast in Ciceronian terms of *inventio* and *dispositio* but preempting to itself all invention, allowing no *loci* to rhetoric as such. This impoverishment of rhetoric in effect set Agricola against Aristotle, although he himself made no issue of being anti-Aristotelian as Ramus was to do.

Agricola had had some effect in England before Ramus' work had matured. His influence is detectable in the *Dialectica* (1545) of John Seton, fellow of St. John's College, Cambridge, and contemporary of Ascham and Cheke, although this work is in essence rather thoroughly Aristotelian. Seton, a Catholic, died in exile in 1567, but his doctrine was kept alive through 1639 by editions of his *Dialectica* equipped with the *Annotationes* which a later Johnian, Peter Carter, had first published in 1563 and which regularly accompanied Seton's text from 1572 on. Like Seton's Latin work, the first logic in English was also basically Aristotelian. This was *The Rule of Reason, Conteyning the Arte of Logique* (1551) by the same Thomas Wilson who was to publish the first full rhetoric in English. Wilson's *Rule of Reason* was plainly intended for the same audiences as his rhetoric, probably the

---

[36] See the entries for these works in Walter J. Ong, *Ramus and Talon Inventory* (Cambridge: Harvard University Press, 1958); for an account of Ramism, see the same author's *Ramus, Method, and the Decay of Dialogue* (Cambridge: Harvard University Press, 1958), as well as Howell, *Logic and Rhetoric in England, 1500–1700.* See also Chapter 7, n. 2 below.

Inns of Court. But neither Wilson's logic nor his rhetoric was so drastically English as the work of a third Johnian, Ralph Lever, whose *The Arte of Reason, rightly termed, Witcraft* (1573, though written about 1550) replaced Latinate terms with forthright Anglo-Saxon formations: witcraft (logic), speechcraft (rhetoric), saywhat (definition), saying (proposition), yeasay and naysay (affirmation and negation), and so on.

Wilson was less Aristotelian than Seton, maintaining with Agricola that dialectic and logic were synonymous. Ramus was more downright still. There was only one art of discourse, he explicitly and contentiously insisted. This was logic or dialectic, which governed all discourse whatsoever, from scientific reasoning through poetry, where the same logic used in mathematics itself was used, only spread rather thin. Logic taught (1) how to find arguments (*inventio*) and (2) how to arrange them (*dispositio* or *iudicium*). These two matters were never the business of rhetoric. The business of rhetoric was twofold: style, which meant for Ramus and his thousands of followers the use of tropes and figures; and delivery, to which Ramus, like most other textbook authors, gave perfunctory acknowledgment but little explicit attention. In logic and rhetoric both, as in many other subjects he wrote on, Ramus enforced an extreme schematic treatment: everything was divided by twos in the famous Ramist dichotomies. Logic had two parts, so did rhetoric. Each of these parts was subdivided into two further parts, each of these dichotomized again, and so on. All the tropes and figures were thus classified in groups of two.

Memory was dropped entirely. Ramus maintained that by using his analytic approach, which followed the "natural" order of things, recall was automatic. The same insistence on analysis gave a special turn to Ramus' use of the places of invention. Like a non-Ramist, to find "arguments" a Ramist went to the headings furnished by dialectic—genus, species, properties, whole, parts, conjugates, and so on—but he characteristically thought of these as implementing a "logical analysis" of a subject, enabling him to draw material out of the subject itself. The Ramist felt less need to rely on the collections of material culled from authors in commonplace books, for he thought of himself as securing his arguments from the "nature of things," with which his mind somehow came into direct contact. Thus he felt he would find arguments against disloyalty by simply understanding disloyalty and "analyzing" its genus, species, conjugates, and the rest, rather than by finding under the headings of the various "places" what had been said about it.

In the second part of logic, judgment or arrangement, Ramus gave attention not only to the proposition and to the syllogism, but to a new arrival on the logical scene, method. This was to become the great bone of contention in the battles between Ramists and "Aristotelians" (Ciceronians), which are referred to ironically by Polonius in *Hamlet* (ii, ii, 208) and by Viola in *Twelfth Night* (i, v, 244) and through countless Elizabethan authors besides Shakespeare. "Method" for Ramus prescribes how to organize larger units of discourse, always by going from general truths to particulars or "specials," except when the audience was unusually recalcitrant, in which case one could betray them into seeing truth by using "cryptic method." Cryptic method moved in reverse, presenting particulars first and proceeding thence to general truths. Ramus triggered the interest in method which came to a head in Descartes. But this "method" was adopted from classroom procedures and rhetorical manuals without any closely reasoned foundation in formal logic.[37] It vaunted orderly sequence, often superficially and sometimes implausibly conceived, over every other aspect of communication and thus directly furthered development of the "plain style." And in its resort to diagrams and other visual models to establish the idea of order—a procedure encouraged both by scholastic logic and by typography—it marked a significant movement away from the world of voice favored by the rhetorical tradition.

With its businesslike stress on method and analysis and its de-emphasis of rhetoric, Ramism appealed largely to the class of rising bourgeois who in England and on the Continent were inclined to embrace Calvinism. It found avid backers in the British Isles. Roger Ascham (1515–1568) in *The Scholemaster* (1570) censures Ramus' anti-Ciceronianism but rates Ramus and Talon apparently on a par with Quintilian. Around 1560 Ramism was espoused by Laurence Chaderton or Chatterton and Gabriel Harvey, both of Christ's College, Cambridge. Cambridge soon became a Ramist maelstrom with Christ's College at its center, although the earliest text of Ramus published in England, the *Dialecticae libri duo* (London, 1574) was edited by a Scot from the University of St. Andrews, Roland M'Kilwain or MacIlmaine (Makylmenaeus).

Editions of Ramus' and Talon's works on the Continent and in the British Isles number nearly 800 (some 1,100 if individual works in collected editions are counted separately); of the *Dialectic* alone over 260 editions have been identified, and of the *Rhetoric* over 160.[38] Between 1574 and 1600 fifteen

---

[37] Ong, *Ramus, Method, and the Decay of Dialogue*, pp. 225–269. See Chapter 7, n. 2 below.

[38] Ong, *Ramus and Talon Inventory;* since the publication of this work, several dozen other editions have come to the attention of the author. See Chapter 7, n. 2 below.

editions of the *Dialectic* and five of the *Rhetoric* had been published in England, but present library holdings make it clear that the Isles were heavily stocked also with copies of Continental printings.

Controversy between Ramists and Aristotelians rocked the Cambridge milieu particularly in the 1580's and 1590's, with Everard Digby and the Ramist William Temple, later provost of Trinity College, Dublin, and grandfather of Jonathan Swift's benefactor, at one another's throats in one controversy, and Thomas Nashe and the Ramist Gabriel Harvey in another. Ramism, however, never became academically respectable on a large scale within the universities. It had an attraction chiefly for schoolmasters or university graduates no longer in residence, and for many of the ambitious commercial class for whom an acquaintanceship with logic was often a status symbol more than a matter of serious scholarly concern. Ramist logic, sometimes epitomized, was often used as "petty logic" to supply the elementary notions of thought structure which training in composition demanded at the pre-university level. Dudley Fenner's condensation in English, published in 1584 and again in 1588 at Middelburg in the Low Countries, evinces even in its title the kind of *simpliste* appeal which Ramism could have: *The Artes of Logike and Rethorike Plainlie Set Foorth in the Englishe Tounge, Easie to Be Learned and Practised* .... At a somewhat higher level Abraham Fraunce's *The Lawiers Logike* (1588) adapts Ramist doctrine in English to legal training. Fraunce was a protégé of Sidney, who himself was to die in the arms of his own Ramist secretary, the older Temple mentioned above. Another related work of Fraunce's, *The Arcadian Rhetorike* (probably 1588), is remarkable for its examples from Sidney's *Arcadia* and from Greek, Latin, Italian, French, and Spanish authors as

well. The real companion piece to this rhetoric is not, however, *The Lawiers Logike* but a further Ramist work of
Fraunce's which remains in manuscript, *The Shepheardes
Logike*.[39] George Downham or Downame, Bishop of Derry,
treated Ramus in more scholarly fashion, publishing an edition of Ramus' *Dialectic* with a commentary, *Commentarii in
P. Rami . . . Dialecticam*, which appeared on the Continent
in six editions, beginning at Frankfort-on-the-Main in 1601,
before its one belated British publication (1669).

Other British writers on Ramist rhetoric, except for Charles
Butler, the grammar-school master and spelling reformer,
come after the Tudor period. Butler's *Rameae rhetoricae libri
duo* probably first appeared in 1593 (first extant printing
1597),[40] going subsequently through many editions, which
eventually drop all mention of Ramus or Talon from the title.
The other English Ramist logicians also belong, like the rhetoricians, to the later seventeenth century, where the most distinguished was of course John Milton, who numbered among
his works an adaption of Ramus' logic, published late (1672)
but apparently done in his younger years.

The sequel to the Ramist "reform" was the development
of various compromises between Ramism and Aristotelianism
or Ciceronianism, largely at the hands of the "Systematics,"
chiefly German polymaths such as the continental theologian-
encyclopedists Bartholomew Keckermann, Heizo Buscher,
and later Johann Heinrich Alsted, the medical writer and
occultist Andreas Libau (Libavius), and the philosophy pro-

---

[39] British Museum Addit. MS. 34361; now edited by Sister Mary
Martin McCormick, P.B.V.M., in a St. Louis University doctoral
dissertation (Ann Arbor, Mich.: University Microfilms, 1968)
scheduled for publication by the Renaissance English Text Society.
[40] For problems concerning dating, see Howell, p. 262.

fessor Clemens Timpler. In England, where the continental Systematics were well known, the chief compromiser between Ramist and earlier logic in Tudor times was Thomas Blundeville in *The Art of Logike* (1599, but written perhaps around 1575). Similar syncretist tendencies are observable in John Sanderson's *Institutionum dialecticarum libri quatuor* (Oxford, 1602). Ramist influence combines with other influence not only in logic but in the complementary works on stylistics as well. Here George Puttenham's *The Art of English Poesie* and Angel Day's letter-writing manual, *The English Secretorie* (1586), though essentially non-Ramist, show some Ramist proclivities in their handling of tropes and figures.

The Systematics did not, however, greatly deviate from the Ramist attitude toward literary performance. If they did not in every case reduce rhetoric to pure stylistics, they did, with the Ramists, consistently make logic the chief determinant of communication, and exploited the Ramist insistence on "method" to produce compendious treatments of any and all subjects foreshadowing modern encyclopedias. The Tudor period, however, ends before either the Ramists or the Systematics could have their full effect. At the opening of the seventeenth century rhetoric stood polarized: Ramist, and to some extent Systematic, doctrine minimized rhetorical display and fostered the plain style favored by many Puritans, while at the other pole a still flourishing Ciceronianism combined with patristic and medieval love of ornateness to produce the lushness met among many writers more or less of the episcopal party.

In sum, the writings on rhetoric during the Tudor age present us with a curiously mixed-up state of affairs. They are mostly in Latin and concerned with Latin expression, only rarely and indirectly adverting to the vernacular. Yet their effect on English is massive, and they merit being looked into by all students of the language. The English-language manuals mark important steps in the development of an English vocabulary adequate for learned expression, they provide samples, often fascinating, of particular turns of expression, and they inform us on the objectives, announced and actual, of Tudor writers of literature. For a show of particular grace, one might single out the works of Lever, Hoskins, Rainolde, and Puttenham mentioned above. Meres' *Palladis Tamia* and the other items in the Bodenham series together with Breton's *A Poste with a Madde Packet of Letters* are doubtless the most colorful pieces.

Works exemplifying the effects of rhetorical training have already been mentioned in limited number. It is not feasible to enlarge the list, nor is it necessary, for to the reader acquainted with the works on rhetoric themselves, almost any literary production of Tudor times is seen to be studded with rhetorical patterns, consciously cultivated, so thickly that to remove the conscious rhetoric would be to demolish the work. Professor C. S. Lewis is quite right in suggesting that our growing knowledge of Tudor views on rhetoric and enjoins the rewriting of literary history, although he himself refrained from the undertaking.

IV. CRITICISM AND LITERARY THEORY

In the Tudor period, criticism in the sense of elucidation of literary works on their own terms hardly existed. The equivalent of criticism must be sought largely in prefaces, disputes, commonplace collections, and theoretical works on education, rhetoric, or poetic which give occasional attention to individual works. Hence writings on "criticism" and literary theory can well be considered together.[41] Criticism and literary theory often became involved with specifically poetic matters, such as versification, but such matters will be touched on only lightly here since they belong rather with a history of poetry.

In literary theory the issue most debated was the worthwhileness of poetry, with some specific attention to the drama. Defenders of poetry had to deal with the general persuasion that poetry was only an inferior form of either rhetoric or logic, and that the imagination tended of itself to be anarchic and needed strict control.[42] In this and related debates, British thinking echoes Italian sources at least indirectly—Daniello, Minturno, Julius Caesar Scaliger, Castelvetro, Cinthio, Pigna—although feeling against Italians ran high and Italian romanticism was frowned on. English writers refer to English works often enough, but seldom at any length, and they take the measure of the English works by the ancient Latin and Greek writings. Vernacular creations "for the common people" such as metrical romances with no patent clas-

[41] The standard treatment of criticism is J. W. H. Atkins, *English Literary Criticism: The Renascence*, 2d ed. (London: Methuen and Co., 1951); see also Vernon Hall, Jr., *Renaissance Literary Criticism: A Study of Its Social Content* (New York: Columbia University Press, 1945), and Ruth Wallerstein, *Studies in Seventeenth-Century Poetics* (Madison: University of Wisconsin Press, 1950). Monographs and articles are legion. An important specialized study is O. B. Hardison, Jr., *The Enduring Monument*. The standard selection of texts is G. Gregory Smith, ed., *Elizabethan Critical Essays*, 2 vols. (Oxford: Clarendon Press, 1904), which is supplemented by the briefer *English Literary Criticism: The Renaissance*, ed. by O. B. Hardison, Jr. (New York: Appleton-Century-Crofts, 1963).

[42] See William Rossky, "Imagination in the English Renaissance: Psychology and Poetic," *Studies in the Renaissance*, v (1958), 49–73.

sical antecedents fare rather badly. Ideals of decorum, balance, and *ne quid nimis* associated with Latin learning are valiantly upheld in principle, although in expressing approval or disapproval of one or another author, writers willingly kick restraint to high heaven: Nashe is "the cockish challenger, the lewd scribler, the offal of corruptest mouthes, the draft of filthiest pennes, the bag-pudding of fooles"[43] to Gabriel Harvey, and Nashe writes of "Songs & Sonets, which euery rednose Fidler hath at his fingers end, and eury ignorant Ale Knight will breath foorth over the potte."[44] But *ne quid nimis* is an elastic rule, and scurrility, provided it had a good rhetorical surface, was seldom disqualified by it. The most excoriating critical ranting, however, is in content sometimes jejune and ingenuous enough, concerned with the far from surprising fact that there were too many poor writers at large.

Everywhere the cry was heard to bring English "into Art," which meant in effect to give it some of the status of the classical languages, with a grammar and all the rest—although Sidney in *The Defense of Poesie* maintained that English was too innately easy to need any such thing. Bringing English "into Art" multiplied objectionable "inkhorn terms" adapted from Latin and Greek, although this borrowing of words with established meanings was commonly easier and more natural and ultimately more successful than fabricating unheard-of English ones as Lever had attempted to do. Conversely, English prosody was often a matter of concern in its deviation from Latin and Greek.

From the first generation of humanists in England later generations inherited a sense of the high purpose of literature

[43] Harvey, *Pierce's Supererogation* (1593), in Smith, ed., II, 267.
[44] Nashe, *The Anatomie of Absurditie* (1589), in Smith, ed., I, 326–327.

and some sense of its mystery. In the circle of Sir Thomas More and Erasmus literature was not at all mere appliqué work on the surface of life—although it could be, half ironically, a "plaything," *nuga*. Literature was utterly central to an integral Christianity, which was conceived of as something positive, not squeamish, free of whining pseudo piety, but also morally uncompromising. Nevertheless, early British humanists have relatively little to show as criticism. The writings of William Grocyn, Thomas Linacre, John Colet, or the Spanish *émigré* Juan Luis Vives, are more prescriptive than interpretive and faced toward classics and Neo-Latin rather than toward the vernacular actuality around them. More's own great vernacular works belong to English literature as such rather than to criticism.

Related to the works of the early humanists are the courtesy books and other educational writings which are treated elsewhere in this history but must be mentioned here because they involve evaluations of literature. Notable among these is Sir Thomas Elyot's *The Boke Named the Governour* (1531), a treatise on the education of the aristocracy. This work stresses the use of selected Latin and Greek literature, especially epic, for moral formation and especially to instill courage. Elyot is aware that the youthfulness (seven to thirteen) of those reading literature demands discreet selection of passages and somewhat summary treatment of Ovid's *Metamorphoses*, but he defends Ovid and others against charges of lasciviousness by urging that they mirror life (rather than by falling back on medieval allegorical interpretation). Roger Ascham's *The Scholemaster*, which was published by Ascham's widow in 1570, two years after his death, and went through a great number of editions continuously into the nineteenth century, enjoyed greater popularity than Elyot's work. Ascham, something of a genial fuddy-duddy who sputters priggishly against the evils of everything Italian, is interested chiefly in Latin style. Toward his native language he expresses a benign, if not very active, interest, tinged with the then ordinary condescension, as when in a letter to Sir William Cecil in 1548 he urges that "even our language allows a man to write in it with beauty and elegance." The influence of courtesy books on critical attitudes persists even at the close of the Tudor age. Edward Blount's *Ars aulica or The Courtiers Arte* (1607), a translation of the work of Lorenzo Ducci, gives attention in some of its thirty-six chapters to proper and effective decorum for discourse with princes.

The polemic matrix of much criticism is again evident in the disputation on poetry, *De re poetica disputatio,* by Richard Willis (or Willes, Wills, Willey) appended together with other scholia to miscellany of Willis' Latin poems published in 1573. Willis defends poetry against the usual charges that it is worthless, demoralizing, lying, debased by association with the stage, and the product of insanity. His Latin answers to these charges are substantially those developed subsequently in English by Sir Philip Sidney in *The Defense of Poesie,* published posthumously in 1595 and called in another edition of the same year *An Apologie for Poetrie.* Although it makes no explicit mention of the erstwhile playwright (and perhaps actor) Stephen Gosson or of the latter's pretentiously Euphuistic attack on plays, poetry, and music in *The School of Abuse* (1579), Sidney's *Defense* disposes of Gosson's arguments by addressing itself to current enemies of poetry generally. It is more effective and graceful than the earlier answer to Gosson by Thomas Lodge (1579) which survives in print but without title page or date. In its organization the treatise is in the strictest sense a *Defense,* an oration of the judicial type, addressed consistently and persuasively to the reader as judge and featuring the proper sequence of clearly indicated parts, from exordium to peroration.[45]

By poetry Sidney clearly understands all writing involving fiction or make-believe, whether this be verse or prose, so that the *Defense* is really a skillful and moving plea for all literature. Following standard academic lines of thought which show the new Renaissance acquaintance with Aristotle's *Poetics*, Sidney's argument is spontaneous and engaging in its style. He treats the value of poetry positively in terms of its antiquity, universality, and high reputation from ancient times, adapting Aristotle's contrast between poets, historians, and philosophers to back his assertion that poets are the best teachers. Poetry exists for Horatian reasons, to teach and to delight (and, Sidney suggests later, to move), but all other purposes are subservient to teaching, understood very broadly. He thinks of poetry as coming from divine inspiration in one place, but shies away from this Platonic view in another. His pivotal neoplatonic theory makes poetry an "imitation" in the sense that it presents things as better than the reality known through history, and, since it serves a frankly pedagogic purpose, provides characters as models for the audience's imitation that are better than those in real life. To the latterly obvious philosophical and Christian objections which could be urged against this view (Christians are not exhorted to imitate poetic heroes and become little Oedipuses but on the contrary to imitate the historical Jesus Christ), Sidney, like his contemporaries generally, does not advert. But throughout, the strong grace of the *Defense* and its eclectic representative value for

[45] See Kenneth Orne Myrick, *Sir Philip Sidney as a Literary Craftsman*, Harvard Studies in English, Vol. XIV (Cambridge, Mass.: Harvard University Press, 1935), pp. 46–83.

sixteenth-century literary theory compensate for its lack of profundity or of original thinking.

Among other things, Sidney's *Defense* is representative of the state of Tudor dramatic criticism. Drama had formed a part of the early humanists' educational program, which proposed the study of Plautus and Terence (Greek drama came into its own in England only in the nineteenth century, although Ascham's *Scholemaster* does rate it above Latin) and fostered academic drama in Latin and even in the vernacular. But interest in dramatic theory or in analyzing dramatic effect in depth is almost nonexistent in Tudor England. The equivalent of dramatic criticism sometimes appears in connection with academic compositions, as in the Dedicatory Epistle to Nicholas Grimald's tragicomedy *Christus Redivivus* (c. 1540), which quite typically discusses drama in terms of rhetorical decorum, but surprisingly defends the mingling of comic and tragic elements. Seneca and the medieval tradition largely determined early sixteenth-century views of tragedy, which is taken to be concerned with the fall of illustrious persons and given a religious interpretation as simply showing the futility of earthly accomplishment—an interpretation which, in view of the complexity of Christian teaching on the temporal order appears patently impoverished today. The more psychological and philosophical Aristotelian theories of tragedy and correspondingly sophisticated theories of comedy known to the Italians [46] are virtually inoperative in sixteenth-century England. There are some faint adumbrations of the three uni-

---

[46] See Marvin T. Herrick, "The Theory of the Laughable in the Sixteenth Century," *Quarterly Journal of Speech*, xxv (1949), 1–16, and the same author's *Comic Theory in the Sixteenth Century*, Illinois Studies in Language and Literature, Vol. xxxiv, Nos. 1–2 (Urbana: University of Illinois Press, 1952).

ties doctrine concerning time, place, and action, but this doctrine becomes effective only in the seventeenth century.

The dispute about the moral value of the stage plays is, however, operative enough. With its Puritan overtones, it involves many more works than those of Gosson, Lodge, and Sidney: Philip Stubbes' *Anatomie of Abuses* (1583) and *Th'Overthrow of Stage Playes* (1599) by John Rainolds of Corpus Christi College, Oxford, are among those of some literary significance.

Although, as has been seen, the tendency to lump poetry with rhetoric at least for pedagogical—and often for theoretical—purposes was strong, there do exist various treatises which take poetry as an art more or less comparable to grammar, rhetoric, and dialectic or logic. These have their own implications in criticism and general literary theory. The suggestion that poetry was some sort of divine gift, the result of a divine indwelling or *enthousiasmos*, helped give it a status distinct from rhetoric even when pedagogical practice and concomitant theory made it only a branch of the latter. George Gascoigne's *Notes of Instruction Concerning the Making of Verse or Rime in English* in his *Posies* (1575), although itself set out at times in macaronic English laced with Latin—"They are *vix* good," or "These things are *trita et obvia*"—champions freedom of range in poetic invention, and is remarkable for circumstantial recommendations concerning meter and rhyme, which involve attempts to describe English phonetics. These attempts show how difficult initially a real description of the phenomena of English speech was: Gascoigne, for example, tries to handle English stresses in terms of grave, acute, and circumflex accents, which he calls *gravis, levis, et circumflexa*.

The relationship of English to Latin and Greek becomes a burning issue in the well-known debate over classical verse forms in English. In four letters written to Edmund Spenser in 1579 and 1580, and published in 1580, Gabriel Harvey discusses the possibility of nonrhymed English versifying based on classical meters, a mode of versifying suggested earlier by Ascham and enjoying for a while a surprising number of eager Thomas supporters and practioners, including Drant, Edward (later Sir Edward) Dyer, Fulke Greville, Richard Staneyhurst, Sidney (who introduces English poetry in classical meters into the *Arcadia*), and, later, the poet-musician Thomas Campion. This ill-advised effort, which interested but did not attract Spenser, was part of a larger program to "illustrate" the vernacular, paralleling in aims that of Ronsard and the Pléiade in France. The aim was to make the vernacular the equal of the classical languages (without, however, doing away with these, for no one seems to have sensed that their currency was transitory).

In *Observations in the Art of English Poesie* (1602) Campion shows some influence of music on metrical theory, particularly in his feeling for duration of syllables in English. Relying heavily on William Lily's Latin grammar (1527), Campion accounted for phenomena in English verse in terms of classical models, allowing about as much artificiality in his reading of English metrical lines as was called for in the accepted way of reading accented Latin words according to quantitative measure. Campion's "inconsistencies" are to be accounted for on these grounds. He recognized both accent and quantity in English, the latter established (quite artificially) by devices often borrowed from Latin, such as an inflexible rule that vowels followed by more than one consonant make inevitably long syllables.[47] Campion shows once more that its impossible for the age to treat English except as some kind of variant of the Latin which was studied in school.

William Webbe shows some enthusiasm for classical hexameters in English in his *Discourse of English Poetry* (1586), a rather exhaustive treatise more notable for lauding the "new poet," Spenser. But the most impressive contribution to poetic theory in Tudor times was the *Arte of English Poesie* (1589), now ascribed with reasonable certainty to George Puttenham (c. 1529–1590), Sir Thomas Elyot's nephew. Prosody, English poets, and even spoken English are treated in this rangy work. Puttenham's notion of the function of poetry in general accords with Sidney's while placing more emphasis than was usual on poetry as affording matter for sheer contemplation, for "solace and recreation."

With these treatises might also be included one by a Stuart which appeared during the Tudor period, *The Essayes of a Prentise in the Divine Art of Poesie* (1584) by the then King of Scotland and future King of England, James, at the time aged eighteen years. The work includes treatment of rhyme, style, "the three special ornaments to verse" (comparisons, epithets, proverbs), genres, and kinds of verse.

The criticism of Thomas Nashe (1567–1601) if it does not mark the end of these academic treatises, at least marks a new departure. Nashe himself is certainly one of the greatest stylists of the language, as a virtuoso probably unequaled until James Joyce. His criticism is part of his virtuosity; far from detached, it is impassioned and partisan. Nashe likes to indulge in witty raillery on grounds that are simultaneously

[47] Jane K. Fenyo, "Grammar and Music in Thomas Campion's *Observations in the Art of English Poesie*," *Studies in the Renaissance*, XVII (1970), 46–72.

personal and stylistic, as when he belabors academic rhetoricians in his Preface to Greene's *Menaphon* (1589):

Give me the man whose extemporall vaine in anie humor will excell our greatest Art-masters deliberate thought, whose inuention, quicker than his eye, will challenge the proudest Rethoritian to the contention of like perfection with like expedition. What is he amongst Students so simple that cannot bring forth (*tandem aliquando*) some or other thing singular, sleeping betwist every sentence? Was it not *Maros* xij. years toyle that so famed his xij. *Æneidos*: or *Peter Ramus* xvj. yeares paines that so praised his pettie Logique? Howe is it, then, our drowping wits should so wonder at an exquisite line that was his masters day labour? [48]

It is evident here that Nashe is not against rhetorical display, of which he himself is past master. Although he declares himself against Euphuism and fine writing of all sorts, he utterly scorns "pettie logique" and the plain style associated with Ramism, whose Puritan practitioners he pillories with more than ordinary relish. His own style is high and low simultaneously, for he ransacks the most idiomatic and colloquial veins of English, not, as has sometimes been said, to return to the idiom of "ordinary life"—no ordinary man could talk the way Nashe writes and live—but to mingle seemingly ordinary expression with the most self-conscious rhetoric which could possibly have been soaked up out of the formal education to which he had been subjected at Cambridge and before.

Nashe's critical observations are scattered through his writings, but are found particularly in his Preface to Robert Greene's *Menaphon* (1589), in the *Anatomie of Absurditie* (1589), and in *Strange News* (1592). His railing is not all

---

[48] Smith, ed., I, 309.

perverse or negative. He gives high marks to the early humanists such as Elyot, Cheke, and Ascham, as well as to Turberville, Phaer, and Golding, but he takes a dim view of the Arthurian romances. Not only Chaucer, but Gower and Lydgate as well, were true ornaments of the language in the past, and Spenser, as well as Peele and Warner, of Nashe's own age.

Other critical writing beginning with the last decade of the century addresses itself to old problems in a somewhat more realistic manner. Such is the case in the Preface with which Sir John Harington equipped his translation (1591) of *Orlando furioso*, in Thomas Campion's *Observations in the Art of English Poesy* (1602), a condemnation of rhyme by a quondam most effective rhyming poet, and in Samuel Daniel's historically sensitive answer to Campion and the quantifying English versifiers in his *Defence of Rhyme* (1603). Joseph Hall attacks current poetry, and some earlier poetry as well, in his *Virgidemiarum or Toothless Satires* (1597–1598), chiefly on moralistic grounds and without adding much to previous discussion.

Francis Meres' *Palladis Tamia* (1598), mentioned earlier, is a commonplace collection with significant critical value. Its section entitled "A Comparative Discourse of Our English Poets with the Greeke, Latine, and Italian Poets" espouses the accepted Plutarchian doctrine that poetry was a propaedeutic to philosophy. It also yields, besides critical evaluation of English poets, biographical data which is otherwise unavailable, but which is also suspect since it is tailored to fit Meres' rhetorical equations between ancients and contemporaries: Meres sets out to match items in the lives of classical poets with corresponding items in English poets' careers. In assembling his materials he works well within the omnipresent rhetorical tradition, since one of his chief sources for information concerning classical antiquity is itself another printed commonplace book, the Latin *Officina* (1520) got up by the sixteenth-century French scholar Ioannes Ravisius Textor for schoolboys. Rhetoric is here producing not only criticism but biography of a sort by feeding on itself.

V. THE CLOSE OF THE TUDOR AGE

The death of Queen Elizabeth in 1603 marks the end of the Tudor age but not the end of the rhetorical tradition. Changes in this tradition had indeed come about during the Tudor reigns, affecting both rhetorical works themselves and the literature conditioned by the teaching of rhetoric. The highly prescriptive, academically oriented works on rhetoric of the late fifteenth and early sixteenth centuries were codified and somewhat deadened through codification, often effected by Ramists. But rhetorical practice kept its earlier vigor and gained suppleness as it was worked out in increasingly self-sufficient vernacular tradition. The literature affected by the teaching of rhetoric—which was virtually all Tudor literature —bore everywhere the mark of rhetorical flair and rhetorical control. But there were subtle changes in the modes of rhetorical operation. The use of the "places" or commonplaces, which fostered lushness and often profundity in style, gnomic "strong lines," and weighty sententiousness, was in many quarters sapped imperceptibly in the late sixteenth century by a newly exclusive passion for "logical analysis" among Ramists and others. The change would not have its full effect until Dryden and after, but the older preoccupation with logic as an instrument of discourse (rather than of private thought), with its accompanying sententious rhetoric, was giving way to interest in a logic of private inquiry and a more tenuous rhetoric (later to be supplanted by a rhetoric of sentiment, passion, and "feeling").

Meanwhile, literary criticism, such as it was, remained largely subordinated to rhetoric. Writing about literature was largely a matter of defending poetry against its accusers, of raising the literary status of English (without, however, the slightest thought of lowering that of Latin and Greek), of propounding one or another more or less rhetorical principle (against stylistic excesses, for or against rhyme), or of more or less scattered remarks on individual works. The place of poetic improved: it began the sixteenth century pretty much as adjunct of rhetoric, but by the century's close achieved a modest independence, at least outside the classroom.

Francis Bacon may serve as a figure with which to close, since he is highly representative of the state of rhetorical affairs at the end of the Tudor age, to which his most active years belong. Bacon's progam for remaking the intellectual world shows not only how the rhetorical way of life was being modified, accommodated to a designedly exploratory and experimental approach to reality, but also how ambivalent such accommodation still had to be. Bacon's great educational work, *The Advancement of Learning* (1605), remains in the midstream of the rhetorical tradition, for it is organized as a classical oration and "proved" by examples. In this work he makes rhetoric one of the three arts devoted to the "tradition" or delivery of understanding, the other two being grammar and "method." But Bacon's "understanding" itself consists, he tells us, of invention, judgment, memory, and elocution or tradition, which last includes style.[49] Here, at the heart of Bacon's

[49] See Maurice B. McNamee, "Literary Decorum in Francis Bacon," *Saint Louis University Studies*, Series A, Humanities, Vol. I, No. 3 (March 1950), esp. p. 48 and the diagram on p. 9.

notion of intellect itself are the five parts of Ciceronian rhetoric again! *Plus ça change, plus c'est la même chose.* Bacon's scheme to provide a new organization different from the older rhetorical one is itself dependent for its basic construction on the older rhetorical view.

In other ways, too, Bacon is inextricably tangled in the rhetorical tradition. His *Essays*, as has been seen, are essentially collections of gnomic commonplaces. He understood poetry in much the same way as many early humanists. It is a play of fancy or imagination, not to be taken too seriously, "feigned history" but with claims less serious than Sidney allowed. Of the "deeper meaning" or allegorical sense of poetry, Bacon was aware, but sceptical. Logic and rhetoric are "the gravest of sciences, being the art of arts, the one for judgment, the other for ornament," he writes in *The Advancement of Learning* (II, Ded., 12). What respect for poetry and for the fictional in general he preserves is kept alive for him by their association with rhetoric and by the common respect for rhetoric which was the heritage of his age. Bacon's voice was indeed a new one in many ways, but it spoke to the opening seventeenth century with the unmistakable—if not always unmistaken—accent of the rhetorical past.

4 { Memory as Art

I

To the unlearned, a recent book by Frances A. Yates, *The Art of Memory*,[1] would appear pedestrian enough in its aims and promise. It undertakes to trace schemes for implementing memory from the ancient Greeks through Cicero and Quintilian, the Middle Ages, and the Renaissance into the Cartesian era. The book is in fact far from pedestrian. It is not only a model of erudition but also certainly a seminal book for vast areas of intellectual and cultural history. More particularly, it opens many new vistas within the rhetorical tradition as this is examined in the present volume, so that it merits special attention here.

With our modern devices for knowledge storage and retrieval (print, encyclopedias, indexes and concordances, photographs, sound tapes, video tapes, computers, and all the rest), it is easy to believe that the studied cultivation of memory was always as sterile as it now appears. Miss Yates shows how far this is from the truth. Earlier ages needed memory desperately. They structured knowledge for memory with the

[1] Chicago: University of Chicago Press; London: Routledge and Kegan Paul, 1966.

result that memory affected the entire intellectual enterprise and the very nature of what man knew or could know.

As her title makes clear, Miss Yates is concerned not with the more purely oral memory skills which have their roots in prehistory and which Albert B. Lord has reported on in *The Singer of Tales* [2] but rather with the memory "arts" or systems or sciences. These were initially part of the art of rhetoric, and like this art itself as a formally stated body of knowledge came into being when a culture still experiencing knowledge in dominantly oral forms ("rhetoric" refers primarily to oratory) undertook to organize knowledge with the kind of reflectiveness which writing makes possible.

The principal basis for all the memory arts, Miss Yates finds, is the imaginative organization of space and of spatially arranged imagery. One visualizes some kind of structure in space made up of recognizable parts standing in fixed relations to one another and then associates what one wishes to memorize with the various parts of the structure. From classical antiquity this has entailed the actual committing to memory of the interiors of entire buildings—and not merely imaginary ones, but real ones, where the parts are inalterably fixed— such as specific temples or law courts or cathedrals, column after column, alcove after alcove (each to be imagined, incidentally, as "well lighted" in the mind's eye). One can then commit to memory, say, a list of the Roman emperors or of the various parts of the virtues of prudence, by associating each in succession with one of the envisioned fixed parts of the building. To facilitate this association, each part of the building is often permanently equipped in the visual imagination with a highly symbolic figure (a sign of the zodiac, a goddess,

[2] Harvard Studies in Comparative Literature, Vol. 24 (Cambridge, Mass.: Harvard University Press, 1960).

and so on) or a combination of such figures, sometimes spectacularly deformed or frozen in a weird or bizarre action to fascinate the imagination. Such symbolically "loaded" figures make the part of the building they inhabit more exploitable, since almost anything or indeed whole clusters of data can in one way or another be readily associated with any of them at will. The individual columns or alcoves or other structural elements thus utilized were known as memory "places" or "rooms," and the entire display—at least by the sixteenth century—was often called a memory "theatre." [3]

One sees here of course the association technique which we all still use in one way or another to expedite recall, but a technique cold-bloodedly organized into a baroque complexity which dizzies the modern mind. For a culture fostering this traffic in images, iconography is not a diversion but a technique needed for managing knowledge and a way of life. Some paragons of mnemonic *sprezzatura* are said to have carried tens of thousands of these places (several whole buildings full) in their heads as a regular part of their thinking equipment. To such persons travel was less culturally than mnemonically enriching, for it could provide whole concatenations of new buildings for memory structures: "To the

[3] For a study of the actual effectiveness of these memory techniques, see Gordon H. Bower, "Analysis of a Mnemonic Device," *American Scientist*, LVIII (1970), 496–510. Bower reports that this memory procedure treated by Miss Yates has been shown by psychological experimentation to be highly effective in all its salient features (enumerated as nine) except one, namely the use of unusual, bizarre imagery, which thus far has tested out as inconsequential. Even here, however, one might suggest that unusual or bizarre imagery—which Bower points out is difficult to define—may in fact mean high-definition concrete imagery. Bower does not go into this possibility, but it suggests itself in connection with what he has on p. 503 about "imaginal associations and their properties."

last of the Roman places you may add the first of the Parisian places," Giordano Bruno counsels (quoted by Miss Yates, p. 311).

In tracing the history of memory arts, Miss Yates shows how the memory systems of the Greek and Roman rhetoricians (already architecturalized in "places" or "rooms") migrated in the Middle Ages from rhetoric to ethics (which interpreted memory as part of the virtue of prudence) and acquired devotional overtones, coming to be associated, not unwarrantedly, with recollection and prayer. With the advent of movable alphabetic type, memory systems divide more clearly than ever before into two classes: the rational (such as Quintilian's and Aquinas' earlier mnemonics and Ramus' post-typographical system), which print will largely exterminate by performing more effectively their essential functions of providing fixed spatial distribution to implement recall, and the magic systems (such as we find, to a degree, in Raymond Lull during earlier times, and later, spectacularly elaborated beyond all belief, in Giulio Camillo and Bruno).

Magic memory systems built on the microcosm-macrocosm concept. Since the microcosm and the macrocosm were cued into one another so that grasp of one automatically involved grasp of the other, some minds could not avoid the inference that a memory structure (that is, an elaborate assemblage of highly symbolic figures distributed strategically in structured space) might well be devised which would pick up the structure common to microcosm and macrocosm, that is, the structure of being itself. The trick would be to arrange the symbols in your memory theatre not arbitrarily but in correspondence with the true forces constituting actuality. The idea of such a system of memory which would be the key to all knowledge haunted Lull and many others and utterly obsessed Camillo and Bruno. It also entered not a little even into Ramus and Descartes, although, as Miss Yates shows, these last two declared allegiance to nonmagic, rational memory. Ramus' declaration was deceptive, since he claimed for his dichotomized "analyses" of everything under the sun the same validity (things really were "this way") that magic mnemotechnicians claimed for their systems. Because Ramus' cues were printed words deployed in abstract geometrical space rather than iconographic images set in architectural space, his system could be made to appear "rational" rather than magic.

Miss Yates opens vistas into areas of intellectual history so vast and diverse as to leave the reader breathless. Modestly and tentatively, but always circumstantially, she shows how the use of memory places tied in more or less directly from antiquity with the rhetorical and dialectical places (*topoi, loci, loci communes,* commonplaces) and rhetorical and dialectical invention, with medieval superclassifications of virtues and vices, with the topography of Dante's *Divina commedia* (one might add, with the plan of Spenser's *Faerie Queene*), with the concept of "theatres" of knowledge and "keys" to knowledge (one might add other images or concepts under which the Middle Ages and the Renaissance subsumed their learning, *speculum, tabula, synopsis, methodus, encyclopedia,* and many more), with the development of iconography and its quick deterioration upon the appearance of alphabetic typography (one might add in particular the weird iconography of Pieter Breughel the Elder, for Miss Yates suggests how memory images encouraged traffic in the grotesque and served also to perpetuate an elaborate pagan heritage through the Middle Ages), with the history of Dominican thought, and all but certainly with the origins of Rosicrucianism and of speculative Freemasonry. One thinks also of possible influence of memory systems on the long-delayed development of movable alphabetic type (Lull had introduced moving letter-symbols into memory) and on the "preludes" to meditation in the Spiritual Exercises of St. Ignatius Loyola and hence on Professor Louis Martz's "poetry of meditation." If all this seems a large order, further study and reflection will certainly warrant making it larger still.

Miss Yates's own conclusions are in fact as sober and judicious as they are startling. Often she might have appealed to still vaster stores of evidence, had she time or space or need. For example, her lengthy discussion of memory "theatres" generally and her few pages on possible connection between memory systems and Shakespeare's Globe (cf. what was said above here of the microcosm, the little "world" or little "globe" and the macrocosm or "great world," "all the world's a stage," etc.) could be elaborated by adducing the late sixteenth-century rash of "theatre" titles for printed commonplace books or other collections of materials of "invention" used by dramatists and other writers. Such would be Robert Allot's *Wit's Theatre of the Little World* (1599), Jean Bodin's *Universae naturae theatrum* (1596, etc.), Theodor Zwinger's huge *Theatrum humanae vitae* (1565, etc.), and so on and on and on—I have myself a collection, assembled quite casually, of dozens of such titles, including theatres of poets, of plants, of history, of calamities, of insects, of machines, of truth and justice, of war and peace. Thomas Heywood's *An Apologie for Actors* (1612) has much material on the theatre as a symbol, too, and Bacon's concept of "idols of the theatre" is probably by no means irrelevant to magic memory systems.

There are certain perspectives which Miss Yates has completely avoided, and probably with reason and wisdom, for what she has been able to exhibit is so many-faceted and rich that further suggestions might only bewilder and confuse the mind. Nevertheless, it is stimulating to think where Miss Yates's scholarship may enable future scholarship to go. There is no doubt that the tendency to develop magic memory systems is related to the fact that memory spaces are typically conceived of as something the individual is *inside of* (a building) or, if visual rather than tactile perception is maximized, as something one peers *into* (a theatre). In either case the space and its contents are felt in one way or another as an extension of oneself. As the macrocosm (the "outside" world) is an extension of the microcosm (myself inside my skin, with special reference to my consciousness), so the memory structures mediate between microcosm and macrocosm and thereby inevitably promise (to a mind such as Bruno's) to be *the* link between the two. Phenomenological analysis would be enlightening here.

II

Study of the art of memory needs further to be connected explicitly with the work of Milman Parry, Lord, and Eric Havelock on oral memory. This work on oral memory would seem to suggest some reasons why Cicero, Martianus Capella, and other memory artists minimize "memory for words" in favor of "memory for things"—their oral mnemonic cultural background was thematic and formulaic rather than verbatim and in part a relict of mankind's original oral, prechirographic culture. Finally, Miss Yates says nothing about the relationship of memory to the Hebraeo-Christian religious tradition, in whose ambit many of the developments she studies took place. This religious tradition differs from all others in being the religion of historical memory, an event-religion, built on remembrance of what God did for His Chosen People and culminating for Christians when the great memory feast of the Passover is transformed and focused for all time in the Last Supper, where Jesus says, "Do this in memory of me." This fact provides positive historical background for much of the Christian attention to memory as well as for the suspicion under which most memory "arts" have ultimately tended to fall among Christians. For in their preoccupation with space and frozen symbols (and with cyclic models of actuality), these arts appear strongly antihistoric—to a limited degree in their rational forms, but massively in their cosmographical magic forms, such as that of Bruno, whom Miss Yates shows here as elsewhere to be the most explicitly anti-Christian of the magic mnemotechnicians.

Miss Yates's study of Ramism as a memory system is fascinating and beautifully executed. She brilliantly ties in some of my own work,[4] as well as that of many others such as Paolo Rossi and W. S. Howell, with her own further discoveries. I wish I had thought of her fine expression (pp. 234–235, etc.), the "inner iconoclasm" enforced by Ramus when he replaced iconographic mnemonics with his "logical analysis," for such "inner iconoclasm" makes clearer than ever before the

---

[4] At one point, however (pp. 233–234), I believe Miss Yates misconstrues me. In *Ramus, Method, and the Decay of Dialogue* (Cambridge, Mass.: Harvard University Press, 1958) I do not maintain that spatial memorization was an entirely new development introduced by the printed book but rather (my p. 79) that with print the possibility of exactly reproducing complicated spatial arrangements of words or other symbols (far harder to transcribe in manuscript than a text) was spectacularly improved, since the most complicated display, once it is locked up in a forme, is just as easy to multiply typographically as straight discursive material. Dichotomized tables such as those of the Ramists can of course be seen in manuscripts, but print gave them ready currency on paper and in the mind to a degree never remotely approximated in a manuscript culture.

alliance between Ramist epistemology and the "outer icon-oclasm" of Calvinism. Microcosm and macrocosm again.

Miss Yates has added a tenth and very important dispute, the Bruno-Dicson-Perkins dispute, to the other nine major Ramist controversies which I document in my *Ramus and Talon Inventory* [5] and in the process has identified several additional persons whose names should be added to my list of Ramists, Anti-Ramists, and Semi-Ramists in the same book. Although her present work does not take up these other controversies or even mention the *Inventory*, it opens the way magnificently to further exploration of material which the *Inventory* catalogs and to a deeper understanding of the total significance of many intellectual developments well past Descartes and Leibnitz, particularly in connection with the rhetorical developments which are the concern of this present book.

[5] Cambridge, Mass.: Harvard University Press, 1958.

5 Latin Language Study as a Renaissance Puberty Rite

I

The reasons why any particular society follows the educational curriculum which it does follow are always exceedingly complex. Because, in being a preparation for the future, it is inevitably a communication of what is available from past experience, education is always primarily a traffic in this experience and only secondarily a matter of theory. The theories concerning the handling of this experience never quite compass the actuality and totality of the experience itself. They are generally rationalizations, afterthoughts, however valuable or venturesome they may be under certain of their aspects.

This is true of education today, and it was true of education during the Renaissance. To be sure, no one bristled with educational theory more than Renaissance man. He had often very definite ideas as to what should be done to produce the proper sort of courtier or soldier or scholar or even ordinary bourgeois. Yet his theories never quite came to grips with everything in the pedagogical heritage.

Such is the case particularly with the Renaissance teaching of Latin. Depending on how much or how little he was influenced by the humanist tradition, the Renaissance educator thought of Latin as bringing students into contact with the ancients, whom Erasmus had declared to be the sources of practically all human knowledge. But quite independently of this theory, the Renaissance educator was also compelled to teach Latin because the books in use, contemporary as well as ancient, were books written in Latin or translated into Latin. These included the books on language and literature, on "philosophy" (which meant, besides logic, physics and what we might best style general science, inextricably interwoven with psychology and snatches of metaphysics), books on medicine, law, and theology, not to mention books on military science, botany, alchemy, physiognomy, geography, and on every other more or less learned subject. This unacknowledged reason for teaching Latin—the fact that to establish and maintain contact with academic and scientific thought pupils had to be able to read it, write it, and think in it—in actuality outweighed all other reasons through the Renaissance period.

This fact also made the teaching of Latin inevitably different from the teaching of Greek or Hebrew, although in the upper reaches of humanist theory these two languages were recommended for study at least as urgently as Latin. The humanists' own encomia of Greek and Hebrew, from Erasmus to Ramus and beyond, together with institutions such as the nominally trilingual colleges of Louvain, Salamanca, and Alcalá, attest the existence of this equal theoretical esteem for Greek and Hebrew and of a desire to implement theory. Yet the Renaissance Greek and Hebrew are sorry failures compared to Renaissance Latin. They produce no perceptible literature at all. When someone in Western Europe, such as Poliziano, writes epigrams in Greek, this achievement—or, perhaps better, this tour de force—is completely overshadowed by the bulk of the same author's Latin writings. And the currency of Hebrew, outside specifically Jewish circles, never even remotely approximated the extremely limited currency of Greek.

As compared with the other "classical" languages, the Latin of the time thus has a viability which is not at all accounted for by humanist theories and attitudes regarding the ancient world. To understand the practices of the Renaissance educator we must look beneath his theories for other things, for the psychological and social drives, for the complex of psychological and social stresses and strains and compulsions to which he is heir and which register in his performance. Here I should like to single out for attention some patterns in the Renaissance teaching of Latin which manifest certain of these complexes and suggest that the Renaissance teaching of Latin involved a survival, or an echo, devious and vague but unmistakably real, of what anthropologists, treating of more primitive peoples, call puberty rites.

II

There is a vast literature on puberty rites, but a brief summary of some of their features will suffice to make the necessary points about Renaissance Latin language teaching and study.[1] Peoples of simpler culture have, almost universally, a

[1] See Hutton Webster, *Primitive Secret Societies*, 2d ed. rev. (New York: Macmillan, 1932), pp. 20–73; A. E. Jensen, *Beschneidung und Reifezeremonien bei Naturvölkern* (Stuttgart: Strecker und Schröder, 1933); Arnold van Gennep, *Les Rites de passage* (Paris: E. Noury, 1909), pp. 93–164; Goblet d'Alviella, "Initiation (Introductory and Primitive)," *Encyclopedia of Religion and Ethics*, ed. by James Hastings (Edinburgh: T. and T. Clark, 1914), VII, 314–319; Charles W. M.

systematic ceremonial induction of adolescent youths into full participation in tribal, as opposed to family and clan, life. These rites have certain more or less well-defined characteristics. The individual being initiated is established in a special "marginal environment" so that the puberty rites are accurately styled by Arnold van Gennep *rites de passage*. The past of the individual is considered to be cut off, and certain excesses—license, theft, arson, violence—are often allowed. This sense of a break from the past may be dramatized, for example, when the home of the boy destined to undergo the rites is invaded by those who are to initiate him and who tear him forcibly from the company of the women, and sometimes physically from the very arms of his mother, who puts up a show of resistance, half conventional and half real. During the period of initiation the boy is made to do many things that are hard, often, it appears, simply because they are hard. In some cases, special taboos are enforced. Thus a boy may not touch his own body anywhere with his hands, but only with a stick—if, for example, he wishes to scratch himself. An atmosphere of continual excitement is cultivated to enlist the youth's interest. As Nathan Miller states it, "Put on edge through ingenious torments, sleeplessness, and nerve-racking frights, the candidate becomes keenly sensitive to the power of

---

Hart, "Contrasts between Prepubertal and Postpubertal Education," in *Education and Anthropolgy*, ed. by George D. Spindler (Stanford, Cal.; Stanford University Press, 1955), pp. 127–145, and the discussion by various persons which follows, pp. 145–162, etc. See also Hutton Webster, *Taboo, a Sociological Study* (Stanford, Cal.: Stanford University Press, 1942), p. 109n. For a brilliant, if somewhat precious and erratic, extrapolation on a theme relevant to puberty rites, see José Ortega y Gasset, "The Sportive Origin of the State," chap. i in his *Toward a Philosophy of History* (New York: W. W. Norton, 1941).

his preceptors and indelible, life-long impressions are made." [2] The role of the preceptor is important, for the puberty rites are essentially didactic, "the chief vehicle to link generations in the transmission of the culture complex." [3] The climax is reached in the inculcation of lessons in tribal law, morality, and tradition. Bushman puberty rites, for example, feature religious dances in which animal masquerades predominate. Over all these presides the belief that the youths must be made by their preceptors to assimilate their lessons the hard way.

Among the Bechuans, the boys in a state of nudity engage in a dance during which the men of the village pummel them with long, whip-like rods while asking such questions as, "Will you guard the chief well?" or "Will you herd the cattle well?"

Needless to say, because they incorporate youth into the tribe rather than into the family, puberty rites involve sexual segregation. The rites for boys are for boys alone. There are comparable rites for girls, but we are concerned with the boys alone here, for, generally speaking, it is boys alone who are taught in Renaissance schools, or who are given a systematic formal education. There are some few rare references to school education for girls in the Renaissance, [4] but commonly

[2] Nathan Miller, "Initiation," *Encyclopedia of the Social Sciences,* ed. by Edwin R. A. Seligman and Alvin Johnson, VIII (New York: Macmillan, 1937), 49–50.

[3] *Ibid.*

[4] See Norman Wood, *The Reformation and English Education* (London: George Routledge and Sons, 1931), pp. 77–78, 181–182; cf. *ibid.,* pp. 3–7, 28, 159 ff. Cf. Carroll Camden, *The Elizabethan Woman* (New York and London: Elsevier Press, 1952), pp. 44–50; Ruth Kelso, *Doctrine for the Lady of the Renaissance* (Urbana: University of Illinois Press, 1956), pp. 58–77, esp. pp. 66, 68, 73 (girls' reading to be in the vernacular); A. F. Leach, *The Schools of Medieval England,* 2d ed. (London: Methuen, 1916), pp. 88–89.

the girls of the time learned what reading and writing they learned outside the schoolroom, in the privacy of the home. The patterns which put in their appearance here are due to the fact that puberty rites are (or were) more urgent for boys than for girls. Girls normally moved directly from their family of origin to their own family of marriage. Young males, however, needed an intermediate stage, away from their family of origin but not yet in a family of their own making, a period to attune themselves to extrafamial existence, which includes among other things, the ceremonial fighting marking the behavior not only of human males but also of the males of most, if not all, other animal species.

Puberty rites are thus ceremonial inductions or initiations of the youth into extrafamilial life which involve a sense of break with the past (a "marginal environment") together with segregation from the family and from those of the other sex, and chastisement under the direction of elders for didactic purposes. Any system of schooling which separates boys from girls and is carried on outside the home will, of course, to a greater or lesser extent involve all these things, with the possible exception of chastisement. And it is common knowledge that in the school from early Greek and Roman times well through the Renaissance, chastisement was definitely involved. Thus any formal education through the Renaissance might well tend to activate the complex of behavior on the part of preceptor and student characteristic of puberty rites, and, indeed, almost any conceivable educational procedure outside the home will to some extent do the same thing. The coincidence of various forms of hazing with schooling everywhere is ample evidence of this fact.

The point of this chapter is that, although there are these general connections between school education and puberty rites, in Renaissance times (and to a great extent through the

Middle Ages, as these led into the Renaissance) the status of Latin encouraged in a special way the development of a puberty-rite setting and puberty-rite attitudes in the educational activity of the time, and, incidentally, that traces of these attitudes can be found in the few places where Latin lingers on the educational scene today. This is thus an attempt to explore certain of the complex social implications of Latin as a learned language.

These social implications were large. For when Latin passed out of vernacular usage, a sharp distinction was set up in society between those who knew it and those who did not. The conditions for a "marginal environment" were present. Moreover, the marginal environment was one between the family (which as such used a language other than Latin) and an extrafamilial world of learning (which used Latin). The fact that the marginal environment was primarily a linguistic one only heightened the initiatory aspects of the situation, for the learning of secret meanings and means of communication is a common feature of initiatory rites. It is through ability to communcate that man achieves a sense of belonging.

III

The cleavage between the vernacular world and the Latin world did not coincide with the division between literacy and illiteracy, but it did coincide with the division between family life and a certain type of extrafamilial life and with a division between a world in which women had some say and an almost exclusively male world. Literacy could be, and frequently was, acquired at home, often under the tutorship of women in the family. But this literacy, which can be distinguished from "learning," was commonly restricted to ability to read and write the vernacular. Schools often prescribed that a boy be able to read and write at least the alphabet as a requirement for admission,[5] for it was the business of the school proper to teach, not reading and writing, but the Latin language. This medieval and Renaissance situation still registers in our vocabulary, where elementary schools are called not reading and writing schools but grammar schools—the "grammar" here referring historically to the teaching of beginners' Latin, which was Latin grammar. This situation meant that, in general, girls, who were educated at home and not in schools, could be quite literate without having any effective direct access at all to the learned world, which was a Latin-writing, Latin-speaking, and even Latin-thinking world. There were only occasional exceptions such as in the Middle Ages the abbess Hroswitha or in the Renaissance the women mentioned in chapter 3, Lady Jane Grey, Margaret More, and Queen Elizabeth—or perhaps Shakespeare's Portia —to ruffle the masculine sense of self-sufficiency. Because their sex was so committed to the vernacular, women could become—as Raymond W. Chambers and others have shown they did become—both a major audience for English literature and some of its chief patrons.

Closed to girls and to women, the schools, including the universities with their own "schools" (*scholae* or classrooms), were male rendezvous strongly reminiscent of male clubhouses in primitive societies. At the top of the academic structure, in the universities, with the exception of doctors of medicine, who at Paris, for example, were allowed after the year 1452 to marry and continue as regents,[6] teachers through the Middle Ages and the Renaissance (and in many universities

[5] For example, the statutes of Canterbury School and St. Paul's School so prescribed in the sixteenth century (Wood, p. 3).

[6] Hastings Rashdall, *The Universities of Europe in the Middle Ages,* new ed. by F. M. Powicke and A. B. Emden (Oxford: Clarendon Press, 1936), I, 446.

much later than the Renaissance) were obliged to remain un-
married so long as they continued active teaching, and this
whether or not they were clerics in the ecclesiastical sense at
all. Peter Ramus, his erstwhile secretary and biographer tells
us, often spoke about marriage but decided to forego it because if he had married he should have had to resign as principal of the Collège de Presles and as a university master.[7]

Somewhat mysterious in its origins and implications, this
specially closed environment of the universities was maintained by a long apprenticeship or bachelorship (common to
medieval guilds of all sorts) terminating in the *inceptio* or
inaugural act of teaching. Today the *inceptio* is echoed really
but faintly in the now wholesale ceremony known by the
mystifying name of commencement, and words surviving on
university diplomas, *periculo facto* or "having undergone the
(requisite) danger or trial," bear witness to the old feeling
that education was an initiation. But in helping to maintain
the closed male environment the psychological role of Latin
should not be underestimated. It was the language of those
on the "inside," and thus learning Latin at even an infra-uni-
versity level was the first step toward initiation into the closed
world. Earlier groups of learned men—the Academy, the
Stoa, the schools at Alexandria—seem never to have achieved
the close-knit, jealously guarded internal organization of the
university. It seems not irrelevant that they did not have a
secret language to nourish their *esprit de corps*.

The humanists, who for various reasons often thought in
terms of a home-centered system of education, were hard put
to find a substitute for the closed male environment of the

---

[7] Nicolas de Nancel (Nancelius), *Petri Rami . . . vita* (Paris,
1599), in Nancel, *Declamationum liber . . . : addita est P. Rami . . .
vita . . .* (Paris, 1600), pp. 58–59.

school. One recalls the embarrassment of Erasmus, More, and Ascham when they speak of rearing a youngster in a home where he would hear the proper use of language at an early age. These educators of course mean the proper use of the Latin language—they are giving no thought to the vernacular at all—and they are visibly nonplused by the fact that this means that the youngster will be in the company of women, since it had proved impossible, even for the humanists, to have homes without women in them. Roger Ascham speaks rather glibly of the way in which Tiberius and Caius Gracchus were brought up in the home of their mother Cornelia, where "the dailie use of speaking were the best and readiest waie to learne the Latin tong." [8] But Ascham here is not merely resorting to humanist piety by preferring a classical example to a current one. He is bowing before historical fact. There were no current examples, and could be none. We can be sure that no English mothers cooed to their children in the language native to the mother of the Gracchi, and thus we find Sir Thomas Elyot more realistically stating, "After that a childe is come to seven years of age, I holde it expedient that he be taken from the company of women, savynge that he may have, one yere, or two at the most, an auncient and sad matrone attending on hym in his chamber." [9]

Sir Thomas pleads here that this arrangement will remove the child from temptations against chastity. However, al-

[8] Roger Ascham, *The Scholemaster*, ed. by Edward Arber, English Reprints, [No. 23] (London, 1870), p. 28. Subsequent references here are all to this edition. See also Lawrence V. Ryan's scholarly modern-spelling edition of the same work (Ithaca, N.Y.: Cornell University Press for the Folger Shakespeare Library, 1967).

[9] Sir Thomas Elyot, *The Boke Named the Governour*, ed. by Henry Herbert Stephen Croft, 2 vols. (London: Kegan Paul, Trench, 1883), I, 35 (Book I, chap. vi).

though this reason might conceivably at times apply with reference to servant girls or other attendants, the separation of the child from his own mother which Elyot seems to envision here, and which families such as Sir John More's practiced (his son Thomas grew up in Cardinal Morton's household), is here generating its own special warrant in humanist educational aims. In cultivating the young boy's ability to speak Latin, women, not being part of the Latin world, were commonly of no use to a child after the age of seven, for this is the age when Elyot and others prescribe that a boy begin to learn and to speak Latin—and, for that matter, Greek as well. The difficulty was that if there were too many women around, the child would speak English, not Latin. He would slip back into the vernacular family circle instead of being forced out already at this tender age into the world of the "tribe," of men. We are faced here with a rather precocious appearance of the puberty-rite situation around the age of seven, but the humanists favored precociousness and promoted it when they could.

Sir Thomas More and others, more realistic, would try to remedy the situation by educating the women of the household, making them not only literate but learned (that is, in Latin). But their efforts would meet with no large-scale success. For some mysterious reason Latin was tied up with schools, and by the time it became accessible to women generally in schools, it had practically disappeared as a medium of communication. Even in its present attenuated form Latin has never been assimilated in the curriculum for girls' schools as it has in certain curricula for boys'. One suspects that something of what it stood for, and in a certain degree still stands for, cannot be assimilated. It is a matter of record that the women students who today matriculate at Oxford or

Cambridge Universities, where some classical tradition remains fairly strong, are almost invariably less well prepared in Latin than the men matriculating from the English public schools. Curricula are the product of complex and fugitive forces, but the forces are real and cannot be gainsaid.

IV

Flogging was a common practice in the schools of antiquity, as we know, for example, from St. Augustine's rueful remarks in the *Confessions* about his own boyhood experiences.[10] The fact that school pupils were all boys of course encouraged rule by the rod. In the Middle Ages not only does this environment and rule persist, but there is evidence that the specifically initiatory cast of the punishment grew more intense and evident. This is made abundantly clear by Leach, who collects stories about the flogging in school of boy aspirants to monasteries which accompanied the early stages of initiation into monastic life, and quotes from Ælfric's *Colloquy* the "highly characteristic" question which Ælfric has his typical master put to his typical pupils: "Are you willing to be flogged (*flagellari, beswungen* or *swinged*) while learning?"[11] To this the boys—in this case not monastic aspirants—answer at once that they prefer flogging to ignorance. The question, answer, and setting suggest the initiation practice among the Bechuans mentioned above. The boy must acknowledge the equation of learning and flogging, and thereby face courageously into learning as into an initiation, something of itself taxing and fearsome.

[10] St. Augustine, *Confessiones*, Lib. I, cap. ix in *Opera omnia*, Vol. I, Patrologiae cursus completus, Series prima (Latina), ed. J.-P. Migne, XXXII (Paris, 1841), cols. 667–668.

[11] Leach, pp. 81–82, 89.

Renaissance educators did not, on the whole, abate the ferocity of medieval or ancient school punishment. Pictures of Renaissance classroom activity, such as Pieter Brueghel the Elder's engraving "The Ass at School," feature bundles of switches as regular classroom equipment. "Advanced" ideas on education did not necessarily entail diminishing physical punishment. Whereas an earlier tradition had, in Erasmus' phrase, tended to regard pupils as merely small-sized men, the Renaissance educator was often quite sensitive to the immaturity of his charges and to the psychology of child education. But for him psychology included the use of the birch. In Thomas Murner's *Mnemonic Logic* (*Logica memorativa*, 1509, etc.), which in an extremely "progressive" fashion purveys the otherwise terrifying logic of Peter of Spain in the form of a logical card game, one of the woodcuts of "cards" features a master holding three bundles of switches.[12] These, we are told, are to suggest the three questions, "What? What kind? and How many?" used in handling enunciations, for, as Murner explains, it is with the aid of the switches that the answers to these questions are extracted from the pupils. Switches serve as mnemonic devices in both the real and the allegorical orders.

It is well known that the Renaissance Jesuit plan of education provided for a *corrector* for the "little boys" (in effect, those still studying Latin) to "keep them in fear," although the plan registers an oblique protest against beating as compromising good teacher-pupil relations, for it provides that this *corrector* never be one of the Jesuit teachers but either a per-

[12] Thomas Murner, *Logica memorativa, Chartiludium logice, sive Totius dialectice memoria; et Nonus* [i. e. *novus*] *Petri Hispani textus emendatus, cum iucundopictasmatis exercitio* . . . (Strasbourg, 1509), fols. Bv^v-Bvi^r.

son specially hired to do the beating or another student.[13] We should not suppose that punishment in Renaissance schools was always mild. Nicolas de Nancel, Peter Ramus' biographer and erstwhile pupil and secretary, a physician who goes into biographical detail with a whimsical clinical objectivity, reports that Ramus, who was a highly successful educator with "advanced" ideas, often punished his pupils in savage outbursts of temper, not only whipping but also kicking them until they were "half dead" (*semineces*) although—and Nancel adds wistfully here, "for this he must be praised"—during all this process he never swore.[14]

However, although Renaissance reliance on physical violence as a teaching device was not new, the connection of this punishment with Latin teaching acquired a greater urgency. This was due to the greater prestige of Latin established by the humanists, but also to an increasing divorce between Latin and extracurricular life and communication. In the Middle Ages, for casual communication between scholars, young or old, Latin was unblushingly vernacularized. Hence the venture into Latin, while a break with the past, was a relatively less violent break. For the humanist, only "correct" classical Latin should be spoken, even by small boys beginning the language. The break with the past thus reached a kind of maximum in the Renaissance, and the sense of the Latin school as a special marginal environment reached its greatest intensity. The break with the past—that is, with the vernacular of one's childhood—was further enhanced by the concurrent growth of vernacular literature and its greater and greater independence of Latin which marked the Renaissance period.

[13] See the documents in George E. Ganss, S.J., *Saint Ignatius' Idea of a Jesuit University* (Milwaukee: Marquette University Press, 1954), pp. 26, 309, 331.

[14] Nancel, p. 60.

v

In the Renaissance the association of violence with teaching takes another special and interesting turn, for the Renaissance educator appears aware of the teaching environment not only in terms of the violence sometimes resorted to on the side of the teacher but also in terms of the courage which he hopes to develop in his pupils. The emphasis seems connected with the tendency of the humanist educator to think of educating his pupil as a whole person. Humanist teachers frequently functioned less as members of teachers' unions or university faculties than as *familiares* or even employees of bourgeois or noble families. Hence they show an interest in the pupil's total upbringing not so often met with in the medieval university, where all pupils were by definition (if not always in actuality) mere apprentices learning the more or less highly specialized teaching trade.

The new interest manifests itself in the many courtesy books and in the various *rationes studiorum*, or works on educational procedure, which were turned out in the humanist tradition and which connect in many ways with the courtesy literature. In this setting, where educational objectives are formulated under the more or less direct influence of well-to-do or noble households, concerned with family tradition and prestige, there flourishes the Renaissance cult of "glory" and there develops the curious interest in the epic poem, together with the typical Renaissance view that such a poem is the highest creation of the human mind and consequently the normally preferred focus of literary (as apart from oratorical) study. By the same token there develops, under the concurrent influence of Plato's *Republic*, a keen interest in courage (which makes the glorious epic hero) as an express objective in the education of boys.

It has not been sufficiently remarked how much Renaissance poetic and other language study finds itself wandering from the consideration of poetry or language to the consideration of courage, or of its opposite, softness or effeminacy. In part this common deviation is undoubtedly due to the fact that in the Renaissance generally poetry tended to be exclusively a matter for education at what we should consider the secondary-school or even the elementary-school level. With our present upper-division courses and graduate courses in poetry and literature, we are likely to forget that the ordinary Renaissance student finished his rhetoric and poetry in his early teens and went on immediately to "philosophy" and shortly after, if he continued his formal education, to medicine or law or theology.[15] On his own initiative or in some more or less special circumstances a student could study literature at an advanced level, and in the later Renaissance, students, in Great Britain at least, tended to linger on in Latin for a longer time, but, by and large, literary studies in the Renaissance were for youngsters. In the mid-sixteenth century Peter Ramus had explained how his students had finished not only rhetoric (together with what poetry was included in this "art") but philosophy as well by the age of fifteen.[16] Rationalizing about the existing situation, Ramus states that poetry is

[15] See Ganss, p. 45. The curriculum and students' ages here outlined may be taken as fairly representative of Continental practice generally, since the Jesuit program of studies was conceived on an international basis and drawn up by pooling international educational experience.

[16] Peter Ramus, *Oratio de studiis philosophiae et eloquentiae coniungendis*, in Peter Ramus and Omer Talon (Audomarus Talaeus), *Collectaneae praefationes, epistolae, orationes* (Marburg, 1599), pp. 248–250; Peter Ramus, *Pro philosophica Parisiensis academiae disciplina oratio*, in his *Scholae in liberales artes* (Basel, 1569), cols. 1019–1020.

taught at a very early age because the logic in it is diluted and thus assimilable by the tender youthful mind, unable to absorb the more concentrated logic of philosophy.[17]

This statement that poetry respects young boys' weakness is, of course, another way of saying that it gets them over the weakness. The Jesuit savant Martin Antonio Delrio a few years later will explain how the lowly humane letters toughen the young boys who suffer from too great tenderness in age and mind, preparing them for the weightier disciplines of philosophy, medicine, law, and theology. He goes on to add that not only poetry, but drama, history, oratory, and literature generally should be studied only by young boys, not by adults, whose sole concern with these things should be to edit texts for boys—Delrio is here apologizing for his own preoccupations, for these remarks of his occur in the preface to his collection or "line-up" (*syntagma*) of Latin tragedies, which turn out to be entirely Senecan.[18] The idea that Seneca is exclusively for children may strike us as amusing and might have seriously upset even the Stoic Seneca himself, but Delrio's views represent one standard Renaissance position, supported chiefly by two considerations. First, in the actuality of the curriculum, if literature was to be studied at all, it had to be studied in the early years of school, for literature was used in the schoolroom chiefly to perfect the boy's competence in Latin so that, as soon as possible, he could move on to philosophy and the sciences. This was not Erasmus' ideal, but then Erasmus' ideal of an education terminating not in philosophy

[17] Peter Ramus, *Oratio initio suae professionis habita* (Paris, 1551), p. 31.

[18] Martin Antonio Delrio, *Syntagma tragoediae Latinae* (Antwerp, 1593), Preface, fols. *3[v], **1[r]. A translation of Delrio's Preface by Richard G. Wittmann is available in typescript at St. Louis University on application to the present author.

and science but in language and literary study, with theology itself cast in a grammatical rather than a philosophical mold, was never effectively realized.

A second consideration moving Delrio would have appealed to Erasmus: Seneca was a stern Stoic moralist and could thus be counted on to make the young boy manly and courageous. At this point we are reminded of the tendency of Renaissance educators to assimiliate to the linguistic portion of the curriculum not only literary works of Stoics such as Seneca or his nephew Lucan, but also more properly philosophical works, such as the *Enchiridion* of Epictetus, which appears in a great number of Renaissance editions, often together with the *Tabula* of Cebes. The somewhat aphoristic character of the philosophy of the *Enchiridion* made it a congenial adjunct of rhetoric, which often cultivated the epigram. But, more than this, its strong moral and ascetical bias fitted the Stoic philosophy to the puberty-rite mentality which we have been considering here as connected with language study. Epictetus' was a toughening philosophy in a way that Aristotle's was not.

The Renaissance humanist could be disturbed by the plausibility of the charge that literature, and poetry in particular, was actually soft or effeminate, so that, being purveyed to youngsters at the very age when they should be maturing in manliness (the puberty-rite attitudes clearly evince themselves here), it actually only weakens him. This is the burden or background not only of Ramus' opinion that poetry has little "logic" in it but also of Gosson's attack on poetry, revealed by his charge, taken up by Sidney, that poetry is "the schoole of abuse." Although Gosson's principal concern is not poetry taught in schools but drama seen in the playhouses, his resort to the school symbol not only in his title but constantly through his argumentation—"I have been matriculated my selfe in the schools [i.e., of the stage], where so many abuses flourish. . . . I should tell tales out of Schoole, and be Ferruled for my faulte. . . . Liberty gives you head [i.e., in the playwright's world, conceived of as a school], placing you with Poetrie in the lowest form" [19]—leaves no doubt that the case for or against drama and literature generally is to be adjudicated in a pedagogical frame of reference: Do these things serve to make boys men (or men more manly)? Sidney works in this same frame of reference when he asserts that he knows *men*—the word is deliberately pointed and is Sidney's own—"that even with reading of *Amadis de gaule* (which God knoweth wanteth much of a perfect Poesie) have found their hearts moved to the exercise of courtesie, liberalitie, and especially courage." [20]

In Gosson and Sidney the connections between poetry, courage (or the lack thereof), and the education of young boys are suggested rather than explicitly dealt with. But in specifically educational treatises connected with the courtesy tradition they come definitely to the fore and show some of the real grounds for the Renaissance educator's preoccupation with the hero and with glory—these grounds being in this case associated with the proper toughening of the youth in his initiation into extrafamilial society.

Thus in Book I, chapters x to xvi, of *The Boke Named the Governour* (1531) where Sir Thomas Elyot treats the scholastic curriculum of his youthful pupil, it is striking that at every juncture where he mentions the age of the boy, he

[19] Stephen Gosson, *The Schoole of Abuse*, ed. by Edward Arber, English Reprints, [No. 3] (London, 1869), p. 24.

[20] Sir Philip Sidney, *The Defence of Poesie*, in *The Complete Works*, ed. by Albert Feuillerat (Cambridge: The University Press, 1922–26), III, 20; cf. *ibid.*, 28.

brings in courage or "corage" for explicit comment.[21] At seven, we are told, the child begins grammar, but not in too great detail, for too detailed grammar "mortifieth his corage" (chap. x). Up to his thirteenth year, "the childes courage, inflamed by the frequent redynge of noble poetes, dayly more and more desireth to have experience in those things that they so vehemently do commende in them they write of" (chap. x). After fourteen, and some study of oratory and cosmography, it is time, says Elyot, "to induce a childe to the redinge of histories; but fyrst, to set him in a fervent courage, the mayster . . . expressinge what incomparable delectation, utilitie, and commodite shall happen to emperours, kinges, princis, and all other gentil men by reding of histories" (chap. xi).

The connection of literature (Latin) with toughness of moral fiber is here explicit, and this toughness of moral fiber goes with physical toughness as well. Thus, says Elyot, "for as moche the membres by movyng and mutuall touching do waxe more hard," physical exercise must be insisted upon for boys, "specially from the age of xiii yeres upwarde, in whiche tyme strength with courage increaseth" (chap. xvi). However, by the time the boy comes to the age of seventeen, a different emphasis must be given, for at this age "to the intent his courage be bridled with reason, hit were needful to rede unto him some warkes of philosophy, especially . . . morall" (chap. xi).

The picture is here complete. By seventeen the child has become something of a man, his courage has been proved and he must now practice what one practices after crossing the threshold of maturity, namely, control. For our present purposes what is of interest is the absolute coincidence in the

[21] All quotations from Elyot are from the edition cited in n. 9 above.

ending of language studies and the ending of emphasis on developing and proving courage. Both mark the ending of a period of initiation. Courage or "corage" (heart-iness, strength of heart) designates for Elyot something definitely connected with the process of maturing, not merely with high spirits, although it would include this. And this strength of heart is communicated by the study of literature—that is to say, of Latin literature (with some smattering of Greek).

It is true that Elyot is interested specifically in educating a "governor," or, as he puts its elsewhere, a "gentleman," one who rules or at least is part of the ruling class of a *respublica*. Still, his program of Latin and Greek studies for his governor-to-be is basically no different from that of Renaissance schools generally, where it would presumably inspire the same kind of "courage" in the sons of merchants and tradesmen as in prospective governors. In showing how the typical ideal Renaissance educational program built around Latin is suited to nobles—the fighting class, who, above all, must pass through the puberty rites ("Will you guard the chief well?" ask the Bechuans)—Elyot is revealing something of the way this program was felt as operating. In books such as Elyot's the humanists set out to show that even the nobles should be educated men—which, from one point of view, means that the humanistic study of Latin was a good and desirable substitute for more barbaric practices of initiation. In this context, how could it be entirely dissociated from such practices?

A cluster of forces sustaining and sustained by the Renaissance cult of the epic hero and of the epic can be seen here. This view of literature as inculcating "courage" both nourishes and feeds on the cult of the hero and his "glory" which the epic fosters. This cult, which affected governors and governed alike, has far-reaching and mysterious roots in human history.

At this point we can only indicate that the position of Latin in Renaissance culture, the way in which this Latin was taught, the things it was supposed to do to the pupil, and the interest in the epic which by the seventeenth century in Western Europe amounts almost to a frenzy are not unrelated phenomena.

It is true also that Elyot's focus on courage in his educational plan is related to a similar focus in Plato's *Republic*, the major source for much that was explicit in the Renaissance cult of courage. However, the point here is not whether or not Elyot has assignable sources but rather where such sources strike root in his thinking—for not everything that Plato said manages to root itself in Renaissance educational theory or practice. What interests us here in Elyot is the association of courage with language study, and in particular with Latin. The study of Greek for Plato's pupil involved no break with the past. For Elyot's pupil, the study of classical languages did. The Renaissance environment for Platonic ideas was different from the original Greek environment.

Moreover, because of the attitude toward the classical languages peculiar to the humanist tradition, for Renaissance boys the learning of Latin represented, like the passage through puberty rites, not only something difficult but precisely a transit from ignorance to tribal wisdom, that is, to the accumulated wisdom of mankind. This wisdom was thought of as stored behind doors linguistically controlled from the inside. "In the Greeke and Latin tong," writes Ascham, "the two onlie learned tonges, which be kept not in common taulke but in private bookes, we finde alwayes wisdome and eloquence." [22] In any generation the wisdom of the past, which is not only the matter communicated to neophytes in

[22] Ascham, p. 117

puberty rites but a major item in all formal education, may be thought of as "situated" somewhere. The only point we are making here is that Renaissance man regularly located this somewhere in linguistic terms.

The connection of the teaching of Latin and of literature with puberty rites is further manifest to us, if it was not manifest to Renaissance educators themselves, when these educators explicitly discuss the problem of physical punishment. In the long dialogue on the pro's and con's of corporal punishment with which Roger Ascham opens his famous educational treatise, *The Scholemaster*, he provides glimpses of issues relevant to our present subject which he never really fully exposes. Some pupils have recently run away from Eton, we are told in the course of this dialogue, "for fear of beating," and the discretion of schoolmasters is called into question because they may flog to punish "weakenes of nature rather than the fault of the Scholer," thus actually driving boys from learning.[23] This seems a clear indication that, whether it should be or not, punishment is felt by some masters as advisable for reasons other than the encouragement of formal learning.

We note further on in the dialogue that Master Mason and Master Haddon vastly enjoy reminiscing about schoolboy escapades (one recalls that in puberty rites the ordinary rules of behavior are often suspended and outlawry is regarded with approval). Master Mason proves "very merry with both parties, pleasantly playing with shrewd touches [trials—i.e., of the schoolmaster's patience] of many cours'd [flogged] boys and with the small discretion of many lewd schoolmasters," and Master Haddon remarks that "the best Scholemaster of our time [we know that he refers to Nicholas

[23] *Ibid.*, p. 18.

Udall] was the greatest beater." [24] Masters Mason and Haddon here plainly speak not as scholars but simply as men who had "gone through" the *rites de passage* and who look back on such experiences, with their aura of lawlessness, as trials which others should perhaps go through not so much for learning's sake as simply to prove their prowess as members of the "gang" and to achieve a sense of belonging. This is a line of argumentation which Ascham, like earnest educators today, does not like, but the fact that it is used and reported testifies to an existing state of mind.

Ascham himself suggests that native ability, not attributable to their experience of Udall's birches, might account for the success of Udall's pupils and leaves no doubt that he himself is against flogging as a device for teaching Latin. He himself does not state that there were other things besides the mere learning of Latin in the back of Reinaissance educators' minds when they beat their boys. Yet the fact that there were, that the flogging served the purpose—unstated, unformulated, but real—of initiating boys into a tough, man's world, as suggested by Masters Mason and Haddon, is curiously confirmed by the example which Ascham himself brings forward to prove that beating is not necessary. The example has become classic. For it is an example not of schoolboy or budding young gentleman, but that of a girl, none other than the young Lady Jane Grey, whom Ascham, to his delight, found one day reading Plato's *Phaedo* while the more boisterous members of her family were out hunting.

Lady Jane was at great pains to explain how nice a person was her teacher, "Master Elmer," by comparison with her strait-laced parents, by whom she was constantly "so sharplie taunted, so cruellie threatened, yea, presentlie some tymes

[24] *Ibid.*

with pinches, nippes, and bobbes." [25] Ascham does not pause to note that, rather than straightforwardly contrasting schooling based on kindness with schooling based on physical punishment, his example really contrasts the romantic world of a maturing young girl with the rough-and-tumble world his society prescribed for young boys. Despite Ascham's attempt to make something else out of his example, what is remarkable about Lady Jane is not that she is not being flogged—Master Elmer certainly could not have flogged her—but that she is studying the classics *instead* of hunting. This suggests that Lady Jane's approach to literature was somehow radically different from that of the ideal Renaissance gentleman, who liked both the classics *and* hunting. Had not Ascham himself written a treatise on the use of the longbow?

The *rites de passage* prescribed for the Renaissance gentleman were to initiate him into an aggressively competitive man's world. For Lady Jane, too, the study of literature was a kind of *rite de passage*, an initiation into a new world ahead and a break with the past. But the breakthrough was at a different point. It opened out upon a pleasant, fanciful, romantic world. As a *rite de passage* the study of literature here meant to a girl something different than to a boy. One made the *passage* to Lady Jane's world precisely by staying away from the hunt, just as the medieval lady, intrigued with vernacular romances, had done. One thinks of the Green Knight's lady in *Sir Gawain and the Green Knight*, or perhaps even of Paolo and Francesca.

I do not wish to pass on the relative merits of the two worlds, that of literature-and-hunting and that of literature-and-Master-Elmer, or to speculate as to where in the dialectic between the two we are at present situated, but only to point

[25] *Ibid.*, p. 47.

out that they can engender a dialectic because they represent different and opposed positions. In view of this fact, however, it seems not entirely irrelevant that *The Scholemaster*, never published during Ascham's lifetime, is presented to Sir William Cecil and to the world by a woman, who writes the preface, Ascham's widow, Margaret. Nor does it seem entirely irrelevant to this dialectic that corporal punishment and the stress on Latin in school have, pretty generally, been disappearing in modern times with the emergence of coeducation.

VI

This study has been a sketch of certain forces at work in the Renaissance attitudes toward Latin, toward literature, and toward education. It could be elaborated indefinitely, and no doubt refined in many ways, by exploiting more and more examples, of which there is certainly "copie" in Renaissance documents. Here we have limited ourselves to samplings from better-known sources, chiefly British. Perhaps further development is worth while, perhaps not. In either event, we can sum up our present conclusions.

First, I have not sought to maintain that Renaissance educators explicitly thought of Latin study as a puberty rite. They had no definable, abstract idea of what a puberty rite is or was—and neither, for that matter, do the primitive peoples whose puberty rites we have taken as a term of comparison. Renaissance educators, like primitive peoples and like ourselves, have no rationalized explanation for everything they do. They do certain things because they feel these things should be done, finding reasons for them afterwards if at all—and, if they are observant and honest, often being surprised at the reasons which turn up on close inspection.

The basic conclusion is that when Latin, in which learning

was encoded, became in the Renaissance more than ever a school language divorced from family life, initiation into Latin became more than ever a *rite de passage*. Thus, when other Renaissance courses were being labeled "methods" and "systems," Comenius finds it natural to describe his course in Latin and other languages as a "door"—*Ianua Linguarum* he entitles his famous textbook. Thus, in a Western society destined to become progressively more humane in its educational procedures, the status of Latin helped maintain the relatively violent puberty-rite setting, a sense of existence on a threshold, within a marginal environment (associated with forced seclusion from the company of women and to a certain extent from one's own family), in an atmosphere of continuous excitement and of that aggressive competition or *aemulatio* which, toned down or outlawed in modern de-Latinized coeducationalism, was a key principle of most Renaissance education.

This complex of attitudes, not new but concentrated with new urgency around language study, helps explain (although I do not wish to suggest that it entirely explains) the frenzied fascination with epic poetry (most of which was in Latin during the Renaissance), with the courageous epic hero (given to war much more than to love-making), with epic theory, and with courage itself, which marks linguistic studies in the period when Renaissance Latin education was having its full effect on society.

Seeing Renaissance Latin teaching in the psychological framework of the puberty rite helps us to explain much in the later trajectory of Latin teaching. In the nineteenth century, when Latin was on its way out as the core subject of the curriculum, educators produced the theory that Latin "strengthened" or "toughened" the mind. This theory, which is still met with today, has been labeled new,[26] and it was new in the sense that earlier educators had not explicitly advanced it. But the complex in which Latin was normally taught had associated the language in a special way with some sort of toughening. Were not nineteenth-century educators, and are not the few twentieth-century educators who repeat their words today, merely giving voice to a vague feeling which has its roots in the psychological setting of the Renaissance Latin school—the feeling that the teaching of Latin, independently of the communication of the ability to read the language (the immediate aim of Renaissance Latin teaching), had somehow to do with toughening the youngster for the extrafamilial world in which he would have to live?

Translated, this means the feeling that a boy's education was basically a puberty rite, a process preparing him for adult life by communicating to him the heritage of a past in a setting which toughened him and thus guaranteed his guarding the heritage for the future. Latin had indubitable connections with the past, and it was hard, indeed all the harder as motivation waned when real use for the language began to wane. This association of Latin with a toughening marginal environment of a puberty-rite type was sufficient to keep Latin in its place as the basic discipline forming the prep school character, with its twin emphases on Latin and physical hardihood (modulated eventually into good sportsmanship).

The perspectives proposed in this chapter are, of course, suggestive rather than complete, but they open the way, I believe, to a better understanding of some curious and important momentums developed by past ideas and practices. And, since it is impossible to study the past without reference to the present, they suggest matter for reflection—forward-

[26] See Ganss, pp. 210–211, 219 ff.

looking, let us hope, rather than nostalgic—concerning the twentieth-century situation. Where are the *rites de passage* for youth today? Does a technological society have any? Should it have any? If so, what should they be?

6 { Ramist Classroom Procedure and the Nature of Reality

The Renaissance is an age particularly rich in educators and educational literature, but nowhere in it is there a figure more profoundly involved in educational theory and practice than Peter Ramus. Ramus' involvement is virtually total. If we consider his life apart from his educational activity, we find very little to consider. Pierre de la Ramée came up to Paris with a limited, elementary education and a driving desire to secure more education. With notable sacrifice he did secure more. He completed the work which made him a master of arts and thereupon committed himself for the rest of his life to purveying learning to others and to improving educational methods by a determined, single-minded effort at devising what he considered suitably organized textbooks backed by attacks on what he took to be ill-organized educational works. His influence on literature and philosophy in England and her American colonies, as elsewhere, was to be largely due to the textbooks with which he and his followers flooded the schools.

In his complete devotion to education, Ramus was true to the central tradition of the university, which, from its medieval beginnings to his time, was doubly committed to teaching. It not only purveyed knowledge, but also, at least in principle if not always in actuality, had insisted that its students in turn teach others. We do well to remind ourselves that early universities were basically normal schools. To complete one's course at Paris, one had to become first a bachelor or apprentice teacher and then a master teacher—of arts, medicine, law, or theology, as the case might be. Statutes had required that all those receiving the master of arts degree honor it by actually teaching for at least two years "unless excused for a reasonable cause," [1] although one must indeed add that "reasonable causes" seem not to have been hard to come by.

Ramus himself, however, did not look for any excuse to avoid educational work. He distinguished himself as a teaching master of arts first at the little Collège de l'Ave Maria and later at the Collège de Presles, where he soon became principal, devoting his time thereafter to supervising and writing more than to direct teaching.[2] Despite his anti-Aristotelian invectives impugning the professional competence of his colleagues—and in some measure even because of these invectives, which won him a great deal of attention—he soon was named one of the first of the newly established group of regius professors (who much later were to be styled the Collège de France) and eventually became known, not without resentment on the part of the others, as the dean of this group of savants.

[1] *Chartularium Universitatis Parisiensis*, ed. by H. Denifle and E. Chatelain (Paris, 1889–97), I, 78; Hastings Rashdall, *The Universities of Europe in the Middle Ages*, new ed. by F. M. Powicke and A. B. Emden (Oxford: Clarendon Press, 1936), I, 409, etc.

[2] Nicolas de Nancel, *Petri Rami . . . vita* (Paris, 1599), in Nancel, *Declamationum liber . . . : addita est P. Rami . . . vita . . .* (Paris, 1600), pp. 17–19, 49–50.

Ramus sacrificed much for his career as an educator. As noted in Chapter 5, since no married man could teach at Paris except on the faculty of medicine, he decided to forego marrying entirely, not because he was not interested in marriage, which his biographer Nancel testifies was frequently on his mind, but quite simply because he wanted to retain his academic position as head of the Collège de Presles. Nancel says that he himself was called Little Ramus (*parvus Ramus*) because of his close association with Ramus as pupil and secretary, but that his own marriage helped destroy the resemblance, shaping his own career differently.[3]

A total commitment to education is manifest also in Ramus' manifold publications. Of his sixty or more extant works, there is not one which, directly or indirectly, is not written for the academic world, and most of them are, in fact, written exclusively for this world as textbooks, commentaries on classical authors (sometimes rather commentaries against them), or reflections on educational conditions or procedures which often proceed as wholesale attacks on Ramus' adversaries, real or imagined. The thirteen or more extant works by Ramus' literary associate Omer Talon (Audomarus Talaeus) are of the same academic cut. Ramus' and Talon's works enjoyed an astounding circulation in the academic world.[4]

A Montaigne or a Bacon was read by many outside educational circles and exercised his influence on education in great part indirectly. Ramus' influence, thanks to his textbooks in particular, was far more immediate and widespread, felt at the very heart of the educational milieu, where devoted followers—almost all from the educational milieu—adopted and adapted his books, reprinted them, annotated them, equipped them with verbose commentaries and then stripped

[3] *Ibid.*, pp. 57–59.   [4] See Chapter 7, n. 2 below.

them of the commentaries again and of some of their own original text as well, reducing the Ramist "arts" of grammar, rhetoric, dialectic or logic, arithmetic, geometry, and to the rest to forms even sparser than Ramus' originals. Teachers in the Ramist and other traditions worried about these changes for generations. In his 1620 English-language adaptation of Ramus' *Dialectic* called *Syntagma Logicum or the Divine Logic*, Thomas Granger remarks:

Too much brevity (as it is a generall complaint of Ramus) causeth many and large Commentaries, and Tractates for understanding and knowledge, and they again bring forth Epitomes for memorie and practise. The former are too prolix for the learned, who have already attained to perfection. The later [*sic*] to [*sic*] briefe for them that either are ignorant, or having some smattering hereof, are desirous to attaine to more skill.[5]

But both inflation and abridgment were planned, if we can believe some title pages, for the highest of pedagogical motives—"the benefit of the student." One suspects that the good of the student was interpreted in terms of publishers' returns—but this commercial aspect of textbook development does not reduce either the number of editions or their educational impact.

It is not surprising that Ramus' most informed and discerning biographer, Nicolas de Nancel, was himself an educator. First Ramus' pupil and then for some twenty years his secretary, amanuensis, literary collaborator, and general understudy, Nancel (1539–1610) had taught in Ramus' Collège de Presles while completing his studies there. In 1562 he accepted a professorship from the King of Spain at the new University

[5] Thomas Granger, *Syntagma logicum, or The Divine Logike* (London, 1620), fols. a2^v-a3^r ("Epistle to the Reader").

of Louvain, returning in 1565 to the Collège de Presles in Paris and taking up the study of medicine. Although he left the University to begin medical practice, he continued his academic interests in his Latin writings. His *Life of Peter Ramus* is written out of his memories and records of the old academic milieu and addressed to Ramus' former pupils.

Thus not only is Ramus' own life devoted to education, but our most definitive record of his career belongs in a special way to the educational world. So, in a somewhat different fashion, do our other two contemporary biographies of Ramus, those by Johann Thomas Freige (Freigius) and Théophile de Banos (Banosius). These men were also educators. Among Freige's numerous Latin works is the *Paedagogus* (1582), a Ramist compendium of all the "arts" for use in elementary instruction and a kind of forerunner of the modern encyclopedia. Théophile de Banos, an exiled French Protestant minister at Frankfort-on-the-Main, was matriculated at the University at Basel.[6] Neither of these biographers was acquainted with Ramus personally at all so well as Nancel. But their sources are no less academic. Where Nancel used records and memories deriving from his association with Ramus at the Collège de Presles as his chief source, Freige and Banos use—often verbatim—autobiographical passages which occur incidentally in some of Ramus' works written for the academic milieu, particularly his *Inaugural Address as Regius Professor* (*Oratio initio suae professionis habita*) and *Defense of Aristotle against Jacob Schegk* (*Defensio pro Aristotele adversus Iacobum Schecium*). Ramus is thoroughly enmeshed in the academic world. It is little wonder that he is a major retrospective source for the history of even the

6 See E. Droz, "Des autographs," *Bibliothèque d'Humanisme et Renaissance*, XIX (1957), 502–503.

medieval universities in Rashdall's famous work on this subject.

II

Ramus has long been identified as an educational reformer, but his activity as a reformer and the nature of his reform need reassessment at present if only because we are able to understand Renaissance educational procedures better today than hitherto. First of all, in our twentieth-century educational activity we have moved away from the procedures somewhat —although not, of course, without learning from them. We have larger perspectives in which to view them. Secondly, and more importantly, a great deal of scholarly work has been done lately on Renaissance educational views and methods, much of it quite circumstantial. The Ramist reform in all fields was, in the view of Ramus himself as well as of others, based on the reconstituted Ramist dialectic, of which the almost equally well known Ramist rhetoric was a correlative. Within the past three or four decades studies of the nature of Ramus' dialectic and rhetoric and their influence in and outside the English-speaking world have been a major preoccupation of American scholars such as William G. Crane, Perry Miller, Hardin Craig, Wilbur Samuel Howell, T. W. Baldwin, Rosemond Tuve, Sister Miriam Joseph, and a host of others, including the present writer. The work of several of these authors has taken up explicitly educational practice itself. Other works concerned more or less directly with Ramist educational practices have appeared also in the French-speaking, German-speaking, and Italian-speaking milieux, although with some few exceptions, notably the studies by Cesare Vasoli and by R. Hooykaas, until the 1960's, Europeans, whether on the Continent or in the British Isles, have not been so directly concerned with Ramus and his followers as Americans have been.[7]

Nevertheless, with all our present detailed research, there is much still to be done concerning Ramus' connection with pedagogical procedures. Ramism has been considered as a movement complexly related to scholasticism, to Agricolan humanism, to typography, to Puritanism and the doctrine of hard work, and to educational institutions generally. What we now need is an attempt to spell out more fully the connections between concrete educational practice and the trajectory which Ramism describes in the sixteenth and seventeenth centuries within Western civilization. The present study is a modest attempt to suggest a few of the little explored connections.

Ramus' position with regard to educational practice can be considered under two headings: institutional organization on the one hand, and on the other curriculum and classroom procedure. As an institutional organizer or reorganizer Ramus himself appears to have achieved a high degree of efficiency. Paris seems to have had a better study program than some other universities, but even Paris left much to be desired, for Ramus complains that of the thousand or more students who, he says, came to hear him and the other regius professors, all but

[7] For a complete bibliography to 1958, see Ong, *Ramus, Method, and the Decay of Dialogue* (Cambridge, Mass.: Harvard University Press, 1958), pp. 375–391. Since the appearance of this bibliography, one should note also J. Moltmann, "Zur Bedeutung des Petrus Ramus für Philosophie und Theologie im Calvinismus," *Zeitschrift für Kirchengeschichte*, LXVIII (1957), 295–318; R. Hooykaas, *Humanisme, science et Réforme: Pierre de la Ramée, 1515–1572* (Leyden: E. J. Brill, 1958); and Wilhelm Risse, *Die Logik der Neuzeit*, Vol. I, *1500–1640* (Stuttgart-Bad Canstatt: Friedrich Frommann Verlag, 1964). Studies of Ramus and editions of his works since 1964 have been too numerous to list here.

two hundred or so should be sent back to the *collèges* to better their basic training so that they could follow the lectures intelligently.[8] Inefficiency of this order is probably endemic to any general educational system. Ramus' own Collège de Presles was noted for its systematic and efficient curriculum,[9] although Ramus' claim that he was turning out students who at the age of fifteen were already full-fledged masters of arts "in fact and not in name only" has to be taken in context.[10] A master of arts could be, and often was, a quite immature youngster in sixteenth-century Europe, as in the Middle Ages.

Discipline at Ramus' Collège de Presles was strict at Ramus' own insistence. We have already noted Nancel's reports on Ramus' daily forays about the corridors and classrooms as well as on the brutal severity of the punishment he meted out to offenders in study or discipline. Whatever Ramus' other achievements, he can hardly be seen as an advanced humanitarian.[11]

Ramus' *Notes on the Reform of the University of Paris* was published in Latin in 1562 as *Prooëmium reformandae Parisiensis academiae, ad regem* without indication of author, publisher, or place of publication, and again in French the same year as *Advertissements sur la réformation de l'Université de Paris, au roy* with André Wéchel's name as publisher, but still anonymously. Only in 1569 was Ramus' name attached to

[8] Ramus, *Prooëmium reformandae Parisiensis academiae* in his *Scholae in liberales artes* (1569), cols. 1075–1076.

[9] Nancel, pp. 18–19.

[10] Ramus, *Pro philosophica Parisiensis Academiae disciplina oratio* (1551) in his *Scholae in liberales artes* (1569), col. 1020; cf. cols. 1011 ff. Nancel says he himself completed the course at the Collège de Presles in "six years or thereabouts" (Nancel, p. 17).

[11] See Chapter 5 above, p. 126.

this "oration" in a collection of Ramus' lectures and orations published at Basel, and not until 1577 did a Paris publisher issue it under the name of Ramus, then five years dead. As its title indicates, this is Ramus' most direct statement concerning needed institutional reforms at Paris. The reforms advocated are reasonable enough and supported by forceful arguments, but they turn out to be rather routine. The number of teachers at the University is to be reduced, the inefficient ones being weeded out, and those retained to be paid out of state funds so that poor students who are capable need not pay exorbitant fees. Along with this, Ramus urges that the "nugatory barbarism" of such authors as Alexander de Villa Dei be replaced with classical authors and that thorny theological disputation be supplanted by commentary on the Hebrew Old Testament and the Greek New Testament. Neither of these lines of thought was in the least original. The regius professors, of whom Ramus had been one since 1551, were already paid out of state funds (if sometimes tardily), so that Ramus was only advocating the extension of an existing policy. And the suggestion concerning theology was quite standard long before 1562 at Paris and all over Western Europe. In his relations, then, with educational institutions as such, Ramus stood for working with and bettering existing arrangements. He proposed nothing notably original at all.

III

Concerning the curriculum and teaching practices that Ramus used or favored, we have information which is rich and detailed. A great deal can be gathered from Ramus' own works, and has been. But we have even more than this. In 1576 Johann Thomas Freige (Freigius), who had met Ramus when the latter had visited Basel and had thereafter become an ardent promoter of Ramus' ideas, had edited the basic

Ramist textbooks all in one volume which he entitled *The Regius Professorship* (*Professio regia*). At the beginning of this book, he presents in good Ramist outline form [12] the complete program which, from Ramus' *Lectures on Dialectic* (*Scholae dialecticae*) and elsewhere, he understood Ramus had given his pupils at the Collège de Presles to enable them to become "perfect philosophers" by the time they turned fifteen. This outline presents in meticulous detail the seven-year curriculum which could produce this result for boys starting at the age of seven with no knowledge of reading or writing. Freige is careful to specify not only subject and procedure but time, often down to the number of hours and the period of the day.

The succession of subjects ran as follows: grammar (Latin and Greek) for three years, and then successively, for one year each, rhetoric, dialectic (or logic), mathematics (purportedly arithmetic, geometry, music, and optics but in reality less than all of these), and physics. This curriculum is to a slight extent Freige's own construction based more or less on Ramus' declared aims, for the later reaches of the curriculum are considerably less developed than the earlier in Ramist textbooks. So far as we know, Ramus did no work on music. He was apparently working on an optics textbook before his death, but it is at present uncertain how much if any of Ramus' own work is included in the Latin *Opticae libri quatuor* or *Four Books on Optics* "written by Friedrich Risner [or Reisner] at the wish of Peter Ramus" which appeared in 1606 edited by Nicolaus Crugius [13] and is the only work on the subject even this closely associated with Ramus' name.

[12] Reproduced in Ong, *Ramus, Method, and the Decay of Dialogue*, p. 261.

[13] See Ong, *Ramus and Talon Inventory* (Cambridge, Mass: Harvard University Press, 1958) pp. 401–402.

The "physics" which Freige lists as culminating the course is Ramus' annotated edition of Virgil's *Georgics*, plus some preliminary additions. This relative disorganization in the upper reaches of the curriculum is not, of course, peculiarly Ramist, for the trivium of grammar, rhetoric, and logic or dialectic had always been more firmly rooted in the European educational tradition than the quadrivium, which was more often a velleity than an actuality.[14] But Freige's picture certainly represents Ramus' general designs.

Freige's understanding of Ramus' classroom procedure divides it into explanation (*explicatio*) and exercise or practice (*usus*). The time allowed for each is meticulously specified, and is the same for all the subjects from grammar through "physics." One hour in the morning and one in the afternoon are assigned to hearing the teacher's explanation, and the rest of the day is broken down into "meditating, studying, and exercise work concerned with what is explained" (*in rebus expositis meditandis, ediscendis, exercendis*). This breaks down into two hours of study, one hour of recitation, and two hours distributed among conversation (*communicandi*), disputation, imitation, and exercise work (*exercendi*).

Of these last four activities, the first two are given no special explanation, and one has the impression, which Ramus' own various writings confirm, that conversation and disputation were subordinate to written work. The favoring of written work over disputation establishes Ramus definitely within the humanist rather than the scholastic tradition, insofar as these can be distinguished from one another. The seeming proportionate indifference to conversation, however, is not a humanist trait. Quite the contrary, for collections of sample

[14] Ong, *Ramus, Method, and the Decay of Dialogue*, p. 138, and references in notes there.

conversations are a regular part of most humanist repertoires from Erasmus on. About this more will be said later, but here one can note that indifference to conversation fits with Ramus' own lack of interest in making conversation and with a deeper attitude toward reality manifest in indifference to proverbial, sententious, or "pointed" expression of any sort.

Exercise (*exercitatio* or *usus*) is divided by Freige, following Ramus exactly,[15] into analysis (*analysis*) and composition (*genesis*). What kind of analysis one uses—grammatical, rhetorical, logical, mathematical, or physical—will depend on what discipline is being taught in the year in question, for we must remind ourselves that this classroom procedure is an all-purpose one. Ramus is explicit about the universal applicability of his analysis to any and all material taught.[16] He conceives of analysis as finding the whole of an art in one or another work representing the subject in question, or as near the whole as possible. Thus in logical analysis, no matter what work one is considering, one finds in it the "arguments," enunciations, syllogisms, and "method"—this last governing structure as a whole, determining the relationship of the separate syllogisms to one another. In grammatical analysis one identifies each word and the relationship of words in a way close to what we today would call parsing. In rhetorical analysis, one identifies the tropes and figures. Arithmetical analysis identifies in a text the use of numbers for counting, geometrical analysis the use of measurements, and so on. The kind of analysis used, we are told by Freige, echoing Ramus, depends on the art being taught, showing how the text under scrutiny accords with the "rules" of the art in question. It is part of

15 Ramus, *Scholae dialecticae*, Lib. VIII, cap. i, in *Scholae in liberales artes* (1569), cols. 191–193.
16 *Ibid.*

Ramist doctrine here that the various arts are to be found "mixed" in virtually any text of reasonable length. From an analysis of the classical authors one can generate not only grammar, rhetoric, and dialectic or logic, but also arithmetic (Do not the historians mention numbers?) and geometry (Does not Caesar use measurements?) and the other arts as well. Of course, in some texts there is a higher concentration of one or another art than in others. Poetry is completely logical, but in it the logic is spread thinner than in orations. This is why looking for an advantageous concentration of the proper material, Ramus himself chooses to derive his "physics," which Freige displays, not from any classical text chosen at random but from Virgil's *Georgics*.

Genesis is conceived of as the reverse of analysis, being always an imitation (*imitatio*) which makes either something to match the text previously analyzed or something more of one's own (*proprium*).

IV

Today in a world which in some ways is ultrascientific and in others ultraromantic, this educational program strikes us as a debasement of both the scientific and the literary. On the one hand everything, even physics, is conceived of as some kind of operation on a text. But on the other hand, the operation does not seem to respect sufficiently the mysterious nature of verbal expression. The literary text becomes for a Ramist a kind of uninspired collection of miscellaneous details from a life which is ordinary and everyday to the point of being humdrum. Something in the relationship between science and reality has been misconstrued. Literary expression is allowed to masquerade as physical reality. And in the process, literature and language have been utterly misrepresented. Literature has been atomized and expression reduced to a mere reassembling of interchangeable parts.

One must acknowledge, however, that in its distressing formalism the Ramist program is not greatly different from many other Renaissance programs of education as these are formally stated by their proponents. Education was carried on at this time, we must recall, in a tongue foreign to all those who used it, Latin, and in acquiring this tongue a certain amount of mechanical drill was the order of the day. Moreover, even after the four years devoted not only to the study of Latin but even to study of Latin in Latin, the upper reaches of the curriculum were still designed for quite young boys, for whom a certain disciplined formalism was helpful.

And yet the Ramist educational procedures can be contrasted at certain sensitive points with other Renaissance procedures. One of the most sensitive points concerns the attitude toward language as manifest by the attitude toward the collections of *sententiae*, apothegms, and the like for the use of students which formed a kind of centerpiece in the educational designs of other humanists, particularly those in the Erasmian tradition. One of the most significant points of difference between Ramus' program and other programs generally is Ramus' failure to provide any such collection for the student to work with. Although Ramus himself uses in abundance "examples" and "arguments" from classical sources, there is no document in the Ramist canon corresponding in any way to Erasmus' *De copia* or to his *Colloquies* which would help supply the student with a fund of expressions and thoughts as "matter" for conversation or writing. Whatever is caught in any expression is, for Ramus, not usable *unless it has first been analyzed*. The passion for analysis militates against Ramus' resort to aphorisms and other "pointed" sayings. This attitude touches rhetoric and possibly dialectic or logic most closely, of course, but it touches other subjects, too. For aphorisms were still being used to teach such things as medicine. They would certainly have provided a more "scientific" physics than an analysis of Virgil's *Georgics*. And yet, by and large, Ramus would have none of them.

The use of all sorts of cullings from literature had been known to antiquity, but had since gained in importance when Latin had shifted from a vernacular language to a learned language. After this shift had occurred, the small talk with which a youngster ordinarily perfects his linguistic skills during the early years of his life had to be supplied artificially. The Middle Ages had known in quantity its florilegia, collections of things to say, and had used these as integral parts of its educational system. The *Disticha de moribus* attributed to an author thought to be named Dionysius Cato is perhaps the best known elementary collection. With the help of typography, the Renaissance had multiplied similar collections of aphorisms, *sententiae*, epigrams, apothegms (generally with no clear-cut distinctions between such genres) or of mere miscellaneous excerpts to an extent which the world had never before seen, and the seventeenth and eighteenth centuries would multiply them more. Ancient collections of pithy sayings such as that of Stobaeus were edited and re-edited, and to them were added new authors by the dozens, all borrowing from one another and from the ancients, too. Erasmus' *Adagia*, Michael Neander's *Gnomologia Graeco-Latina*, collections by Conrad Wollfhart (Lycosthenes), Jan Gruter (Gruterus), are significant samples of the genre. Recently the work of DeWitt T. Starnes, Ernest William Talbert, and other scholars has shown how dependence upon these collections and on various dictionaries of classical quotations and lore was inculcated in the educational practice of the time, and T. W. Baldwin's two massive volumes have traced much of Shakespeare to its commonplace sources in such works. These collections and dictionaries are published throughout Western Europe in quantity from the sixteenth century on. A recent study of Renaissance concepts of the commonplaces [17] and another concerned with the use of aphorisms [18] have shown in detail how much Renaissance writing, Continental as well as British, was nurtured by the schoolboy's indoctrination in sententious, witty, pointed sayings. Lack of ability or of occasion to make small talk in Latin was in a way compensated for by this cultivation of gnomic utterance and by other efforts at furnishing a *copia verborum et rerum*, as has been detailed in Chapter 2 above.

We cannot, of course, be sure that Ramus or those who taught in his Collège de Presles never used collections of sayings or quotations. Indeed, there would be a specific place in Ramist dialectic or logic where such items could be introduced as arguments from "testimony," from something that someone says. But Ramists took a rather dim view of this sort of argument. It was definitely peripheral, "inartificial" in the sense that it fell outside the domain of the art of dialectic strictly understood since the sayings of other persons were adventitious to the real state of affairs.

The Ramist antipathy to collections of pregnant sayings has deep roots which extend not only to educational procedures

[17] Sister Joan Marie Lechner, O.S.U., *Renaissance Concepts of the Commonplaces* (New York: Pageant Press, 1962).

[18] Sister Scholastica Mandeville, Ad.PP.S., *The Rhetorical Tradition of the Sententia; with a Study of Its Influence on the Prose of Sir Francis Bacon and of Sir Thomas Browne* (Ph.D. dissertation, St. Louis University, 1960; available on microfilm, University Microfilms, Ann Arbor, Michigan).

but wind themselves into Ramus' own personal habits of life. Nancel records specifically that Ramus exhibited an indifference or positive aversion to apothegms, *sententiae*, and all such things. When he used such modes of expression, he used them ineptly, for they lacked wit and appeared careless or affected, quite unlike Cicero's adroit turns.[19] This attitude Nancel connects with Ramus' tendency to be thinking always of "grave" and "serious" matters (*gravia et seria*), implying that in Ramus' mind apothegmatic, sententious, or "pointed" utterance was frivolous. We should say today that Ramus' attitude shows, among other things, that Ramus was a singularly unimaginative person, as Nancel himself makes clear in other places, too.[20] This is an important fact, for it is of some moment that certain sectors of the sixteenth- and seventeenth-century world could take to heart the message of so profoundly unimaginative a man and make this message the center of their educational systems.

Nancel expatiates on Ramus' lack of interest in sententiousness and wit, which he obviously takes to be a regrettable lacuna in Ramus' outlook at a time when wit was considered in many quarters to be a close associate of wisdom. In his wry, clinical way he tells the story of how Ramus, when confronted by some German autograph hunters with a selection of emblems (*emblemata*)—apparently the sort of illustrations with mottoes and explanations which one finds in Alciati, but possibly merely a list of apothegms—chose the one with the motto which had always been his favorite, "Unremitting labor conquers all" (*Labor improbus omnia vincit*), and copied it out in his own hand for his German admirers. Nancel's own wit in his choice of this episode should not be missed. Here

---

[19] Nancel, p. 62.

[20] He notes Ramus' singular lack of success in lecturing on poetry (*ibid.*, pp. 22–23, 32).

was an apothegm which downgraded all apothegms, however pregnant with wisdom, in favor of hard, and presumably dull, work. Ramus' choice of this *sententia* strikes Nancel as of a piece with Ramus' performance in the company of others. He was able to hold the attention of a large audience in a public address but quite incapable of getting along in a smaller social gathering because he could not make small talk and in ordinary conversation "could scarcely express himself even with great effort and exertion and puffing."

At this point Nancel's clinical diagnosis of the situation becomes intensely interesting in the questions it suggests. "This sort of thing," he reports, "is common among the French, even the most learned, because they are not at all accustomed to use Latin in ordinary human affairs." One wonders how many persons who were the product of this Latin-centered education of the humanists ever really did use Latin in their ordinary human affairs. There is no end of books as Sir Thomas Elyot's *The Boke Named the Governour*, which prescribe that a boy learn to speak Latin from the age of about seven on, and no end of school statutes requiring the use of Latin by the boys even during times of recreation, and no end of phrase books (among non-Ramists) to further this use. But there are indications that Latin speaking "in ordinary human affairs" was often a mere velleity. Even in Paris itself the University was divided into "nations" along lines which were roughly linguistic, and the individual colleges, often established for students from a particular locality, tended to be dominated by one or another vernacular—Ramus' Collège de Presles by boys and teachers from near Soissons, the Collège de Navarre by Iberians, and so on. Many things militated against the use of Latin, but conditions such as these seem to have guaranteed a certain tolerated neglect.

In noting the failure of his fellow Frenchmen to use Latin

in ordinary daily affairs, it is conceivable that Nancel was using as a term of comparison the Italians, who presumably would be more adept at using Latin for small talk than the French were. And yet in the somewhat spectacular encounter he records between Ramus and a group of Italian diners, we find Ramus leaving in a huff because the Italians were ignorant not only of French but of Latin as well.[21] One suspects that Nancel's accusations against Frenchmen applied also to other nationalities. Erasmus and others of the great humanists, and Nancel himself, can be imagined resorting to Latin for their ordinary conversation as they moved around Europe from Italy to Germany and to the Netherlands and France and England. Such professional educators were out to set a good example. But the ordinary product of the Latin-centered education of the times can be less readily imagined as doing so. The linguistic situation was a strange one, and increasingly difficult for us to reconstruct, when persons of no matter what nationality and mother-tongue were educated only in Latin (with a sprinkling of Greek and sometimes a glance at Hebrew) in the schools. By 1655 in England John Wharton conjectures, in his *New English Grammar*, that after leaving secondary school only one boy in a hundred has any need at all for the Latin which he spent virtually all of his elementary and/or secondary school years in studying.

v

Whatever the exact dimensions of the neglect of Latin in ordinary daily affairs, the changed and changing status of the language is not irrelevant to Ramus' indifference toward collections of sayings and toward the wit which the use of such collections fostered. Such sayings—aphorisms, proverbs, *sen-*

[21] *Ibid.*, p. 62.

tentiae, and the like—tend to focus wisdom in the word in ways which were uncongenial to the Ramist mind and which reflected the tendency to restrict the actual use of Latin more to writing and print and to the classroom, where writing bulked larger than in prehumanist times.

The use of collections of sayings has a complex history,[22] which has been reviewed earlier in Chapters 2 and 3. From ancient and medieval times sayings had served to "amplify" or expand upon a theme. Ramus, like other Renaissance teachers of rhetoric and logic, was necessarily concerned, if not directly with wisdom as such, certainly with giving his pupils something to say. But he chose not to view this as amplification, which was a rhetorical procedure. To expand upon a subject a Ramist turned not to rhetoric—which provided in Ramus' scheme only "ornament" and some slender help in oral delivery—but to logic. In logic he ransacked the "places," commonplaces or "topics," which he called "arguments." Here he found matter which was relevant to the subject in hand and which he arranged in statements and syllogisms, availing himself of "method" to give the whole its proper over-all organization. He turned to rhetoric only to make the whole appealing by decorating or ornamenting it with suitable figures of speech and suitable delivery.

This is the picture which we see when we consider the Ramist arts of dialectic and rhetoric in themselves. When, however, we turn to classroom practice, where all the arts are to be mingled in *usus,* we see a somewhat different pattern. Here analysis becomes paramount. By analysis the student acquires the necessary familiarity with the "matter" he must use in expressing himself. Analysis is the activity which pre-

[22] See also Lechner and Mandeville, nn. 17 and 18 above, and bibliographies there.

cedes and makes possible *genesis* or composition, saying something about a subject, enlarging upon it.

Ramists are of course not the only ones who insist upon analytic study of texts. Virtually all Renaissance educators demand some analytic work of one sort or another, under one or another name. But for the Ramist, analysis assumes exceptional importance because it is seen as the process providing material for one's own discourse. Without a stock of "sayings" to spill off his lips, a Ramist avails himself of textual analysis to find something to say. With this insistence on analysis, it is not surprising that the term analysis itself is for a period associated with Ramist views particularly: sixteenth- and seventeenth-century writers speak of "the logical analysis of the Ramists" in a way which gives us to know that they feel that analysis is something which Ramists have more or less successfully pre-empted. When Samuel Butler satirizes the Puritan in *Hudibras* he is advertising the Puritan's Ramist background in stating

> He was in *Logick* a great Critick,
> Profoundly skill'd in Analytick.

The tendency of a Ramist not to exploit utterances unless he has first analyzed them has two implications which bear upon the educational situation and classroom practice. First, it fixes attention on the written word rather than the spoken word, for analysis is primarily an exercise conducted upon a written text. Indeed, it is hard to see how it would be feasible without writing. This special fixation upon what is inscribed rather than upon what is spoken makes human expression less a conveyance of a truth or of wisdom and more an object upon which one performs an operation. Ramist analysis strengthens the tendency to regard the word as a thing.

Secondly, when Ramus and his followers replace the use of collections of *sententiae*, aphorisms, and other sayings of the sort which make their way into commonplace books, with analysis as a device for finding "matter" for discourse, they shift from a word-wisdom to a kind of classroom-wisdom. Ramist analysis forces the pupil to process all his mental possessions through some art or curriculum subject before he puts them to use. If an apothegm or a proverb or an aphorism should by any chance come to mind, before one uses it one had best write it down and analyze it—grammatically, rhetorically, logically, mathematically, or "physically." What it "contains" is what comes out of the analysis, not what it actually says before it is analyzed. In this spirit Johannes Piscator will undertake to do a "logical analysis" of every book in the Bible to see what truth each "really" states—which will be the truth Piscator analyzes out of it, not what it originally says. To this mind the sense that utterances can somehow touch mysterious depths which analysis can never quite fathom (without itself opening still greater depths) is of course lost. All statement is flat, plain, and if it is not this, it is deficient as statement.

Seen in these quite valid and central perspectives, Ramism might seem merely quaint, perhaps artistically lethal, but of no great importance. Yet its great spread will hardly allow us to regard it as educationally insignificant. As a matter of fact, it has educational significance of the headiest sort, for it implies no less than that it is the "arts" or curriculum subjects which hold the world together. Nothing is accessible for "use," that is, for active intussusception by the human being, until it has first been put through the curriculum. The schoolroom is by implication the doorway to reality, and indeed the only doorway. It is true that Ramus would not quite subscribe to so bald a statement. He makes much of a kind of "natural" grammar and rhetoric and logic.[23] But Ramus is far from the most consistent of thinkers. The implication is there, for every educator to read and take comfort from, that in the last analysis the curriculum is all. Ramus is indeed a pedagogue's pedagogue. In his work *On Ramist Studies* (*De studiis Rameis*, 1597), Caspar Pfaffrad is true to his master's teaching when he says, "*Formal* education brings man to his *natural* perfection."

[23] Ong, *Ramus, Method, and the Decay of Dialogue*, p. 8, and *passim*.

7 | Ramist Method and the Commercial Mind

One of the persistent puzzles concerning Peter Ramus and his followers is the extraordinary diffusion of their works during the sixteenth and seventeenth centuries. The general pathway of this diffusion has been well known since Waddington's *Ramus* in 1855. It proceeds chiefly through bourgeois Protestant groups of merchants and artisans more or less tinged with Calvinism. These groups are found not only in Ramus' native France, but especially in Germany, Switzerland, the Low Countries, England, Scotland, Scandinavia, and New England. Perry Miller's work, *The New England Mind: The Seventeenth Century* (New York, 1939), is still the most detailed study of Ramism in any such group. Many merchants and artisans were becoming much more influential socially and were improving themselves intellectually, and Ramism appealed to them as they moved up. Ramus' works thus enjoyed particular favor not in highly sophisticated intellectual circles but rather in elementary or secondary schools or along the fringe where secondary schooling and university education met. For the most part, despite a persistent but uneasy currency in some German university circles,

Ramism never established itself firmly anywhere as a university movement.

But to note this well-marked general pathway of diffusion is not to explain Ramus' popularity. Scholars such as Pierre Mesnard, Neal W. Gilbert, and Wilhelm Risse have declared themselves still puzzled by the quantity and popularity of works by Ramus and his literary associate Omer Talon.[1] These works, as noted in Chapter 3, run to nearly 800 known and identified editions represented by extant copies, or, if one numbers separately works published together, to some 1,100 editions.[2] Almost all of these appeared between 1543 and about 1650. These figures represent only editions of works written by Ramus himself and, in much smaller numbers, by Talon, who died before Ramus after writing in collaboration with him for some years. They do not include

[1] Pierre Mesnard, review of Walter J. Ong, *Ramus, Method, and the Decay of Dialogue* (Cambridge, Mass.: Harvard University Press, 1958) and *Ramus and Talon Inventory* (Cambridge, Mass.: Harvard University Press, 1958), in *Bibliothèque d'Humanisme et Renaissance*, XXI (1959), 568–576; Neal W. Gilbert, review of the same volumes in *Renaissance News*, XII (1959), 269–271; Wilhelm Risse, review of the same volumes in *Deutsche Literaturzeitung*, LXXXI, No. 1 (1960), cols. 7–11.

[2] Ong, *Ramus, Method, and the Decay of Dialogue*, p. 5, where the count is based on Ong, *Ramus and Talon Inventory*. In the latter work, p. 14, I have conjectured that "in all the countries together where Ramism appears there is perhaps a total of about two hundred further printings which might qualify as editions or adaptations of one or another of Ramus' and Talon's works" and which do not appear in the *Inventory*. Since publication of these two works, co-operative readers, among whom I must mention particularly Dr. Wilhelm Risse of Cologne, have supplied me with additional entries to the number of nearly thirty at present, a few of these being rare Scandinavian editions, and I have located a good many more myself. I plan an enlarged edition of the *Inventory*.

any works about Ramus written by others or any of the countless works manifesting Ramus' influence which were published during this same period.

My own attempts to explain the appeal of the Ramist view of knowledge and of the educational process have focused in part upon the shift in sensibility marked by the development of typography. This shift brought Western man to react to words less and less as sounds and more and more as items deployed in space. Printing made the location of words on a page the same in every copy of a particular edition, giving a text a fixed home in space impossible even to imagine effectively in a pretypographical culture. Printing thus heightened the value of the visual imagination and the visual memory over the auditory imagination and the auditory memory and made accessible a diagrammatic approach to knowledge such as is realized in the dichotomized tables which often accompanied the typographical treatment of a subject at the hands of Ramists and, to a lesser extent, of their contemporaries. Typography did more than merely "spread" ideas. It gave urgency to the very metaphor that ideas were items which could be "spread."

My attempts to explain in part the currency of Ramism have also relied heavily upon certain pedagogical conditions and procedures which grew up because of the youth of most students. But I am quite aware that these explanations, whatever their value, do not account for everything in the spread of Ramism, for they stop short of showing why Ramus' works appeal to certain groups of persons more than to others. Supplementary explanations have been suggested. Gilbert conjectures that Ramism perhaps appealed to the Protestant ethic by providing a way of training students "to make useful contributions to the betterment of man's estate in the shortest possible time." [3] Or perhaps Ramism served more particular religious needs. Ramus' development of Rudolph Agricola's argumentative logic to supersede Peter of Spain's less argumentative logic came at a time, Professor Howell believes, when argumentative logic was much in demand for religious controversies, and lost favor when a new demand for a logic of inquiry was met by Descartes.[4]

These suggestions have value, but, I believe, need still further supplementation. Ramists were not any more articulate about bettering man's estate than were those, such as Francis Bacon, who held Ramists in relative scorn. And, while Ramist logic was used by some religious controversialists, it was passed over almost completely unnoticed by others. Calvinists took to it most freely, but Catholics, Anglicans, most Lutherans, and many others who were in the heat of the religious fray paid it little heed and less respect. Indeed, its greatest currency was among masters in secondary schools, who were often far removed from actual religious controversy.[5] Although it is true that they may have been governed by some concern with such controversy, the proponents of Ramism produced with Ramist "method" relatively little theology but rather works adapted for lawyers (as those by Johann Thomas Freige and Abraham Fraunce), arithmetics (by Heizo Buscher), and, often influenced by currents outside Ramism as well as by Ramus himself, polymath compendia presaging our modern encyclopedias (Freige's *Paedagogia* or Alsted's encyclopedic works). Theologians influ-

---

[3] Gilbert, p. 271.

[4] Wilbur Samuel Howell, "Ramus and the Decay of Dialogue," *Quarterly Journal of Speech*, XLVI (1960), 90.

[5] Ong, *Ramus and Talon Inventory*, *passim*.

enced by Ramism can be militant defenders of their own Protestant religious views, but when they are, the most prolific of them, such as Johannes Piscator or William Ames, tend to be systematizers and consolidators of lines of defense rather than debaters, and some, such as Thomas Granger, who published an English adaptation of Ramist logic for the use of divines, are remarkably free from contentiousness. Moreover, the collections of theological *loci* which are the armories for religious dispute lie outside the Ramist tradition, even when a Ramist such as Alsted did publish such things.

I do not mean to suggest that Ramist logic and rhetoric did not serve disputatious purposes, for they all too obviously did. But their use in disputation was subordinate to their use in education generally. Whereas Ramist logics are seldom advertised as serving the purposes of debate, they frequently manifest an express concern with simplification for pedagogical purposes. Many editions, such as those prepared by Marcus Friedrich Wendelin, state explicitly on their title pages their usefulness for "youthful beginners" (*tirones adolescentes*).[6]

One of the difficulties in accounting for the appeal of

---

[6] *Logicae institutiones tironum adolescentium captui accommodatae . . . a Marco Frederico Wendelino* (ed. novissima, Amsterdam, 1654). Among the many other editions advertising their pedagogical value on their title pages one might note: Beurhaus's "pedagogical dialectic," *De P. Rami Dialectica praecipuis capitibus disputationes . . . et . . . comparationes: . . . paedagogiae logicae pars secunda* (Cologne, 1588), . . . *pars tertia* (Cologne, 1596); Freige's Ramist logic for beginners, *Logica ad usum rudiorum in epitomen redacta* ([Nürnberg?]: typis Gerlachianis, 1590); the editions published by Wéchel and Antonius "with the commentaries omitted for the sake of studious youth," *Dialecticae libri duo, nunc in gratiam studiosae inventutis absque commentariis in lucem editi* (Frankfort-on-the-Main, 1591; Hanau, 1598; Hanau, 1600); etc.

Ramism is its complex relationship to the society which gave it birth and surrounded it. Elsewhere I have attempted to define Ramism as

at root a cluster of mental habits evolving within a centuries-old educational tradition and specializing in certain kinds of concepts, based on simple spatial models, for conceiving of the mental and communicational processes and, by implication, of the extramental world.[7]

We can no longer be satisfied with a view which presents Ramus and his followers as revolutionaries whose aims can be defined in terms neatly opposed to a group of other academic persons, his enemies.

Ramism has little which is *absolutely* distinctive of itself. Most things which originate in the Ramist camp—for example, the concept of "logical analysis," which seems certainly to put in its first appearance among Ramists—soon prove so congenial to other "systems" that they are quickly assimilated by them and, like logical analysis, seem always to have belonged to non-Ramists as well as to Ramists. What is really distinctive about Ramism is not one or another such item, but its special concentration of items, most of which can be discerned individually outside Ramist circles.[8]

II

Some new leads concerning the appeal of Ramism to the social groups espousing it were opened a few years ago by R. Hooykaas of the Free University of Amsterdam in his excellent book *Humanisme, science, et Réforme: Pierre de la Ramée, 1515–1572*, published at Leyden in 1958. Hooykaas

[7] *Ramus, Method, and the Decay of Dialogue*, p. 8.   [8] *Ibid.*

treats Ramus' works without special attention either to the educational milieu or to the dialectic which, as "the art of arts, the science of sciences, possessing the way to the principles of all curriculum subjects," [9] tied in the Ramist outlook with the educational milieu. Instead, Hooykaas' chief interest is the commercial world of the artisans and merchants in whose schools the Ramist curriculum took such strong hold. He traces a connection, carefully and with appropriate nuances, from the "literary empiricism" of Ramus and other humanists to the "scientific empiricism" of Bacon, which he finds congenial to the bourgeois mind. The connection is established largely through Ramus' interest in thinking in accord with "nature," his emphasis on classroom exercises, the lip service he pays to "induction," his interest in an astronomy without "hypotheses" as manifest in his correspondence with George Joachim Rheticus, and his devotion to mathematics as a practical subject.[10] Ramus' commitment to the ideals of the bourgeois world, shown for example in his desire to found mathematics on the practice of bankers, merchants, architects, painters, and mechanics, is interpreted by Hooykaas as favoring a more empirical and experimental approach to science on the grounds that these last-named occupational groups were dealing with new problems and techniques which tended to encourage more empirical and experimental approaches to knowledge.

Hooykaas makes less of Ramus' talk about "induction" than of his enthusiasm for *usus*, that is, classroom practice or exercise, in establishing the relationship of Ramus' educational

[9] This definition of dialectic opens the *Summulae logicales* of Peter of Spain, against whom Ramus and other humanists rebelled. But the paramount role assigned by Peter to dialectic or logic was not by any means contested by Ramists.

[10] Hooykaas, *Humanisme, science, et Réforme: Pierre de la Ramée, 1515-1572* (Leyden: E. J. Brill, 1958), especially pp. 20–32, 64–90.

aims and procedures to bourgeois culture. The break with older ways both among burghers generally and among Ramists in particular consisted more in an interest in pupil activity than in anything we should recognize today as experimentation or "induction." These points Hooykaas makes are valid and follow recent lines of thinking in discerning a certain intellectual fertility in the meeting of the artisan and academic minds during the sixteenth and seventeenth centuries. But how far does a stress on *usus* explain Ramus' special appeal? Insistence on practice or exercise is hardly distinctive of Ramism. The humanists commonly exhibit the same emphasis, and Ignatius of Loyola found this stress so general at Paris that he recognizes it as the great difference between the educational procedure at Paris (*modus et ordo Parisiensis*) and that in other universities he knew.[11] Ramus, it is true, was known as the *usuarius*, but this sobriquet was apparently given him in mockery not because he alone talked of *usus* but because he made such an issue of *usus* during a hearing before the Parlement, at which his opponents give the impression that he was resorting to mere histrionics, since *usus* was a procedure to which none of them was particularly opposed.[12]

There are, however, other grounds besides those examined by Hooykaas which confirm his belief in the connection between the bourgeois mentality and the Ramist outlook. If the artisans' and the burghers' interest in induction and what Ramus styles *usus* is at times equivocal, their interest in manufacturing and commerce is patent and easy to deal with. I

[11] George E. Ganss, S.J., *Saint Ignatius' Idea of a Jesuit University* (Milwaukee: Marquette University Press, 1954), p. 30.

[12] Adrien Turnèbe, *Disputatio ad librum Ciceronis de fato* (Paris, 1556), fol. 27; Nicolas de Nancel, *Petri Rami . . . vita* (Paris, 1599), in Nancel, *Declamationum liber . . . : addita est P. Rami . . . vita . . .* (Paris, 1600), p. 36.

should like to suggest that one of the major features of Ramism is its tendency to reduce knowledge to something congenial to the artists' and burghers' commercial views. In particular, Ramus' central and controlling concept of "method," which Hooykaas largely passes over, gives evidence of what can legitimately be called a kind of intellectual commercialism. By this intellectual commercialism I do not mean a tendency of Ramus or his followers to be venal. I mean rather that in a veiled, unwitting, but nevertheless real fashion Ramus tends to regard the knowledge which he purveys in his arts as a commodity rather than as a wisdom.

Ramus lived in an age when the instruction of the universities was being organized still more efficiently with an eye to a "consumer culture." By comparison with the older university mind, the humanist mind was "consumer oriented" to no small degree. Rudolph Agricola, Erasmus, Melanchthon, Sturm, and their congeners tended to be not representatives of a teachers' guild, as university masters were, but to be freer agents of the family in the preparation of its younger members for life in society. Typically, therefore, they were pupil-oriented, interested in the kind of person whom they were producing quite as much as in professional attitudes toward the subjects that they were teaching. The works on educational methods generated by humanism show the overpowering tendency to view knowledge in terms of those to whom it was purveyed.

There is an obvious relationship between this mentality and the mentality of a commercial, merchandising world, where goods had to be thought of in terms of operations with a view to possible users or consumers. Ramist "method" makes it possible to think of knowledge itself in terms of "intake" and "output" and "consumption"—terms which were not familiar to the commercial world in Ramus' day, of course, but which do refer to realities present within that world.

III

The "method" which Ramus proposes is central to the entire Ramist operation and to the movements which develop in the wake of Ramism. Nevertheless, the influence of Ramus' concept of method is not easy to trace with absolute precision. The concept itself is somewhat protean or chameleonlike, at times losing outline when it is applied—as Ramus insisted it should be—to every conceivable subject and to some which may strike us now as inconceivable. Basically, however, Ramist "method" is a way of organizing discourse. In common with most of his contemporaries, and despite the attenuated rôle he assigned to rhetoric, Ramus assumed that the typical form of discourse was the oration, which, moreover, set out to "prove" something. To avail oneself of Ramus' method, after first discovering "arguments" which will prove what one wishes to say, one organizes the arguments into enunciations and syllogisms, and then, to relate the syllogisms to one another, one avails oneself of the principles of "method." [13] These principles are utterly simple: one proceeds from the better known to the less known, and this, Ramus insists, means always that one proceeds from what is general to what is particular.

Proposing his method as applying to all discourse whatsoever and assuming that oratory is discourse in its basic form, Ramus further assumes that the ultimate purpose of all oratory and of all discourse is to teach. Here he is following a

---

[13] For a full explanation of Ramist "method" and the relevant texts in Ramus' works and the works of others, see my *Ramus, Method, and the Decay of Dialogue*, esp. pp. 225–269.

line of thought feeding out of classical antiquity, but he is also yielding to and indeed strengthening the prejudices developed in the massive pedagogical tradition of the universities.[14] The process of teaching he refers immediately to the classroom. His favorite example of the use of method is its use to organize material presented in a course in grammar. The fact that this material is hardly comparable to that in an oration, that it is not "found" by running through the "arguments" (which non-Ramists usually called the places or commonplaces of invention), that it is not organized syllogistically, and that it is very difficult to see how the declension of nouns would be more "general" than the conjugation of verbs, or vice versa, makes no great difference. All this and more does not deter Ramus from his insistence on a "one and only method," from general to particular, applying first to curriculum subjects but also to all other discourse as well.

There is, of course, a very loose sense in which such a concept of method can apply to at least some portions of a book on grammar as well as to certain other instances of discourse. But Ramus strains this sense beyond all endurance. He makes only one concession to his rule of proceeding from the general to the particular. This is his "method of prudence" or "cryptic" method, which is used by orators and poets particularly. The "method of prudence," however, turns out to be simply the usual method running backwards. It allows some particulars to appear before generalities. Such "cryptic method," moreover, is used only in dealing with recalcitrant, unusually ignorant, or otherwise ill-disposed audiences, and for reasons which, it must be avowed, Ramus is at a loss to explain with any convincingness at all.

One of the curious features of Ramist method is that, al-

14 *Ibid.*, pp. 149–167. See also Chapter 6 above.

though it is only a subordinate part of Ramus' dialectic, which rules all discourse, and the last part at that, it somehow or other envelopes the entire field of discourse. This it does because it governs the larger units of expression as wholes: a letter, a curriculum subject, an oration. When one has discovered all one's arguments and "disposed" or arranged them in enunciations and syllogisms, "method" thereupon determines how these are to be assembled for the total speech or other presentation. The "one single method" of Ramus purportedly could organize anything, even, as it turned out, the Christian religion itself, which is methodized in Ramus' posthumously published *Commentaries on the Christian Religion* (*Commentariorum de religione Christiana libri quatuor*, 1576). Ramists took pride in the stubborn simplicity of their one-method views. Eight years after Ramus' death, his follower William Temple (1555–1627) is able to distinguish with horror no less than 110 different senses in which his anti-Ramist opponent Everard Digby attempts to employ the term "method," urging the ease of the purportedly single-method Ramist approach over this chaotic assortment of Digby's.[15] Nearly a hundred years after Ramus' death his followers are still sure of the complete serviceability of Ramist method in the face of no matter what intellectual crisis. Jan van Aelhuysen explains in his 1664 edition of Ramus' *Dialectic* that, just as dialectic is concerned with both being and nonbeing, so method is concerned not only with the arts but with absolutely everything whatsoever (*omnium omnino rerum*) which the poet or the orator or the historian or the philosopher wishes to teach (that is, to deal with in any way), even though

---

[15] William Temple, *Admonitio Francisci Mildapetti* (Frankfort-on-the-Main, 1589), pp. 97–100; this work was first published in 1580 at London.

such persons do at times stray a little from the way which they should follow.[16]

In part, this purportedly universal Ramist method represents literary humanism run wild. At first blush, it is true, the cold-blooded, analytical, diagrammatic, and even mechanistic view of communication which Ramus' concept of method implies makes any connection with the purely literary implausible. Ramists were, as a matter of fact, the least literary of people. They tended to be dry-as-dust analysts, polymaths with a penchant for classification. Nevertheless, it is a historical fact that Ramist method has its source in rhetoric. In *Ramus, Method, and the Decay of Dialogue* I have spelled out the way in which Melanchthon, Sturm, and Ramus had all found treatments of "method" established first in a rhetorical context, in the works of Hermogenes and in letter-writing manuals, and how they simply transplanted this originally rhetorical method into their manuals of dialectic or logic and with no further ado styled it logical method from this point on. The transplantation took place between 1543 and 1547, and is all the more evident because all three authors had originally published treatments of dialectic or logic without this section on method which they tacked on later.[17]

Back of Melanchthon's, Sturm's, and Ramus' concept of method there lies nothing like any rigorous logical inquiry. "Method," when they picked it up, signified a rule-of-thumb organization serving essentially practical rhetorical purposes —helping one to meet a situation in which one was called

[16] Ramus, *Dialectices libri duo, cum notis variorum quas colligit . . . I. van Aelhuysen* (Tiel, 1664), pp. 170–171 (Lib. II, cap. xviii).

[17] I am indebted to Dr. Wilhelm Risse of Cologne for pointing out to me that there are indications of interest in method to be found in Melanchthon rather earlier than I had suspected, although I have as yet been unable to follow up his valuable leads.

upon to say something. It did not properly belong to a logic of inquiry or a logic of speculation at all. For pedagogical reasons such a concept of method was transferred to logic, being rigidified in a mechanical sort of way in the process. This operation just before the mid-sixteenth century appears to have been the maneuver establishing the tradition that a treatise on method belonged in logic—as the age of Descartes was simply to assume. One has the impression that the assumption was an egregious blunder based on ignorance of intellectual history. The careful thinking which was supposed to underlie the conviction that method was a matter of strict, formal logic had in reality apparently never been done.

IV

The affinity between Ramist method as just described and the practical, down-to-earth outlook of burghers and artisans shows not as an explicit, declared relationship but rather in terms of a state of mind. Ramist method presented a view of the intellectual world congenial to the dealer in physical wares. The merchant and artisan were concerned with commodities, with things visible, definite, and things moreover demanding itemization and inventory. The peasant dealt with things too, but in a way closer to nature and less involved with account keeping. With the burgher, control achieved by a tally of wares was much more important. Studies exemplified by John U. Nef's *Cultural Foundations of Industrial Civilization* (Cambridge: The University Press, 1958) have shown the massive account keeping and the trend toward statistical analysis of operations which marked the development of commerce through the late Middle Ages and on past the Renaissance. Edgar Zilzel has traced connections between the growing capitalist economy and the development of the quantitative method in the physical sciences.[18]

The bourgeois world and the philosophical and religious worlds met often in the sixteenth and seventeenth centuries, of course. Burghers and artisans did not always neglect the quest of wisdom, and still less the quest for union with God through Christian piety. But the world with which they habitually dealt day by day was somewhat less than neoplatonic. It was concrete, resistant, sensible, and, if one was to make a profit, it had to be kept in a certain physical order. When a mind accustomed to such a world turned from it to the unfamiliar and misty realms of learning, Ramist method could be highly reassuring. It enabled one to deal with the imponderables which haunted the world of learning and to deal with them on a basis more definite and orderly than had hitherto been available. If the definiteness was sometimes illusory and the order superficial or even irrelevant to the subject in hand, Ramism would of course not survive—as it did not survive. But learning was such a mystery that it would take some time before the deficiencies of any moderately plausible mode of facile organization could become evident to the minds to which the organization had great initial appeal.

In a way it is true that Ramist method appealed primarily to a desire for order, not to a desire for experimentation. First of all, it was a method of organizing discourse, not of working with things. And secondly, even in the organization of discourse, it did not serve the purposes of either narration or description, which are paramount for dealing with experimentation. In its neglect of narrative and description in our

[18] Edgar Zilzel, "The Sociological Roots of Science," *American Journal of Sociology*, XLVII (1941–42), 544–562.

present sense of these terms and its concentration upon oratory and "teaching," upon moral or practical issues between men, Ramist education was like most education of the times. In the ordinary pedagogical tradition, which tended to consider history, poetry, and even letter-writing as forms of oratory, it was unthinkable to write about an individual "objectively." The whole purpose in speaking or writing of anyone, as one can see in so standard a manual as Aphthonius' *Progymnasmata*, was to praise or blame him. Even in treating of inanimate objects, training in cold, objective, uncommitted description is by and large simply not a part of formal education in the arts of language at least through the seventeenth century. Consequently it is very difficult to find a real connection between any pedagogical system, based as these generally were upon training in the language arts, and an experimental frame of mind. The pedagogy of the time bore hard on personal relations.

Nevertheless, within this framework the Ramist approach did favor a kind of objective view. By moving all questions of organization from rhetoric to dialectic and by concentrating there on "method," Ramus takes what might be called an itemizing approach to discourse. Personal issues tend to pale in this climate, as discourse itself becomes highly diagrammatic in concept. Ramists did not train for objective description such as might have served an experimental approach to reality or even for narrative such as might have served Bacon's not too clear proposals for "histories" of things, but they did inculcate an attitude toward discourse itself which made of discourse a kind of thing. Ramus helped make it plausible to dream of itemizing discourse and even thought itself. At this point his attitude becomes of great interest to the artisan and merchant mind.

v

Although Ramus presents method as part of dialectic or logic, the appeal of his concept of method to the artisan and merchant classes becomes particularly understandable when we consider it in connection with printing. Printing was developed by artisans and promoted by enterprising commercial interests—most of our knowledge of Gutenberg and his associates comes, significantly, from records of lawsuits. Whereas manuscript production had demanded copyists who belonged more to the world of letters than to the artisan's world, printing demanded large numbers of artisans. The compositor had to have some minimum literacy, but he did not have to be able to pronounce, much less to understand, the words which he was called upon to set in type. He was one more remove from the spoken word than the scribe had been. Basically his work was a transaction not in words but in local motion, for printed books were the products of an assembly line—perhaps the earliest real assembly line. This fact established a close connection between printing and the other crafts. At the same time, the growth of the book trade together with the increased production of reading matter which printing made possible gave book production a greater commercial importance after typography than before. We have long known that printing was a commercial enterprise, but our knowledge of the size and details of the enterprise is growing continually, thanks to detailed case studies such as that by Robert M. Kingdon.[19]

Printing, however, involved not only the craftsmen and

19 Robert M. Kingdon, "The Plantin Breviaries: A Case Study in the Sixteenth-Century Business Operations of a Publishing House," *Bibliothèque d'Humanisme et Renaissance*, XXII (1960), 133–150.

merchants, but the academic world as well. Indeed, it brought the crafts and commerce into direct contact with the world of learning more than ever before.[20] Across Europe in the sixteenth century we see persons and entire families directly involved in learning or teaching on the one hand and in publishing activity on the other. Johann Sturm in the Rhineland and Paris, Amerbach at Basel, and the Manutii in Italy, or the Rastells in England are typical examples.

The mass-production methods utilized for bookmaking made it possible and indeed necessary to think of books less as representations of words serving for the communication of thought and more as things. Books came to be regarded more and more as products of crafts and as commodities to be merchandised. The word, living human speech, is here in a sense reified. Even before the advent of typography a marked reification of the word had been begun by the medieval terminist logicians, and elsewhere I have discussed in detail the psychological connections between terminist logic, the topical logic which succeeded it in the humanist age, and the development of attitudes toward communication favoring typography.[21] The terminist logic was indeed still represented at Paris in Ramus' youth or not long before by persons such as Juan de Celaya, John Dullaert, and John Major, and even later by Ramus' own defender Jean Quentin. But the terminist tradition had favored the reification of the word for intellectual purposes. This impulse to reification had come from the

[20] See Curt F. Bühler, *The University and the Press in Fifteenth-Century Bologna*, Texts and Studies in the History of Mediaeval Education, No. VII (Notre Dame, Ind.: University of Notre Dame Press, 1958).

[21] Walter J. Ong, *The Barbarian Within* (New York: Macmillan, 1962), pp. 68–87; *Ramus, Method, and the Decay of Dialogue*, pp. 307–314.

academy. When we view typographical developments from the burghers' standpoint, we see another type of drive to reification joining forces with the first. If the logicians wanted to hypostatize expression so as to subject it to formal analysis, the merchants were willing to hypostatize expression in order to sell it. Ramus' publisher André Wéchel, whose firm printed at least 172 different editions of one or another work of Ramus' or Talon's, shows a keen sense of Ramus' books as tangible realities, subject to inventory. He writes of them not as records of what has been said, as efforts at communication, but quite frankly as "bear cubs," which he says this admirable author licked into shape through successive revised editions—much to Wéchel's profit, it might be added, and not too secret delight.[22]

VI

At this point Ramus' concept of method becomes particularly interesting because of the fact that this standard description of method is highly reminiscent of printing processes themselves, so that it enables one to impose organization on a subject by imagining it as made up of parts fixed in space much in the way in which words are locked into a printer's form.

In Ramus' standard description of method he imagines the rules, definitions, and divisions of grammar as having been "found" by means of a kind of hunting expedition, as having been properly "judged" (that is, put into enunciations), and as

[22] Wéchel's preface in *A. Talaei Rhetorica, P. Rami praelectionibus illustrata* (ed. postrema, Paris: A. Wéchel, 1567). Ramus' adversary Jacques Charpentier (Carpentarius) also had referred to licking his own writings into shape, but before publication rather than after. See his *Descriptio universae naturae pars posterior* . . . (Paris: Gabriel Buon, 1564), fol. *iij^v ("Carpentarius lectori").

being thereupon written out on little slips of paper. The proper spatial arrangement of these slips, the more general first and the more particular last, comes next, he says, and calls for "method," which, indeed, is the only part of dialectic or of any other art or science capable of telling how to arrange such slips of paper.[23] The association of method with writing is here evident, but even more evident is the similarity between the process of "methodizing" these slips and the process of setting up type taken from a font. In each instance, the composition of continuous discourse is a matter of building up discourse by arranging pre-existing parts in a spatial pattern.

The effect of Ramus' concept of method upon the typographical format of his own textbooks has been discussed elsewhere.[24] In general, as these books evolve, they become more organized for visual, as against auditory, assimilation by the reader. Paragraphs and centered headings appear, tables are utilized more and more until occasionally whole folio editions are put out with every bit of the text worked piecemeal onto bracketed outlines in dichotomized divisions which show diagrammatically how "specials" are subordinated to "generals." This increased sophistication in visual presentation is not restricted, of course, to Ramist writings, but is part of the evolution of typography, showing clearly how the use of printing moved the word away from its original association with sound and treated it more and more as a "thing" in space. Nevertheless, in Ramist works this sort of evolution throws

[23] Ramus, *Dialectici commentarii tres authore Audomaro Talaeo editi* (1546), p. 84—a work certainly by Ramus; see Ong, *Ramus and Talon Inventory*, no. 3. This description is repeated over and over again in later editions of Ramus' works on dialectic.

[24] Ong, *Ramus, Method, and the Decay of Dialogue*, pp. 311–312.

light on the Ramist concept of method, for words which admit of neat diagrammatic or semidiagrammatic presentation are those which are methodized properly, whereas those which resist such presentation are not effectively methodized. In principle, every subject properly treated by a Ramist admitted of being diagrammed on bracketed dichotomized outlines of the sort which editors such as Freige or Samuel Sabeticius [25] use to present Ramus' works. It should be noted that, while such outlines are not unknown in the pretypographical manuscript tradition, in this tradition they do not admit of the ready multiplication which printing makes possible. In manuscripts, diagrams are much more laborious productions than straight text, for manuscript copying only with great difficulty controls the position of material on the page. Typographical reproduction controls it automatically and inevitably. Ramist method is psychologically of a piece with typographical reproduction.

It is telling in this connection that Ramus' hostility to the pedagogical works on which he comments is generally directed against pretypographical productions and is uniformly grounded on their lack of "method." If Ramus did indeed defend his reputed anti-Aristotelian thesis, *Quaecumque ab Aristotele dicta essent commentitia esse*—for the story of his having defended such a thesis, as I have attempted to show elsewhere,[26] may be only a legend and not a fact—he quite evidently meant this to say not that Aristotle was untruthful (a common interpretation of the reported thesis) but rather that Aristotle's material was poorly organized, not properly controlled by "method." In other words, it was unsuited for the Gutenberg era. Even Aristotle, however,

[25] Ong, *Ramus and Talon Inventory*, no. 40.
[26] Ong, *Ramus, Method, and the Decay of Dialogue*, pp. 37–41.

could conceivably be reformed by method, if one thought him worth the effort, and in the Ramist age there are in fact several examples of attempts to methodize him. Jacobus Martinus was one of those who set out to do just this, arranging Aristotle's logic with its common or general principles first and following these with its proper or special principles.[27]

VII

The accommodation of subjects to the spirit of the age by organizing them according to "method" was, I have so far suggested, an accommodation to the artisan and merchant mind in so far as it reified the field of learning and discourse. In the highly spatialized Ramist concept of dialectic, the mysterious realm of knowledge was reduced to something one could manage, almost palpably handle. The arts and sciences could be viewed as a mass of "wares." But there was another aspect of Ramist method which gave it appeal to the bourgeois groups among whom it was a favorite. It was very much like account-keeping. The merchant not only deals in wares, but he has a way of keeping record of them which levels all sorts of items to the pages of an account book. Here the most diverse products are mingled on an equal footing—wool, wax, incense, coal, iron, and jewels—although they have nothing in common except commercial value. To handle a merchant's wares in terms of his account books, one need not trouble oneself with the nature of the wares. One has to know only the principles of accounting.

Ramist method offered the vision of a world of knowledge leveled in much the same businesslike way. A young boy

[27] *Iacobi Martini . . . Logicae Peripateticae per dichotomias in gratiam Ramistarum resolutae libri duo* (Wittenberg, 1614). In point of fact, this book tends to make Aristotle out as a Ramist.

taught the principles of method could feel assured in advance of control of any body of knowledge which might come his way. If he were intelligent, he might set out to organize all human knowledge according to some simple principles of method, confident even before he approached a new field that he would know quite well how to deal with it. This is the state of mind which produces polymaths in the Ramist tradition such as Johannes Piscator or Johann Heinrich Alsted. When one reads Alsted's huge encyclopedias one after another—for he produced several complete encyclopedias of his own—one cannot but be impressed with the order and sweep of the presentation. Only if one begins to read the text does one begin to wonder how much of this organization is concerned with really communicating anything at all vital about the material purportedly under consideration. And if one attempts to read aloud, one sees immediately that the effect is like that of reading an account book. The presentation of the material is thoroughly dependent upon the visualization of its various divisions and subdivisions. A methodic pattern is imposed adamantly upon the subject at hand, no matter what it is, not in order to understand it but to level it so that everything is of a piece with everything else, much as monetary values are assigned to different sorts of goods so that they can be "balanced" against one another.

The tendency to look for an identical format for any and all knowledge is not new with the Ramists. The human mind always likes to simplify. But Ramists carried this tendency farther than most. Moreover, in doing so they took advantage of a current situation which made methodization and systematization particularly gratifying. There is no time when method and system are more valued than when one is faced with utter chaos. In the sixteenth century, and still more in the seventeenth century, knowledge was growing to an extent which dismayed even the most sanguine. How could one organize the new developments in knowledge or the burgeoning older fields such as geography and history which were asserting themselves more and more in the curricula? How could one even think of them as what we today call "subjects"? This term had other meanings, and the term "art" was more or less sacrosanct, pretty well pre-empted by the standard subjects forming the base of the arts course, namely, grammar, rhetoric, and dialectic. The huge literature regarding the interrelationship of the sciences inherited from past ages was growing apace. Disputes raged as to what was the "matter" and what the "form" of the various branches of knowledge, and concepts concerned with organizing the various branches of learning floated confusingly about—concepts such as *technologia*, which the Ramist Johann Heinrich Bisterfeld says is "the art of reducing terms to the liberal disciplines." [28]

In this situation the idea of method was a godsend. In the face of growing knowledge, it gave courage. Any branch of knowledge, old or new, could be thrown together in any fashion whatsoever and called a "method," as indeed all branches were. Before Ramus began to publish, the trend had started with titles such as Gemma Frisius' "method" of practical arithmetic (*Arithmeticae practicae methodus*, 1540, 1544, etc.), or even Erasmus' earlier "method" of true theology, *Ratio seu methodus perveniendi ad veram theologiam* (Strasbourg, 1521), and it continued through hundreds of other titles during the next three centuries. The list of "method" books cannot be elaborated here, but a cross section

[28] J. H. Bisterfeld (Biesterfeld, Bisterfeldius), *Bisterfeldius redivivus* (The Hague, 1661), II (Part II), 37–41.

of some of the early titles can be found in Neal W. Gilbert,[29] where it becomes apparent that the term is at first used almost exclusively for pedagogical works.

These treatises which style themselves "methods," like the others styling themselves "systems" or "syntagmata" or "apparatus" or "ideas" or "analyses" (which is often translated in English by "resolves") of this or that subject, would still repay a great deal of investigation. It is quite impossible at present even to conjecture how many hundreds of pedagogical and parapedagogical works in the sixteenth and seventeenth and eighteenth centuries are built around such conceptions. It will be sufficient here, however, to note simply that all these approaches to knowledge are approaches congenial to persons who habitually deal with reality in terms of accounting rather than in terms of meditation or wisdom. "Method" is an early step in the procedures which encode knowledge in a neutral, leveling format, reducing it to bits of information such as those which will eventually make their way into electronic computers. From this point of view, Ramus' influence is felt not in terms of experimentation but in terms of calculators and business machines. Many rising commercial groups of course were prevented by various inhibiting factors from becoming interested in Ramism, but in those large groups where Ramism was adopted, the appeal of Ramist method as a somewhat cryptic but very real kind of accountant's approach to knowledge—and indirectly to reality—helps explain the extraordinary fascination which Ramist works had for the artisan and commercial mind.

[29] *Renaissance Concepts of Method* (New York: Columbia University Press, 1960), pp. 233–235.

8 | Swift on the Mind: Satire in a Closed Field

I

It is with Jonathan Swift as it is with most essayists whose pronouncements spring from impulses more strategic than scientific. The frames of thought in which the observations are set are often of more significance than the observations themselves. In Swift's case, certain of these frames of thought, intimately connected with the milieu in which Swift moved, can yield valuable insights into his own mind and style and can account for certain of his attitudes toward the human mind and its functions—perhaps most crucially of all for some remarks of his concerning the vexed question of poetry.

In common with most men of his age, Jonathan Swift put great trust in "reason." Thirty years ago,[1] Gordon McKenzie

[1] Gordon McKenzie, "Swift: Reason and Some of Its Consequences," in *Five Studies in Literature*, by B. H. Bronson *et al.*, University of California Publications in English, Vol. 8, No. 1 (Berkeley: University of California Press, 1940), pp. 101–129. Citations of Swift's works throughout the present study are made from the following editions: (1) *Prose Works of Jonathan Swift, D.D.*, ed. by Temple Scott, 12 vols. (London: G. Bell and Sons, 1897–1908), referred to as *Works*. (2) *Correspondence of Jonathan Swift, D.D.*, ed. by J. El-

showed that this term "reason" in Swift most often designates not a process of arriving at truth but some sort of apprehension of truth independently of any process which may be involved. The notion of a process of movement through premises to the grasp of a conclusion—ratiocination as against the simple act of understanding or fully grasping a truth—is not denied in the usage of Swift and his contemporaries. It is simply pushed aside or not attended to at all. Reason is simply the ability to get at truth—however this is done—truth, "clear, pure, and complete." It comes to something like "right intuition." [2]

Swift felt the compulsion to establish a simple unity in reality—in the operations of the human mind as elsewhere. To a great extent the "difficulties of complexity and variety were anathema to him, whether or not they had any degree of organization." [3] Under the urge to realize such unity, Swift, with many of his contemporaries, not only falls back upon a greatly abridged notion of reason but bolsters this notion with that of "common sense." An echo of the protest against complicated philosophical maneuvers too difficult for most men to follow, this "common sense" is often equivalent to "reason" further simplified by dint of a more popular title.

"Common sense" is here the open door to the "right intuition" of reason. It provides that, at least in those matters

---

rington Ball, 6 vols. (London: G. Bell and Sons, 1910–14), referred to as *Correspondence*. (3) *Poems of Jonathan Swift*, ed. by Harold Williams, 3 vols. (Oxford: Clarendon Press, 1937), referred to as *Poems*.

[2] McKenzie, pp. 104, 109. As McKenzie has suggested, the tendency to reduce the concept of reason or ratiocination in this summary fashion is recurrent throughout human intellectual history as we know it.

[3] *Ibid.*, p. 114.

which concern the practicalities of life, the conquests of reason may be realized by any sane human being. The use of the term "reason" built up the assumption that the human intellect came upon truth in one fell swoop. The term "common sense" conveyed the assurance that competence in executing the simple but necessary operation was to be met with in everyone always. The consistency and unity warranted in human intellectual operations was further warranted by its being viewed as widespread.

Buoyed up by his trust in reason and common sense, Swift's confidence in the potentialities of the human mind remains always essentially unshaken. As he once told his biographer Deane Swift, he never could understand logic, physics, metaphysics, natural philosophy, mathematics, "or anything of that sort"—anything of the sort demanding close thinking.[4] Swift never ceased to ridicule such things. But through all his ridicule his serene confidence in reason and common sense remains unaffected. The realm of the intellect is basically unified and consistent.

Like most of his contemporaries, Swift never carried quite out into the open his notion of reason and common sense, nor his eagerness for immediately apprehensible unities in place of complexities, which to him were only annoying. These things existed not as any direct formulations but rather held their shape in terms of the general contour which Swift's thought had been given by its milieu.

Insofar as one's concepts are given form by being knocked about in one's milieu, as Swift's so largely were, they tend to acquire a consistency derived not from any closely studied and consciously worked out interrelations, but rather from the fact that they roughly fit into the same basic aggregate

---

[4] Deane Swift, *Essay upon the Life, Writings, and Character of Dr. Jonathan Swift* (London: Charles Bathurst, 1755), p. 30.

of patterns. Further examination of Swift's conceptualizations, which the present study undertakes, indicates that his ways of conceiving psychological operations beyond those of reason and common sense receive a rough but very significant kind of organization in terms of such a group of patterns.

The kind of patterning which they thus receive, it would seem, gives a common rationale to certain of Swift's most characteristic observations concerning various operations of the psychological organism—observations of his which seem otherwise to resist being gathered to any common ground. Among the remarks of Swift affected by such a rationale would be some of his observations on the function of poetry discussed some years ago in another connection by Herbert Davis.[5]

II

In describing the state of scientific speculation in the seventeenth and eighteenth centuries, Alfred North Whitehead has called specific attention to the importance of the conviction, which he derives from the medieval theological tradition, that there is an order in things which promises that there will be results attainable by scientific investigation.[6] He also has underlined the importance of the notion of the ideally isolated system which was employed by Galileo and Newton in their physics and which played so definitive a role in the formation of the contemporary mentality and that of succeeding ages.[7]

[5] "Swift's View of Poetry," in *Studies in English by Members of University College, Toronto*, collected by Malcolm W. Wallace (Toronto: University of Toronto Press, 1931), pp. 9–58.

[6] *Science and the Modern World* (New York: Macmillan, 1925), pp. 18–19.

[7] *Ibid.*, p. 66. Although Professor Whitehead does not trace it so far, one might suggest that this latter mentality also very likely has medieval roots—in the kind of thinking cultivated by Duns Scotus

The first of these factors in scientific speculation has an evident connection with Jonathan Swift's compelling desire to establish unities in things. This desire of his is due to the general conviction, like that which enters into scientific work, that there is an order in things, plus a personal disposition to impatience which demands that the order be got promptly in hand and things thus brought to heel immediately. But the second factor, the notion of the ideally isolated system, has an even more immediate relevance. The unity with which Swift wished to invest things was not any kind of unity but a blunt and downright kind. Swift cultivates some sort of elemental delimitation, a determination—unformulated again—to keep his thinking well cut back, to rein in this Pegasus which is the intellect if ever it should want to have its head and venture on a high flight. Such flights, Swift is persuaded, are not necessary. He is convinced that the answers one is going to come out with anyhow will be cast in terms which are thoroughly manageable, which will leave the mind in no straits, stretched on no tenterhooks. There is a readily discernible affinity between Swift's characteristic bluntness, his desire to do away with the frills, ruthlessly to consider all

---

and other nominalists or near-nominalists. Perhaps it is of some significance that in the world of Swift and his contemporaries, Scotus and not St. Thomas Aquinas represented scholasticism. See, for example, *A Tale of a Tub*, in *Works*, I, 56 ("complete abstract of sixteen thousand school-men, from Scotus to Bellarmin"); *Battle of the Books*, in *Works*, I, 164, where it is Scotus who is paired with Aristotle; *ibid.*, p. 173, where Aquinas is mentioned, but only between Scotus and Bellarmine. Scotism has a strong appeal to the mind oriented toward things at the mechanistic, physical-science level: see Béraud de Saint-Maurice, *Jean Duns Scot: Un Docteur des temps nouveaux* (Montreal: Thérien Frères, 1944), esp. pp. 159, 169, 300 ff. Scotistic conceptualization in terms of a multiplicity of substantial forms bears a general analogy to conceptualizations of Swift's discussed in this chapter.

appendages and complexities as irrelevancies, and the tendency to seek truth in terms of an ideally isolated, thoroughly simplified system.

The ideally isolated system was used chiefly in the Newtonian mathematical physics—for example, in the formulation of Newton's laws of motion. The system was conceived of, or at least interpreted, quantitatively. This fact that the mentality encouraged by the Newtonian milieu operated largely numerically, geometrically, quantitatively, has considerable relevance in Swift's case. Conceptualization in terms of an isolated system which tended to become geometrical and diagrammatic affords a framework for an immediate and abrupt simplification which is quite in accord with Swift's thought. If truth comes in this way—in delimited chunks which need be considered only quantitatively—"reason" is indeed most reasonable and "common sense" readily suffices. Naturally, one will be blunt in what he has to say, for that is the way reality is—blunt, chunky, chopped up in little pieces, frills removed, all the edges well in sight.

At this point, then, two questions might be asked. First, how far was Swift in contact with the philosophico-physical speculation of his day? And secondly, in his observations on the workings of the mind, does his imagery and mode of conceptualization give any evidence of connection with this speculation, particularly with the notion of ideally isolated systems conceived of spatially or quantitatively?

III

Swift was not much affected directly by the study of physics—this was one of the subjects he never could understand [8]—but he plainly enough was not a stranger to the intellectual climate in which the recent philosophico-physical

[8] Deane Swift, p. 30.

speculation grew. References in his own works make this clear.

Francis Bacon was considered by Swift to be the beginner of the new philosophy.[9] Swift had personally annotated his copy of Bacon's *Opera omnia*.[10] He cites Bacon often, and there are frequent echoes of Bacon throughout Swift's writings, such as that in the spider and bee episode in the *Battle of the Books*.[11] Bacon is considered by Swift one of the staples having place in a woman's education as well as in a man's.[12]

Swift's attention to Hobbes is not less noteworthy. From his earliest years he was acquainted with *Leviathan*.[13] Besides a two-volume set of Hobbes's philosophical works and a copy of his *Elementa de cive*, two copies of *Leviathan* are listed in the catalog of Swift's books sold at auction in 1745.[14] One of the copies of *Leviathan* is annotated in Swift's own hand. There is, moreover, definite record of a separate set of notes on *Leviathan* found in Swift's study, although these are now lost.[15] Swift cites Hobbes frequently.[16]

Hobbes is much more shot through with the mechanistic and geometrical than is Bacon. He regarded the geometrical

[9] W. Gückel and E. Günther, *D. Defoes und J. Swifts Belesenheit und literarische Kritik* (Leipzig: Mayer und Müller, 1925), p. 63; *Battle of the Books, Works*, I, 177–178.

[10] Harold Williams, *Dean Swift's Library: With a Facsimile of the Original Sale Catalogue* (Cambridge: The University Press, 1932), p. (15), no. 627.

[11] *Works*, I, 167–168; cf. Francis Bacon, *Novum organum*, i, 95, in *Works of Francis Bacon*, ed. by James Spedding, Robert Leslie Ellis, and Douglas Denon Heath (London: Longmans and Co., 1864), I, 201. For other citations see Gückel and Günther, pp. 55, 63.

[12] "My Lady's Lamentation," *Poems*, III, 855–856.

[13] Williams, p. 31.

[14] *Ibid.*, pp. (5)–(13), nos. 153, 202, 255, 506.

[15] Williams, pp. 26, 31.

[16] See Gückel and Günther, pp. 58, 72, 88.

method of demonstration as the true scientific method and liked this sort of conceptualization, which so readily associates itself with some of Swift's kinds of imagery: "When a man *Reasoneth*, hee does nothing else but conceive a summe totall, from *Addition* of parcels; or conceive a Remainder, from Substraction of one summe from another." [17] It is noteworthy that Hobbes conceptualizes not only in terms of addition and subtraction, but explicitly in terms of "parcels" as well. The picture at the beginning of *Leviathan* is worth consulting. It shows the Commonwealth as a huge individual built up of an aggregate of tiny individuals—a kind of geometrical assemblage. This picture strongly caught Swift's fancy. He makes use of it at times explicitly and at times in oblique reference.[18]

The works of Newton himself seem to have been too technical to be of much interest to Swift, although a copy of Newton's *Philosophiae naturalis principia* from Swift's library is listed in the 1745 sale catalog.[19] This work Swift had not annotated. Still, Swift is by no means entirely ignorant of Newton's theories. As master of the mint, Newton had been involved in the Wood coinage project in 1724, and when two years later Swift strikes at Newton in retaliation, he does so in terms of Newton's law of attraction. In Glubbdubdrib he has Aristotle remark that the law of attraction, "whereof the present learned are such zealous asserters," is only a passing fashion.[20]

Toward Locke, Swift was especially sympathetic. Of the works of Locke, the Swift library sale catalog of 1745 men-

[17] Hobbes's *Leviathan*, reprinted from the edition of 1651 (Oxford: Clarendon Press, 1909), p. 32; cf. p. 33.

[18] *Battle of the Books, Works*, I, 201; *A Tritical Essay upon the Faculties of the Mind, Works*, I, 291.

[19] Williams, p. (9), no. 326.

[20] *Gulliver's Travels, Works*, VIII, 207.

tions only *Tracts Relating to Money, Interest, and Trade*,[21] but Swift knew others of Locke's works. Speaking of them in general, but especially of the *Essay on Human Understanding*, while regarding as "dangerous" Locke's doctrine that there are no innate ideas, Swift maintains that Locke is a man to be read: "People are likely to improve their understanding much with Locke."[22] The concreteness of Locke's sensationalist philosophy had a great appeal for Swift, who shows traces of Locke throughout his writings and conspicuously spares Locke in his attacks on philosophers in general.[23]

Swift's interest in speculative writing is roughly in inverse proportion to the strictly metaphysical pretensions of the writing. Swift knew Descartes in a general way well enough to give him intelligent mention, but the sort of thing in the French mathematician engaging Swift is rather Descartes's quasi-mechanistic conceptions than his more abstruse methodology. In the spider-bee episode there is a merely implicit and general reference to Descartes's methodology,[24] but shortly thereafter something more mechanistic out of Descartes is joyfully pressed into service to do Descartes to death. He is made, not to think himself out of existence, but rather to whirl around "till death, like a star of superior influence, drew him into his own vortex."[25]

The tenor of Swift's comment on Berkeley, together with

[21] Williams, p. (8), no. 300.

[22] *Remarks upon a Book Intituled "The Rights of the Christian Church," Works*, III, 113–114.

[23] *Battle of the Books, Works*, I, 172–174. See Kenneth MacLean, *John Locke and English Literature of the Eighteenth Century* (New Haven: Yale University Press, 1936), p. 9.

[24] *Battle of the Books, Works*, I, 168–170.

[25] *Ibid.*, p. 178; the same imagery is again applied to Descartes by Swift in *A Tale of a Tub, Works*, I, 116.

the paucity of his comment, shows that Swift suspected that Berkeley was philosophically important and yet found himself hard put to follow Berkeley's thought. Swift had every opportunity to make much of Berkeley's work. He knew Berkeley well personally and even undertook to circulate his writings, styling him a "very ingenious man and great philosopher." [26] Berkeley's *Alciphron or the Minute Philosopher* stood in a two-volume edition on Swift's shelves with three others of Berkeley's works, but Swift annotated none of these books by his friend. [27] When he does mention Berkeley's philosophy, it is only to write to Gay expressly that he himself finds Berkeley "too speculative." [28] Swift's reaction to Malebranche was much the same. Malebranche's *Recherche de la vérité* was among Swift's books, but the copy is unannotated, and when Swift sees fit to mention Malebranche, it is clear that he finds the French philosopher uncongenial. [29]

Swift seems to have had some immediate acquaintance with Bernard de Fontenelle's *Entretiens sur la pluralité des mondes*, which had appeared in 1686 and which did so much to popularize the work of Galileo, Descartes, and Copernicus. [30] At Fontenelle's popular level, science was something more to Swift's taste than at the level at which Galileo or Newton themselves purveyed it.

One would expect from a general view of Swift's situation and circle of acquaintances what this brief survey of his reading confirms: namely, that Swift was fairly well in touch with the intellectual currents of his day at the point at which

[26] *Journal to Stella, Works*, II, 456.

[27] Williams, p. (7), nos. 271, 273–275.

[28] *Correspondence*, IV, 295.

[29] Williams, p. (3), no. 72. See "The Dean's Reasons for Not Building at Drapier's Hill," *Poems*, III, 900.

[30] Gückel and Günther, p. 97.

they connected with the scientific or philosophical world. It is not necessary for our present purposes to establish exact correspondences. Indeed, such correspondences, one could safely hazard, do not in any detailed way exist. It is sufficient here to note enough contact to give an initial probability to the view that some of the intellectual forces that were in the air, some of the perhaps unconceptualized but operative assumptions and frames of thought—what one can call the current "myths"—which were at large within the field of scientific inquiry, would not be foreign to the frame of thought of Jonathan Swift.

Among these current myths, Swift seems most at home when they are given a mechanistic or geometrical turn. His interest in Bacon, Hobbes, and Newton contrasts with his comparative coolness to Descartes (except in his more mechanistic speculations), Malebranche, and Berkeley.

IV

Swift's interest in the more mechanistically conceived speculation corresponds to a propensity of his to employ mechanistically conceived imagery and conceptualizations. This propensity, which shows itself elsewhere, gives point to some of Swift's remarks concerning the workings of the human intellect. Indeed, in the psychological field the propensity is crucial and quite highlighted. Precisely because they are so forcefully obtruded and enter so violently upon the psychological scene, Swift's downrightness and bluntness, inevitable in a field such as political discussion, here give one to believe that they reveal less the state of the matter under discussion than the predispositions—the myths—which possess the author's mind.

Swift likes to reduce psychological operations and situations immediately into spatial or local-motion components. Position is often made to play a decisive role. A madman is described by Swift as "a person whose intellectuals were overturned, and his brain shaken out of its natural position." [31] Or again, as a person "whose natural reason had admitted great revolutions" [32] (the political reference is perhaps in the ascendancy here, but the notion of whirling, dizziness, and the like is certainly operative as well). Wotton's brain is "half overturned" by one of the whelps of the goddess Criticism which she has thrown into Wotton's mouth.[33]

It is by their local position, following a process of liquid stratification, that various operations of the mind are conceived of as differing from one another, "Wit, without knowledge, being a sort of cream, which gathers in a night to the top, and, by a skilful hand, may be soon whipped into froth; but, once scummed away, what appears underneath will be fit for nothing but to be thrown to the hogs." [34] It is interesting to observe how in this passage Swift reduces a distinction of quality (cream vs. milk) to a difference in position. Separation is effected not by assessing various characteristics but simply by removing what is in the topmost place. (Swift wryly cautions here that there is a type of brain "that will endure but one scumming.")

In the title itself of *A Discourse Concerning the Mechanical Operation of the Spirit*,[35] Swift seizes upon a local motion as the vehicle of his satire. The extremes to which he carries his exaggeration here as elsewhere in instancing mechanical reasons for various phenomena shows the vigor with which he

[31] *A Tale of a Tub, Works*, I, 112–113.
[32] *Ibid.*, p. 113; cf. *ibid.*, p. 118.
[33] *Battle of the Books, Works*, I, 177.
[34] *Ibid.*, p. 160.   [35] *Works*, I, 189 ff.

had appropriated the mechanistic imagery. From the notion that the intellectual anarchy of the "fanatics" is traceable to something's being out of place ("imagination hath usurped the seat" of the brain [36]), Swift moves to a further conception: the difficulties are traceable to the local motion of vapors ("you perceive the vapours to ascend very fast" [37]). The quilted caps of the fanatics, by blocking the passage of the spirit out of the head through the normal channels of perspiration, force it to issue from the mouth.[38] Or finally—Swift cites this as "the opinion of choice *virtuosi*"—the brain is itself atomized, conceived of as "a crowd of little animals," and its various products—poetry, eloquence, politics, and so on—are differentiated in geometric terms.[39] The animals bite the capillary nerves, and hexagonal morsures produce poetry; circular, eloquence; conical, politics; and so on.

The whole passage is plainly, besides an oblique and ambiguous satire of Hobbes, an adaptation—how serious need not concern us here—of the atom philosophy of Democritus and Epicurus for the treatment of psychological material. The sharp teeth and claws with which Swift equips his little animals are plainly nothing other than the hooks with which Democritus had equipped his atoms, for not only are they used by the animals to cling to one another, but Swift expressly refers to these animals' "hamated [Latin *hamus*, hook] station of life."

Swift was plainly conscious here of what he was operating with. In *A Tale of a Tub* he burlesques Epicurean mechanistic atomism by applying it by name to mental operations:

[36] *A Discourse Concerning the Mechanical Operation of the Spirit, Works*, I, 198.

[37] *Idem*.   [38] *Ibid*., p. 201.

[39] *Idem*. Cf., opposite the title page in Hobbes, *Leviathan*, the frontispiece to which Swift here expressly refers.

Epicurus modestly hoped, that, one time or other, a certain fortuitous concourse of all men's opinions, after perpetual justlings, the sharp with the smooth, the light and the heavy, the round and the square, would, by certain clinamina, unite in the notions of atoms and void, as these did in the originals of all things.[40]

Again, in *A Tritical Essay upon the Faculties of the Mind*, Swift returns to the Epicurean atomism as applicable to man.[41] His attention to it is there oblique. He introduces it only to reject it. But it does have a fascination for him.

Other instances of a like sort can be adduced almost without limit. In *A Voyage to Laputa* Swift is preoccupied with attempts to reform language by geometrical means: polysyllabic words were to be sliced down to monosyllables for the betterment of the language,[42] and Swift describes and illustrates with a diagram a machine which turned out books of philosophy, poetry, law, mathematics, or anything else on purely mechanical principles—one needed not to think but only to turn cranks.[43] Writing with less buffoonery, Swift will still work extensively in veins of geometrical imagery:

Others, again, inform us, that those idolators adore two principles; the principle of good, and that of evil. . . . What I applaud them for is, their discretion, in *limiting* their devotions and their deities to their *several districts*, nor ever suffering the liturgy of the white God *to cross*, or *to interfere* with that of the black. Not so with us, who pretending by the *lines and measures of our reason*, to *extend* the dominion of one invisible power, and *contract* that of the other, have discovered a gross ignorance in the natures of good and evil, and most horribly *confounded the frontiers* of both. After men have *lifted up* the throne of their divinity . . .

[40] *Works*, I, 116.   [41] *Ibid.*, p. 291.   [42] *Ibid.*, VIII, 192.
[43] *Ibid.*, pp. 190–192.

after they have *sunk* their principle of evil to the *lowest centre*
. . . I laugh aloud to see these reasoners, at the same time, engaged in wise dispute about certain *walks and purlieus* . . .
seriously debating, whether such and such influences *come into*
men's minds *from above,* or *below.* . . .[44]

Swift did not invent the common mode of conceiving heaven
as above and hell as below, but he obviously reveals in the
notion and protracts it with relish. It forms a happy setting
for what is distinctly his own importation here: "the *lines* and
measures of our reason." Sextants and triangulation will have
their office, too, within the human mind.

But one need multiply instances no further. The present
point has been sufficiently made: in his concern with psychological operations, Swift tends frequently to conceive of
them in a mechanistic or geometrical fashion—a fashion at
once gross and vigorous. He favors immediate reduction of
complex issues in terms of position or local motion. He
satirizes mechanistic thought largely by simply exaggerating
it, not by proposing an alternative. His own modes of thinking
often suggest no alternative, for they are strongly mechanistic
or diagrammatic themselves, if not always spectacularly so.
Swift was moving in the current of mechanistic thought so
strong in his day.

Consciously or subconsciously, he feels this current of
thought as a central item in human existence. It is significant
that when he pictures the Emperor of the Lilliputians as a
great patron of learning, he makes it explicit that the learning
accorded the patronage is mathematical and mechanical.[45]
Swift indeed points to such learning here in an ambiguously

[44] *Mechanical Operation, Works,* I, 199–200; italics mine.
[45] *Gulliver's Travels, Works,* VIII, 25.

satirical gesture. But his very pointing certifies to the fact that mathematical and mechanical learning as such had caught Swift's attention sufficiently for him to regard this learning as a quite self-contained field of activity—separate, for example, from that of metaphysics (or, at any rate, of what he considered metaphysics). The Brobdingnagians, Swift observes, have a mastery of the former kind of thinking but can get nowhere with the latter.[46]

Even when one thinks that he is about to make his way out of this current of mechanistic thinking, Swift will move back into it again. In one place he pokes fun at the popular image of the soul and body pictures in a rider-horse relationship. But he does not work himself clear of the gross imagery. He merely observes that sometimes it is the soul which is the horse and the flesh the rider.[47] The issue he leaves stated more or less in terms of position. The question is who is on top. In a vein vaguely suggesting this, he had defined enthusiasm as "a lifting-up of the soul, or its faculties, above matter." [48] Swift's frequent preoccupation with magnification and shrinking runs in the same mechanistic current. In Gulliver's visits to Brobdingnag and to Lilliput, moral issues are reduced in a way to spatial terms when the repulsiveness of humanity is revealed not by subjecting man to highly cultivated perceptions, but by presenting some such thing as the human skin or ordinary human activities magnified until their mere massiveness weights the reader with disgust, or shrunk till spatial insignificance becomes so striking as to do for moral insignificance too. It is no accident that the readiest symbols for Swift are Lilliputians and giants.

[46] Ibid., p. 140.   [47] Mechanical Operation, Works, I, 204.
[48] Ibid., p. 194.

v

The type of imagery which has here been pointed out in Swift is, of course, not the only type one finds in him. Neither is it a type used only by him or by his age. But its currency during his age becomes highly symptomatic.

Some sort of delimitation or definition of field is inseparable from any scientific investigation, or indeed from any sort of human knowledge. Human knowledge proceeds by abstraction, and the more scientific it becomes, the more it must eliminate from its attention all but the precise material which the science in question regards.

As applied to the ideally isolated system of the Newtonian physics, the economy of limitation takes on a very concrete aspect. All sorts of removal techniques begin to put in their appearance. In mathematical or metaphysical science, isolation of the material to be considered is effected differently. The mathematician, for example, once on his speculative way, need return no more to tinker with the exterior world.

In the world of physics which Galileo and Newton were laying open, such was not entirely the case. There was constant return to examination of phenomena, and, in connection with this examination, the need more and more to exert oneself to eliminate the irrelevant from one's observational field—to control the experiment. One had to cast about for means of removing the influence of heat, or of air currents, or of other interferences. Intellectual activity became more and more involved with the development and exercise of removal techniques.

The effect of the preoccupation with the physical and mathematico-physical sciences on the general mentality of the age is difficult to calculate in any detailed fashion. Some effect it certainly did have. Quite conceivably, the relatively greater appeal at this time of systems proposing some sort of universal compassing of all human knowledge is connected with the delights of conquest enjoyed within the thoroughly delimited systems of Newtonian physics. Of all ages, this was one of those most preoccupied with schemes either to lay out plans for all human knowledge (Francis Bacon), or to compass all possible human knowledge in one methodology (Ramus, Descartes, Hobbes), or even in one set of textbooks as well (Ramus).[49] Language itself came to be envisioned as a field so completely delimitable that its operations could be gathered under legislative control by the Academy in France or by its correlatives in other countries. (Swift himself, as is well known, was one of those interested in "fixing our language for ever." [50])

How far the notion of delimitation applied first to the physical sciences or to the whole field of human knowledge need not concern us here. The delimitation techniques of the physical sciences took root in a mentality which had obviously been somehow prepared in advance, but which was further fertilized by the fruits of the physical sciences themselves.

In terms of the resulting intellectual climate, a theorem can be constructed which will account in part for much of what Swift has to say concerning the operations of the human mind: Swift tended to conceptualize issues according to an elemental spatial pattern. Typically, he liked to conceive of a field of some sort isolated from all outside influence and to

[49] Walter J. Ong, *Ramus and Talon Inventory* (Cambridge, Mass.: Harvard University Press, 1958).

[50] *A Proposal for Correcting, Improving and Ascertaining the English Tongue, Works*, XI, 15.

maintain the conditions of such a field by enforcement of various removal techniques. He liked to do this especially when there was a question of reform involved.

It is noteworthy that Swift is so often preoccupied with boxes, countries immune to all foreign contact, islands marooned not only in the sea, but, far more effectively, in the air. Sometimes there is isolation within isolation: in the isolated country of Brobdingnag, Gulliver himself is isolated within his little box or cage, just as he had been isolated within their country by the Lilliputians, who held him under duress.[51] The issues are rather complicated here. Neither the Brobdingnagian nor Lilliputian isolation, nor Gulliver's own personal isolation, is conceived of as isolation within a perfectly functioning system. Still the isolation is of high psychological import. On the destruction of his box at the hands of the rescuing sailors, Gulliver delivers himself of a sad little eulogy, recalling his happy days within its confines and within the confines of Brobdingnag.[52]

Swift's over-all intent in his use of diagrammatic images is frequently ambiguous. Although he is often evidently satirizing something beyond the images, generally the images or models themselves are included among his objects of satire: in *The Mechanical Operation of the Spirit* Swift satirizes, among other things, the very attempt to reduce spiritual activity to mechanics. But the instances here adduced show that despite a feeling of hostility toward mechanical models, he was not himself so innocent of spiritual or psychological mechanics as he might have wished to be—perhaps by fault of invention, for he appears to have been unable to find alternatives to the view he satirized. Even though an object

[51] *Gulliver's Travels, Works*, VIII, 115 and *passim; ibid.*, pp. 20 ff.
[52] *Ibid.*, p. 149.

of satire, the spatial projection or diagrammatic model often remains an indispensable part of Swift's own thinking. And if such models are probably in one way or another indispensable in all thinking, they are elsewhere not commonly so obtrusive as they are in Swift. Perhaps the remorselessness of Swift's irony is often due to his inescapable dependence upon that which he is satirizing.

As a satirist, Swift is largely occupied with evils and their reform. Within the framework of thought described here, reform is accomplished typically by the simple enforcement of isolation. (The ideally isolated system of the Newtonian physicist was supposed to work with an absolute precison, for the vagaries of quantum physics were happily unknown.) Under the spell of the myth of the ideally isolated system, one is tempted to prescribe as remedies simple removal techniques. If there is something wrong—a "bug" in the system— one does not try to control it but to remove it, lift it out.

Applied to the remedy of evils, the myth of the ideally isolated system thus generates a subordinate myth which we can call the myth of asepsis. The problem of reform is no longer a problem of qualitative adjustment within a system— such as, for example, one might propose for a citizenry in propounding an intelligent educational system in great detail. The problem is one of purging—of getting the germ out and keeping it out. This, says the myth of asepsis, will effectually achieve betterment of things.

Diagnosis and prescription according to the general lines of this myth are readily discernible in Swift time and time again. Swift seldom argues energetically and convincingly for reform by adjustment of qualitative elements within a system. The basis of his tremendous rhetorical appeal is typically more spectacular. It is the cry of the dissatisfied spectator in the grandstand, too aroused for the tediousness of analysis and a subsequent reorganization of attack. "Throw him out!"—this is the formula, impressive, urgent, and delightfully relieving to pent-up emotion, which forms the core of much of Swift's rhetorical argument and is the source of much of his vigor. Stated in a single trisyllable instead of in three monosyllables, the cry is more scientific and quite as familiar: "Sterilize." It is the universal elemental remedy of a mechanistically driven civilization.

Brought to bear on the area of intellectual operations in man, Swift's passionate desire to discern a unity in things, plus his habit of conceiving of things in terms of ideally isolated systems, helps explain some of his most typical remarks concerning the function of poetry. These remarks treat poetry in terms of the myth of asepsis. Obviously to Swift, there is something wrong in the general field of reason and common sense—the kingdom of the mind. What is more natural than that one right matters by a technique of removal, of asepsis? True to what we should expect, Swift hits on a means to this asepsis. Poetry is the means: it is a throwing off of filth from the human mind.[53]

The asepsis motif is carried even further. The whole commonwealth is to be subject to it, kept clean and sterile by the isolation of the poets themselves. Asepsis itself is to be rendered aseptic. Those who need to subject themselves to the practice of personal intellectual asepsis are to do so in a specially isolated place, in a Grub Street. Lack of such a place

has been attended with unspeakable inconveniences: for not to mention the prejudice done to the commonwealth of letters, I am of opinion we suffer in our health by it. I believe our cor-

[53] *A Letter of Advice to a Young Poet, Works*, XI, 107–108.

rupted air, and frequent thick fogs, are in a great measure owing
to the common exposal of our wit; and that with good management our poetical vapours might be carried off in a common
drain, and fall into one quarter of the town, without infecting
the whole, as the case is at present, to the great offence of our
nobility, and gentry, and others of nice noses. When writers of
all sizes, like freemen of the city, are at liberty to throw out
their filth and excrementitious productions, in every street as
they please, what can be consequence . . . ? [54]

This sort of thing is of course not all Swift has to say about
poetry. But it is typical, and in the general framework of his
thinking, it makes some sense. To guarantee the field of reason
and common sense, in the face of threatening evil, that ready
unity and coherence which he felt indispensable, Swift here
comes to see poetry as a way to better the human mind.
Strangely enough, so far he is under the same compulsion as
those who conceive of poetry as a revivification or an education or a refinement of perception. With poetic theory of
this sort, at first blush, his views would seem to have
nothing in common. But the mode of operation of poetry in
Swift's view takes its shape not only from an initial tendency
to see poetry as bettering the mind, but also from Swift's
typical way of representing the mind itself—as something like
an isolated, closed field, perfectly functioning in so far as
interferences are removed.

Such a field does not need revivification or education or
refinement. If it is contaminated, as the field of the human
mind is—for experience shows something is wrong here—it
does need an exit. Evil is to be ousted and the aseptic state
recovered. As the sector of the field least amenable to common sense, most deviously connected with reason, poetry

[54] *Idem.*

was the natural choice. It serves the turn well. It is a quite plausible exist.

Our sketch here suggests a qualification of the opinion, to which many would subscribe, that Swift did not take poetry seriously—not so seriously, for example, as did Pope.[55] He did not take it seriously in the sense that he is inclined savagely to satirize it—yes. In this sense, Swift takes seriously very few things indeed. Moreover, he does not assign it a high-sounding role. Did he take it seriously in considering that it performed a function demanded by the human mind in one of his favorite ways of representing this mind? The answer can only be: Indeed he did.

[55] Davis, pp. 11–14.

9 Psyche and the Geometers: Associationist Critical Theory

I

The temper of eighteenth-century criticism is like the compunction recommended by Thomas à Kempis, more readily felt than defined. Attempts to deal with this criticism have operated with such theorems as Locke's sensationalist philosophy, Berkeley's idealism, various forms of authoritarianism, Cartesian so-called "anti-authoritarianism," the controversy between ancients and moderns, and various other formulas. But, from whatever point it has taken its departure, recent study has brought out more and more the importance, at least strategic and symptomatic, of the contemporary associationist psychology in determining the critical climate. "It may be questioned, indeed," writes Walter Jackson Bate, "whether any philosophical or psychological doctrine has since permeated critical thought in so great a degree as did that of the association of ideas at this time." [1]

Although it seems plain that associationist doctrines were somehow connected with the growing ascendance of the

[1] Walter Jackson Bate, *From Classic to Romantic: Premises of Taste in Eighteenth-Century England* (Cambridge: Harvard University Press, 1946), p. 96.

physical sciences, the details of the connection are puzzling enough. The common denominator of "empiricism" which is supposed to relate the physical sciences and associationism is far from adequate and is even misleading. It is true that associationist psychology talks at some length about the particular in its concern with the sensory origins and manifestations of knowledge. But, while this interest contrasts with Johnson's and Burke's preoccupation with the "general" in art,[2] it does not contrast notably with the preoccupations of a great many earlier psychologies. There is an interest in the sensory origins of knowledge in much medieval philosophy; yet, for all that, such philosophy does not invite comparison with the physical-science mentality in the way that associationism certainly does.

The embarrassing truth is that the associationists would often rather talk about observational procedure than practice it. Associationist critics are not markedly more empirical than an Aristotle or a Quintilian or an Elizabethan such as Puttenham or a neoclassicist such as Fénelon. If anything, they implement their discussion with fewer concrete instances. They are interested in the particular, to be sure—but often in terms of the fine-spun theory they can weave into it; for, in the last analysis, they are among the most viable of die-hard theorizers. What is more, they sometimes know they are. By way of apology for his doctrinaire critical procedure, Henry Home, Lord Kames, sees fit to protest:

I am extremely sensible of the disgust men generally have at *abstract speculation;* and for that reason I would avoid it altogether, were it possible in a work which professes to draw the rules of criticism from human nature, their true source. There is

[2] *Ibid.*, pp. 93, 106, 128–129.

indeed no choice, other than to continue for some time in the same track or to abandon the undertaking altogether. Candor obliges me to notify this to my readers, that such of them whose aversion to *abstract speculation* is invincible, may stop short here; for till *principles be explained*, I can promise no entertainment to those who shun thinking. . . . The foregoing *speculation leads to many important rules* of criticism.[3]

The nature of the principles and speculation from which Kames's three-volume *Elements of Criticism* proceeds is plain enough. In Volume I, the first chapter, from which this apology has been quoted, is entitled "Perceptions and Ideas in a Train."

In the light of Kames's protestation, there is some reason to look to the nonempirical rather than to the empirical aspects of the physical sciences to help account for the all but palpable similarity between the associationist and the physical scientist. Fortunately, one does not have far to look. The physical sciences were progressing through the eighteenth century chiefly in close collaboration with the least empirical of all sciences, mathematics. In terms of the mathematical mentality and the kind of conceptualization it encourages (the projection of problems in function of quantified and spatial imagery), the present study seeks some central clue—though, of course, not the only one—to the relationship between the physical sciences and associationist criticism.

II

Although associationism was given its initial vogue under the auspices of David Hartley (1705–57), an Englishman, and

---

[3] Henry Home, Lord Kames, *Elements of Criticism* (Edinburgh: A. Millar, 1762), I, 33–34; italics mine.

although Addison early sponsored its application to criticism,[4] associationism achieved widespread critical currency chiefly under such Scottish writers as Henry Home, Lord Kames (1696–1782), Adam Smith (1723–90), Alexander Gerard (1728–95), William Duff (1732–1815), John Ogilvie (1733–1813), Archibald Alison (1757–1839), and others, from whom it fed back into England. Such writers provide ample evidence of thinking carried on under the spell of quantified or spatial tableaus.

The imagination, writes William Duff, assembles "ideas conveyed to the understanding by the canal of sensation, and treasured up in the repository of the memory, compounding or disjoining them at pleasure."[5] It enjoys, he adds in the same place, a "plastic power of inventing new associations of ideas, and of combining them with infinite variety." The components of Duff's description are symptomatic. "Ideas," it will be noted, are subject to a good bit of local motion. They are "conveyed" over a "canal." They are hitched one to another. Variety seems to be a matter of different hitching sequences. "Invention," or the creative process, becomes a process of lining up "ideas" in hitherto unfamiliar concatenations.

The tendency to plot psychological operations geometrically was already markedly in evidence in Hobbes and Locke,[6] but it is overpowering among the Scottish associationists as a whole. Alexander Gerard, who also explains invention in terms of "associating power" and describes it as "ex-

[4] Martin Kallich, "Association of Ideas and Critical Theory," *ELH*, XII (1945), 308.

[5] William Duff, *An Essay on Original Genius: And Its Various Modes of Exertion in Philosophy and the Fine Arts, Particularly in Poetry* (London: T. Becket and P. A. DeHondt, 1767), pp. 6–7.

[6] Kallich, pp. 291 ff., 303 ff.

tensive comprehensiveness" of imagination,[7] says that poetry is "a *complication* of beauties."[8]

The geometrical plotting is not always static but often becomes, at crucial points, a geometry of local motion, as in a typical passage in Adam Smith's essay *Of the Nature of That Imagination Which Takes Place in What Are Called the Imitative Arts:*

That train of thoughts and ideas which is continually passing through the mind does not always move on with the same pace, if I may say so, or with the same order and connection. When we are gay and cheerful, its motion is brisker and more lively, our thoughts succeed one another more rapidly, and those which immediately follow one another seem frequently either to have but little connection, or to be connected rather by their opposition than by their mutual resemblance. . . . When we are melancholy and desponding . . . a slow succession of resembling or closely connected thoughts is the characteristic of this disposition of mind.[9]

Once he has stated it, Smith exhibits the usual virtuosity in manipulating his theme. The passions with an affinity for music, the "musical passions" (grief, humanity, compassion, joy, etc.), are passions whose trains move at regular intervals. Thus, by regularity of spacing, they are differentiated from

[7] Alexander Gerard, *An Essay on Taste: With Three Dissertations on the Same Subject by M. de Voltaire, M. d'Alembert, and M. de Montesquieu* (London: W. Strahan and T. Cadell, 1759), pp. 173 ff.

[8] *Ibid.*, p. 82; italics mine.

[9] *The Works of Adam Smith* (London: T. Cadell, 1811–12), V, 284–285. It is perhaps not necessary to mention that Smith, now more famous for his economics, was in his own day equally well known for his rhetorical and literary studies. In his professorship of logic, his first professorship at Glasgow, Smith had spent most of his time not on logic but on rhetoric and belles-lettres.

the nonmusical or less musical passions (furious anger, malice, envy, etc.), whose periods are all irregular."[10] Significantly, Smith maintains in the same breath that the "natural tones" of the musical passions are "clear and distinct" and thereby adapted to measured repetition. Here, as elsewhere, the Cartesian insistence on clarity and, more particularly, on distinctness comes into its own. It meets the demands of a geometric and mechanistic world.

Archibald Alison's concept of the fine arts introduces a more complicated and fascinating graph: a train of ideas with fancy hard on its heels. The fine arts, Alison explains, which are those concerned with sublimity and beauty, address themselves to the imagination by starting a train which "busies" our fancy in its "pursuit."[11] Alison's train is methodically assembled. It is "a regular or consistent train of ideas of emotion."[12]

These and the other critics in the associationist tradition seem quite convinced that their geometric plottings and forthright analogies with mechanistic local motion were designed to revamp the critical thought of centuries. Although the analogies were themselves perhaps as old as the eldest of psychologies, suddenly everything turns on them. Lord Kames, bravely starting his three-volume *Elements of Criticism* with the chapter already noted, "Perceptions and Ideas in a Train,"[13] is symptomatic. This was the deepest interior of the spider, out of which all was to be spun.

Even in John Ogilvie, who, because of his reserved attitude toward associationism, stands somewhat apart from most of

[10] *Ibid.*, p. 276.

[11] Archibald Alison, *Essays on the Nature and Principles of Taste* (London: J. J. G. and G. Robinson, 1790), pp. 1–2.

[12] *Ibid.*, p. 55.   [13] Kames, I, 21–46.

his contemporary Scottish philosophers and critics, geometry leaves some traces. Ogilvie describes the imagination as a faculty which "strikes out happy imitations," but soon modulates into a more acceptable key by noting that, to do so, it forms "new and original assemblages of ideas." [14] Later we are told by him that it effects unusual associations.[15]

III

Associationism, which, even as a detailed theory, goes back at least to Aristotle,[16] rose to the ascendancy which it enjoyed in the eighteenth and early nineteenth centuries on a wave of mathematical and mechanistic interest whose force it would be difficult to exaggerate. It is today something of a commonplace that, at least from the early seventeenth century on, men had been fascinated by the quest of a single method which would act as a universal solvent for all possible knowledge and that they tended to make this method a mathematical or geometric one.[17] Cartesianism has been ably interpreted as the outcome of such a quest, occasioned in part by Descartes's success in joining two fields of mathematics through analytic geometry.[18]

[14] John Ogilvie, *Philosophical and Critical Observations on the Nature, Characters, and Various Species of Composition* (London: G. Robinson, 1774), I, 11.

[15] *Ibid.*, p. 106.

[16] See Albury Castell, *Mill's Logic of the Moral Sciences: A Study of the Impact of Newtonism on Early Nineteenth-Century Social Thought* (Chicago: University of Chicago Libraries, 1936), pp. 11 ff.

[17] See Louis I. Bredvold, *The Intellectual Milieu of John Dryden* (Ann Arbor: University of Michigan Press, 1934), esp. pp. 50 ff.

[18] Jacques Maritain, *The Dream of Descartes*, tr. from the French by Mabelle L. Andison (New York: Philosophical Library, 1944). For a discussion of the influence of Locke, Newton, and allied writers on the poets see Marjorie Hope Nicolson, *Newton Demands*

The force with which the universal-method mentality could move into the field of criticism can perhaps be better appreciated when one observes its inroads into the even more abstruse field of theology in such a treatise as *Theologiae Christianae principia mathematica*, published in 1699 by Newton's friend, the Rev. John Craig. The Rev. Mr. Craig—the Scottish name would read well on the associationist roster—is under a pious enough compulsion, as he explains in his "Praefatio ad lectorem," for he feels that, since nature is governed by geometry (note the unqualified Newtonian assumption here), geometry must lead to God. Most persons would doubtless agree that geometry, like everything else, must somehow lead to God. The question, of course, is how. By geometrical means, says Craig's blooklet.[19]

Craig, who is so confident about his way of imputing Euclidean geometry to the world of physics, is even more ready to have his favorite science replace metaphysics or ontology. In its brash naïveté, his book is perhaps the *ne plus ultra* of monomethodology. Here historical probability and the pleasure principle are marshaled geometrically by dint of axioms and diagrammatic demonstration. At the end, everything is at *AB*'s, and *xy*'s, and the reader is confronted with page after page of geometric designs calculated to confound atheism and deism. Racked on this sort of frame, little wonder if proofs for God's existence enter here on dark days.

The prestige of the quite elementary geometrized mech-

---

the Muse: Newton's "Optics" and the Eighteenth-Century Poets (Princeton: Princeton University Press, 1946); and cf. W. K. Wimsatt, Jr., *Philosophic Words* (New Haven: Yale University Press, 1948), for related phenomena in Dr. Johnson.

[19] *Theologiae Christianae principia mathematica* (London: Johannes Darby for Timotheus Child, 1699), pp. v–viii.

anism, so spectacularly in evidence here, which underlay the representations of reality common to associationism and its entire environment, could come about because of the way mathematics is related to the physical sciences. When mathematics and the physical sciences are combined, as they were to produce the new science, it is mathematics—the more abstract science—which takes the dominant role. This is in the nature of things. Weight, for example, can be considered in numerical terms, subjected to the discipline of numbers. But there is no point in considering numbers in terms of gravity, in speculating as to whether 6 is heavier than 2. Boyle's law, concerning properties of gas, admits of mathematical formulation, but, if in such a law physical properties or qualities can be processed mathematically, it is impossible to process mathematical concepts by means of physics. One does not physically heat or cool mathematical quantities, although one can mathematically measure physical heat and cold. That is to say, there is a mathematical physics in a way in which there is not a physical mathematics. Quality is in a way interpretable quantitatively, quantity is not interpretable qualitatively.

It is in a body of thinking such as this, dominated by mathematics, that a gross mechanism gains plausibility; for, although it is applied to the physical world, the kind of thinking represented by such mechanism—at its worst the crude atomism which interprets reality by dint of jostlings and even of hooks —has more affinity with mathematics than it has with the physical sciences. It often represents the encroachment of mathematics on the physical sciences. Newtonian physics, for all its serviceableness, emerges at this very period as a monument to the tendency of mathematical quantity to obtrude itself in other than mathematical fields. Because quantity abstractly considered is indefinitely divisible, classical physics had proceeded as though physical forces themselves were indefinitely divisible. Only much later, with the development of the quantum theory and the discovery of Planck's constant, was this encroachment of mathematics exposed—by empirical methods. Forces in the physical world simply did not have the indefinite divisibility which is a property of abstract quantity in mathematics. It was illegitimate to proceed here as though one were dividing quantity when one was rather dividing something only measured by quantity. One should not jump, without looking, from the level of geometrical abstraction to that of physics. Thomas Aquinas and others had pointed out a difference in abstractive modes or levels; [20] but, until someone rediscovered or effectively reasserted the lesson of this difference, geometrizing mechanism was carrying the day. Even when not expressly acknowledged, Descartes's ambition to make quantity, not existence, the decisive reality was subconsciously operative, as in many quarters—in a Democritus or an Epicurus or a Pythagoras—as it always had been and as it continues to be to the present day. Quantity remains the most tractable, and hence the most inviting, of the abstractions made by the human mind.

In view of its contemporaneous encroachments elsewhere, it is not surprising to see how openly quantified thinking makes its way into associationism. In the Lockian epistemology, it will be recalled, the primary sensibles, such as extension and local motion, were assigned an objectivity which was denied to other sensibles, such as color or heat, less amenable to mathematical processing.[21] Stated somewhat bluntly, this view comes close to holding that what is not quantity, simply is not, whereas what is quantity, simply is. Quantity and being

[20] *Summa theologica* i, q. 85; cf. i, qq. 40, 77.   [21] Bate, p. 98.

tend to become interchangeable. Adam Smith equates substance with the Newtonian solidity and refers this solidity to extension; [22] and, where a more dispassionate view might see algebra as a species of language, Hartley takes a stand more flattering to quantity. "Language itself," he observes, "may be termed one species of algebra." [23] The Royal Society's mathematical prose was having its inning. The whole field of intellectual activity was coming to be a department of mathematics.

As represented in its various exponents, associationism is often espoused for quite various purposes. The Rev. John Gay, to whose "Preliminary Dissertation concerning the Fundamental Principle of Virtue or Morality" (1702) Hartley traces his interest in associationism, in a rather gingerly fashion introduces associationism into his tract in the hope of refuting Hobbes and the self-interest school and of trimming down the excessive number of internal senses commonly postulated at the time.[24] In 1749 Hartley himself uses association somewhat differently, developing it in connection with his doctrine of vibrations and the theory that the difference between pleasure and pain was one of mere degree—that is, that it was reductively quantitative.[25] In 1726 Kames is putting association to another use as the keystone of an approach to literature. But, for all their divergent interests, associationists tend in common to view reality as coming in quantitative chunks, whether

[22] "Of the External Senses," *Works*, V, 336; cf. 334–335.

[23] David Hartley, *Observations on Man, with Notes and Additions by Herman Andrew Pistorius*, 4th ed. (London: J. Johnson, 1801), I, 280. This work was first published in 1749.

[24] Gay's treatise is published as the introduction to William King, *An Essay on the Origin of Evil*, tr. from the Latin [by Edmund Law] (London: W. Thurlbourn, 1731), see pp. xxii, xxxiii, *passim*.

[25] Hartley, I, 5, *passim*.

"ideas," which, it is insisted, must be clear and distinct, or feelings or sensations or all these together. They seem to rest content with an explanation which does little more than suggest some warrant for considering these bits of reality properly juxtaposed, according to the demands of the particular problem in hand. Indeed, at least from the time of Hartley, who traces his theories to Newton and Locke, associationism is quite explicit in regarding itself as an outgrowth of mathematical physics.[26] At the close of the period here treated, Dr. Thomas Brown of Edinburgh could write with confidence of "physics, whether of matter or of mind." [27]

IV

To consider associationism more in terms of the modes of quantified conceptualization it encouraged than in terms of "empiricism" calls for some re-examination of the very real historical connection between associationist criticism and that concern with individuality which makes so much of the singular (person or experience) and generates the subjectivism coincident with the romantic movement. This subjectivism, ably discussed by Mr. Bate,[28] had to do more with devious psychological compulsions shaped by the view of reality here being treated than with a straightforward pursuit of new observational or "scientific" methods.

Hobbes, to whom so much in associationism traces, pictures the senses as moving out to counter external objects moving in toward the organism. This way of representing things is at once helpful and impoverishing—helpful insofar as there may

[26] *Ibid.*
[27] Thomas Brown, *Lectures on the Philosophy of the Human Mind* (Edinburgh: James Ballantyne for W. and C. Tait, 1820), II, 270.
[28] Pp. 128 ff.

be some sort of real analogy here and impoverishing in so far as one is encouraged by the analogy to consider psychological fact in terms of mechanistic analogy *and no further*. In Hobbes this latter determination is rather obvious. All mental activity, Hobbes says, is to be pictured in terms of local motion, for motion determines all mental activity, and "besides Sense, and Thoughts, and the Trayne of thoughts, the mind of man has no other motion." [29] Precisely for his erection of the association of ideas, with all its implications, as the sole law of "connexion among our ideas," Hobbes was to merit the praise of Hazlitt, who at the same time censures Coleridge for crediting so important a revision of psychology to Descartes or others.[30]

With this sort of outlook, it is almost inevitable that Hobbes would assert, as he does, that a work of art is simply an orderly succession of ideas and that it is one in which speed plays a major role.[31] This notion, which becomes a commonplace, shows the matter of poetic organization reduced to a kind of jigsaw puzzle. The truncated view put forward here can be seen in its plenary form, as so many similar things can be seen, in James Mill, who, with a savage logic learned from associationism and brooking absolutely no interference, states bluntly that predication itself—that is, any conceivable human statement whatsoever—does no more than mark the order of trains of thought. Without explaining how it ever got there, Mill fiercely attacks the "implication" of existence

[29] *Leviathan*, introduction by A. D. Lindsay (New York: E. P. Dutton, 1950), p. 20 (Book I, chap. iii); see Kallich, pp. 292 ff.

[30] William Hazlitt, "Coleridge's Literary Life" [review of Coleridge's *Biographia literaria*], *Edinburgh Review*, August, 1817, reprinted in *Collected Works*, ed. by A. R. Waller and Arnold Glover (London: J. M. Dent, 1902–6), X, 143.

[31] Kallich, pp. 297–298.

which is bound up with predication and which, according to him, has only "produced the wildest confusion." [32] Because it defies his mechanistic processing, existence has become a crime. It is significant that James Mill's notions and the notions which his son, John Stuart Mill, derived through him from associationism have been characterized as unproductive for the precise reason that they did not take enough into account the data of experience.[33]

The tendency for discussions of "color," long a standard rhetorical accouterment, to drop out of the picture at this time is no accident. It is not that the associationists had found a substitute which was better. Color was simply a nonquantitative analogy. Vibrations would do instead.

v

The growth among the associationists of such subjectively weighted concepts as "sympathy," "genius," and "taste" has been treated ably and at length in Bate's *From Classic to Romantic*. This growth, however, shows curious retrogressions not unrelated to the habits of geometrized conceptualization being discussed here. Often enough, in the very act of developing the new concepts, the mind dominated by such habits tended to explain away these same concepts in terms of the very mechanized view which they were meant to supplement.

[32] James Mill, *Analysis of the Phenomena of the Human Mind* (ed. by Alexander Bain, Andrew Findlater, George Grote, and John Stuart Mill (London: Longmans, Green, Reader, and Dyer, 1869), I, 161, 175.

[33] Castell, pp. 88 ff.; Oscar Alfred Kubitz, *Development of John Stuart Mill's System of Logic*, Illinois Studies in the Social Sciences, Vol. XVIII, Nos. 1–2 (Urbana: University of Illinois Press, 1932), pp. 34 ff., 248 ff., 300.

When Adam Smith begins his *Theory of Moral Sentiments* with a protracted treatment of sympathy, he is doing more than honor the established vogue of benevolence. He is introducing surreptitiously a psychological and metaphysical question which had been glossed over by much associationist discussion. The associationist trains had proved versatile enough: they had linked not only idea with idea but, if one can make the distinction come off, thoughts with ideas,[34] perceptions with ideas,[35] and any or all of these with feelings and emotions, especially with pain and pleasure.[36] But, eclipsed behind this elaborate display, there lay a perennial problem of all psychologies, that of correspondence or conformity between what was in the individual consciousness and what lay outside it.

Smith's discussion of sympathy is, among other things, a heavily—if innocently—disguised consideration of this conformity, that is, of the problem of knowledge. He centers attention on concordance of one person's feelings with another's, coming thus by a specialized and somewhat devious route to the problem of representation, which in other philosophies had been handled on a broader footing in terms of "form." The question as to how one thing can represent another is narrowed in Smith to the question as to how one feeling can represent another.

All the ins and outs of Smith's theory need not concern us. What is relevant here is the deflection of the problem as far as possible into associationist lines of discussion. Such deflection was not entirely new, of course. Earlier, drawing on Locke's notion of "simple" and "compound" or "mixed" ideas, Gay had maintained that "men, unless they have their

---

[34] Smith, "Nature of Imitation," *Works*, V, 284.
[35] Kames, I, 381; Hartley, *passim*.   [36] Hartley, *passim*.

compound Ideas . . . made up precisely of the same simple ones, must necessarily talk a different language." [37] Here Gay is plainly feeling for a solution to the problem of communication in terms of configurations or trains.

Gay's interest is only a passing one. Smith's is head-on and crucial, for his whole theory of moral sentiments turns on his view of sympthy with which he begins his treatise. He notes, first, that the correspondence of one person's feelings with another's in what is called "sympathy" comes about through the imagination. I imagine the same things happening to me as are happening to you, and, given the proper conditions, sympathy results.[38] In availing himself of the imagination, Smith has focused the discussion on the associationist assembling faculty. In Duff this serves to effect new combinations,[39] and in Gerard its role is much the same,[40] for it is set against the memory. Smith sets it against the senses, noting that the senses register our own feelings, whereas the imagination enables us to reproduce the feelings of others. His discernment is simple enough and basically sound. Smith is only saying that, when you sympathize with a man who has hurt himself with a hammer, your thumb isn't being hit, too.

But many crucial questions are left untouched by Smith. The most crucial regards an operation represented by Smith as a spatial maneuver. After achieving correspondence of feeling, to generate sympathy I am supposed to set myself in your "place." That this is a common mode of speech and that it roughly covers some sort of psychological reality anyone will concede. But what does this operation comprise? How are the

[37] Gay, "Principle of Virtue," in King, p. xii.
[38] *Theory of Moral Sentiments, Works,* I, 6–10.   [39] Duff, p. 7.
[40] Alexander Gerard, *An Essay on Genius* (London: W. Strahan and T. Cadell, 1774), pp. 100–101.

notes both of otherness and of identity, with the strange tension between them, constituted and maintained? No doubt, Smith does not aspire to handle such matters, and one can have no quarrel with his objectives. The only point is that his discussion of sympathy feeds back into preoccupation with the associationist imagination and that unsolved problems are laid to rest under little heaps of spatial images.

As though he feels that here he has not settled the problem which is his disguised concern, Smith revives the matter of resemblance or conformity in another and still more complex dress in the subjects of self-approbation and self-disapprobation. His account of these is a kind of involution of his account of sympathy. In sympathy, our imagination represents another to us as ourselves. In self-approbation, it represents the self as someone else (who, in turn, is being represented as the self).[41] This sort of thing can be turned back on itself indefinitely. It is curious evidence of the general drift in associationism toward subjectivity discussed by Mr. Bate. In the sea change which the philosophy of knowledge here undergoes, the final subjectivity is indeed rich and strange: theory itself delights in being involuted.

The notions of genius and taste are more intimately related to associationism than are discussions of sympathy. They put in their appearance as supplements to associationist theory, only to have this theory explain them away. A genius, William Duff explains, is so constituted as to have not only imagination to assemble ideas and judgment to counteract the rambling and volatility of the imagination, but also taste.[42] This last is "that internal sense, which, by its own exquisitely nice sensibility, without the assistance of the reasoning faculty, dis-

[41] *Theory of Moral Sentiments, Works*, I, 189–190 and 3–4.
[42] Duff, p. 11; Duff is here adapting Cicero *De oratore* iii. 50.

tinguishes and determines the various qualities of the objects submitted to its cognizance; pronouncing, by its own arbitrary verdict, that they are grand or mean, beautiful or ugly, decent or ridiculous." [43] This is a fairly representative treatment, although elsewhere genius itself is regarded as the third component which makes for outstanding artistic work, while taste is made into a faculty of the spectator or auditor corresponding to genius in the artist. The arbitrariness assigned by Duff to taste is placed by other authors somewhere in the field bounded by taste and genius jointly.

The basically diagrammatic set of associationist thinking is nowhere better revealed than in its by-products of genius and taste and its way of speaking about them. These two "senses" are by-products, thrown off at the periphery of associationist thinking, because they are too directly concerned with questions of value, with the good and the bad, to have place in the associationist field. Their decisions are indeed arbitrary—unaccountable from the point of view of the associationist mind.

Thus it is that in the associationist tradition the person of genius or taste, however much he may be praised or even emulated, becomes more and more a person apart. "A man of Genius," observed Duff, "is really a kind of different being from the rest of his species." [44] The rest of the species are those whom associationism believes it can explain. This is standard associationism, derivative in some way, no doubt, from the *je ne sais quoi* of Montesquieu's *Essay on Taste*, but far more advanced. Duff's explanation is reiterated in Thomas Brown's description of the poet as a man whose trains of association are

---

[43] *Ibid.*

[44] William Duff, *Critical Observations on the Writings of the Most Celebrated Original Geniuses in Poetry* (London: T. Beckett and P. A. DeHondt, 1770), p. 339.

marked by constitutional *difference* [45] and re-echoed in John Stuart Mill's dictum that the orator is heard but the poet *over*heard.[46]

However, here again associationism sets about liquidating the view designed to supplement it. Given a little time—and of this he has plenty in his four volumes of *Lectures on the Philosophy of the Human Mind*—an associationist such as Brown inevitably produces an explanation of the inexplicable. Genius, at first heralded by him as quite "different," will be made to bow off the scene in a spatial equation. Brown's word of dismissal is as crushing as it is simple: the difference between a poetic genius and a mediocre poet is that in the mind of the mediocre poet the trains of association run as they do for ordinary men, but in the mind of the poetic genius they run *backward*.[47] *Sic transit gloria.* In much the same fashion Gerard flirts with the notion of genius as something quite exotic to the associationist milieu. But life *is* humdrum after all, and, when Gerard puts the exotic stranger to work, he finds him nothing but a quite ordinary job, geometrically construed: genius is to order the confused heap of materials collected by fancy and to set "all the members in that position which it points out as the most natural." [48] John Stuart Mill's "mental chemistry" was another and later attempt to salvage the mechanistic view by supplementing it with something transmechanistic.[49] But, like Brown's and Gerard's offerings,

[45] Brown, II, 272 ff.

[46] John Stuart Mill, "Thoughts on Poetry and its Varieties," *Dissertations and Discussions: Political, Philosophical, and Historical* (New York: Henry Holt, 1874–82), I, 97; see also Chapter 10 below.

[47] There is a "change of the direction of the suggesting principle" (Brown, II, 274).

[48] *Essay on Taste*, p. 174.

[49] John Stuart Mill, *A System of Logic* (London: Longmans, Green, Reader, and Dyer, 1872), II, 439 (Book VI, chap. iv, § 3).

Mill's mental chemistry was destined to be consumed by the mental mechanics it was supposed to transcend and redeem.

VI

The geometrized and mechanized system of imagery which dominated associationism makes itself felt not only in the matters of sympathy, genius, and taste but in the more classic notions of judgment and reason. A full study of eighteenth-century "reason" in its connections with quantified or spatial conceptualizations has still to be made. Only some slight indication of significant shifts in emphasis can be given here.

Judgment had been classically connected with the question of truth as such in the general philosophical tradition in which any statement or proposition was designable as a judgment. In this tradition, judgment or discernment was made, in the last analysis, a matter concerning the subject-predicate relationship. Among the Scottish critics and their contemporaries, however, judgment tends to be assigned functions more suggestive of those of a traffic officer. Judgment or discernment comes here, in the last analysis and in its deepest meaning, to be a matter of guaranteeing order in trains of association so that poetry can be differentiated from mere random assemblages of "ideas." Two ways of guaranteeing this order were early foreshadowed in the divergent approaches to associationism taken by Hobbes and Locke and discussed by Martin Kallich.[50] Hobbes assumes that the governing principles of associationism—contiguity in time and place, cause and effect, resemblance (these are variously enumerated in other authors besides Hobbes and roughly trace to Aristotle)—lead necessarily to order. Locke, on the other hand, assumes that association can lead to disorder. In the Lockian view a supplementary

[50] Kallich, pp. 290–315.

ordering principle must be added, most especially when one is faced with the need to assert an order as tenuous and complicated as that of poetry.

The supplementary role is assigned to judgment by the authors generally, and, in the realignment occasioned by the associationist tendency to view order in terms of trains and mechanistic constructs judgment becomes a kind of supervisor charged with detecting and shunting off irrelevancies. Judgment, says Duff, points out homogeneity or discord in the ideas collected by the imagination,[51] the homogeneity and discord being not that of subject and predicate but that demanded by associationist trains.

It is to this sort of function, among others, that "reason" is assimilated. Indeed, functioning in the associationist economy, judgment and reason often become nearly indistinguishable. Gerard enumerates four "intellectual" powers: sense, memory, imagination, and judgment.[52] What others often do at least implicitly, he, with Ogilvie, does explicitly: under the last power—judgment—he expressly includes reason.

VII

Interpretation of psychological phenomena by analogy with mechanistic operations and geometric diagrams, we can note in conclusion, is not without warrant and has always been a part of the methodology necessary or accessory to psychology. Everything depends on how much is made of the analogies—whether, either implicitly or explicitly, they are regarded as quite adequate or whether their analogical character is kept constantly in mind. In the writers here considered, the quantified, spatial, geometrized concepts are ac-

[51] *Essay on Original Genius*, pp. 8–9.
[52] *Essay on Genius*, pp. 27, 32; cf. Ogilvie, I, iv.

cepted pretty much at face value. Embarrassing questions are seldom asked. The reader is infrequently, if ever, reminded that "ideas" do not actually trail after one another over a spatial field and that the conjunction of such things, which no one has ever hoped to touch or see, can bear only a remote and tenuous resemblance to any sort of train.

But to point out the distortion which excessive geometrization can induce in it is not to say that associationist criticism was valueless. It bore much fruit, not only in indirectly encouraging the kind of effective whimsicality found in Sterne but perhaps chiefly in drawing critical attention to the function of the periphery of consciousness and of allusion in poetry and in fostering the study of literary and ideological history. Without associationism, much of Coleridge's best criticism would not have been. In the compact and complex, but logically loose, organization of poetry, associationism has opened many a rich vein.

Few tools, however, are used without mishap. Everyone will own that through the associationist period some sort of "dissociation of sensibility" had had or was having its effect. As a further conclusion to this study, it might be suggested that this dissociation came about not so much in connection with an overevaluation of thought as in connection with a kind of undervaluation, despite all the fine talk about "reason." The Royal Society's ambition to promote a style "bringing all things as near the mathematical plainness as possible" was highly symptomatic. In this milieu the much-commended "reason" tended to be an intellect *manqué*—circumscribed, often ingeniously enough, or ingenuously enough, by a mathematical horizon.

There is no longer much enthusiasm for picturing the

eighteenth-century sensibility as swinging, early in the century, simply and single-mindedly toward intellectual abstraction and later, by way of simple reaction, to romantic emotion. Disregarding the points at which this representation and the evidence do not too well agree, one might reasonably ask why work produced at the mid-point in such a period did not have just the right mixture of mind and emotion. The present study suggests why there could be no such point of equilibrium in the pendulum swing at all (and thus how another crude mechanistic analogy fails). For the poles between which such a pendulum could have moved were not so neatly opposed, and the pendulum could not move so simply from one to the other. The mind or "reason" toward which the pendulum first moved would have been not full intellectual (as against emotional) life but rather intellect hypnotized by the plausibility of pure quantity, failing to be itself in the full life by right its own, the life not only of mathematics and mathematical physics but of other knowledge as well. The tendency at this period so to hypnotize the intellectual powers remains always only a tendency, to be sure. It is never entirely successful. But its effects are unmistakable.

Some of these have been rehearsed. One more particularly deserves to be instanced, for its shows how the quantifying drives of the age were both affecting and registering a movement of sensibility away from the auditory to the visual, away from the sonorous world where rhetoric had held forth since antiquity to the observational world of modern science. Imitation has long been a capital subject in poetic and critical theory. As elaborated by Aristotle, the imitation theory involves a delicate set of principles which produce, among other things, the teasing conclusion that music, and that apart from

words, is the most imitative of the arts.[53] In the hands of
Adam Smith, imitation is dissociated from the oral-aural
world which music implies, and reconstrued so as to refute
Aristotle's conclusion. Instrumental music is the one art which
is not imitative. It arouses one's own original feelings, not
"sympathetic feelings," which are the "reflected disposition of
another person." [54] Smith's treatment of imitation is involved
in lengthy speculation, but, beneath the elaborate guise of
sympathy and feelings and originality, the basic mentality
reveals itself. Smith's favorite way of conceiving imitation is
geometric. He likes to think of it primarily in terms of one
carpet pattern repeating another.[55] Sonority has been disqualified in favor of diagram.

[53] *Politics* v (viii). 5. 1340a18. It has been pointed out that Aristotle
and the Greeks who share this opinion with him are apparently
thinking of music as such, apart from words (S. H. Butcher, *Aristotle's Theory of Poetry and Fine Art* [London: Macmillan, 1923],
pp. 128–131).

[54] "Nature of Imitation," *Works*, V, 287–288.   [55] *Ibid.*, p. 243.

10 J. S. Mill's Pariah Poet

To an age given to prying open psychological secrets and casting up the accounts of other men's minds, John Stuart Mill's painstaking reports on himself in his *Autobiography* offer possibilities for the most part still strangely unexploited. Matched against his own psychological states as disclosed in his own story of his life, certain of his typical pronouncements reveal themselves as highly symptomatic and lay open both his mind and central issues at stake to an astonishing depth. Mill's carefully pursued ratiocinations and the results to which they lead him are often more revealing than he knew.

In particular, this is true concerning his dictum that eloquence is heard, poetry overheard—a dictum quoted often enough in connection with Mill but so far never really measured against the movements of his mind. The dictum in itself is remarkable in that it prescribes poetic results to be achieved not so much by the poet as in spite of him. The poet is to function by accident, by indirection. He sits outside human society—talking to no one, singing only to the moon. His mission is fulfilled only in so far as the venturesome make their way a little toward his seclusion and install themselves as eavesdroppers at his essentially private soliloquies.

Mill's attitude toward the poet here has connections with his own case history which are both symbolic and more than symbolic. The poet as pariah, as outlaw, is a symbol of Mill himself, whose youthful nervous breakdown with its accompanying sense of isolation had terminated with a drastic revision of his attitude toward poetry. But in the very act of functioning as a symbol for Mill himself, this same poet emerges also as a logical consequence of Mill's earlier intellectual position. He provides an interpretation of the workings of Mill's mind and of the milieu in which Mill's mind was formed.

I

Like many other positions taken by Mill, his stand concerning poetry can be described largely as an attempt to bring into conformity with Hartleian notions of the nature of man—in which Mill had been drilled by his father—aspects of reality which for one reason or another impressed him as urgent. In the case of poetry, the urgency was connected with the "mental crisis" which Mill underwent at the age of twenty. It is shortly after this crisis that the most important of his pronouncements concerning poetry appear, the essays "What is Poetry" and "The Two Kinds of Poetry," in the *Monthly Repository*, respectively for January and October, 1833.[1] The dictum that "Eloquence is *heard;* poetry is *overheard*"[2] occurs in the first of these essays.

[1] Reprinted together, with slight revisions, as "Thoughts on Poetry and Its Varieties" in John Stuart Mill, *Dissertations and Discussions: Political, Philosophical, and Historical* (New York: Henry Holt, 1874–82), I, 89–120.

[2] *Ibid.,* p. 97.

The general nature of Mill's mental crisis or nervous break-down is well enough known.[3] It was connected with the circumstances of his rigorous and precocious intellectual training at the hands of an intelligent father determined to force his eldest son into the mold he had prepared for him—that of a young liberal who could think for himself. Identifying the crisis with an overpowering feeling of futility and purposelessness, of complete deorientation, Mill traces it to a stunted emotional development connected with his associationist training.

My course of study had led me to believe, that all mental and moral feelings and qualities, whether of a good or of a bad kind, were the results of association; that we love one thing, and hate another, take pleasure in one sort of action or contemplation, and pain in another sort, through the clinging of pleasurable or painful ideas to those things, from the effect of education or of experience. As a corollary from this, I had always heard it maintained by my father, and was myself convinced, that the object of education should be to form the strongest possible associations of the salutary class; associations of pleasure with all things beneficial to the great whole, and of pain with all things hurtful to it. This doctrine appeared inexpugnable, but . . . I now saw, or thought I saw, what I had always before received with incredulity—that the habit of analysis has a tendency to wear away the feelings.[4]

Mill dates his recovery as beginning from an experience which reassured him that he was still capable of emotional response.[5] The experience was occasioned by reading, and

[3] *Autobiography* (New York: Henry Holt, 1874), pp. 132 ff.

[4] *Ibid.*, pp. 136–137; see also *ibid.*, pp. 132–146.

[5] *Ibid.*, pp. 140–141. In "The 'Mental Crisis' of John Stuart Mill," *Psychoanalytic Review*, XXXII (1945), 86–101, A. W. Levi con-

Mill henceforward turns more and more to poetry. He had indeed read poetry after a fashion before, but, with the Hartleian bias against "sentiment," had been schooled to discount its emotional aspects. Now, however, from that of relative indifference his attitude changed drastically, and, as periodical articles only recently identifiable as Mill's [6] make more evident than heretofore, he began to insist that poetry and art were necessary to develop the feelings.[7]

In view of the fact that Mill felt so strongly the sad neglect of the feelings or emotions in the education which he had been given,[8] his intellectual maneuvers at this period become

---

cludes that Mill's mental crisis or nervous breakdown "was brought on primarily by the repression of his hatred and death wishes against his father," who, since Mill entertained also a great esteem for him, was indeed a center of conflict in Mill's life—as the *Autobiography* makes abundantly clear (*passim*). Levi notes that the emotional experience from which Mill dates the start of his cure was one in which, after a period of complete emotional stagnation, he "was moved to tears" at reading a passage in the *Mémoires* of Jean-François Marmontel concerning the death of Marmontel's father and Marmontel's assumption of his place as head of the family. The passage in Marmontel is certainly a psychiatric gold mine. Mill's case here, the classic instance of the breeding up of one Utilitarian by another, is interesting if only for the terrifying authoritarian procedure involved.

[6] Ney MacMinn, J. R. Hainds, and James McNab McCrimmon, eds., *Bibliography of the Published Writings of John Stuart Mill*, edited from his manuscript with corrections and notes (Evanston, Ill.: Northwestern University Press, 1945), pp. vii–xiv.

[7] John Robert Hainds, "John Stuart Mill's Views on Art," *Summaries of Doctoral Dissertations . . . Northwestern University*, VII (1939), 9–13.

[8] See Mill's article "Bentham," published during this same period (*London and Westminster Review*, August, 1838), reprinted in *Dissertations and Discussions*, I, 355–417. Mill felt his emphasis of feeling to be, in a way, a reversion to the "sentimentalist" position from

extremely interesting. Mill feels compelled somehow to escape from the narrowness of his previous training. But to do so is not at all easy. It is noteworthy that his own explanation of his nervous breakdown just quoted, together with the larger context in which the explanation is set, is itself couched in terms of strict Hartleian associationism and that the very position from which he wants to revolt was one which to Mill somehow "appeared inexpugnable."

Mill was becoming aware of the difficulties of his associationist and utilitarian training, but, like many another man who has seen the weaknesses of his early training, he found himself approaching these difficulties in terms of the very position he wanted to criticize. His reproach of Hartleianism and Benthamism is a little Hartleian and Benthamite masterpiece. So thoroughly had his father drilled him to one point of view that even conceiving of another became for Mill a major struggle.

Mill's two essays on poetry, which set forth what was to be pretty well his permanent view,[9] thus are in a sense a resolution of a curiously ingrown personal struggle. The problems which Mill poses, "What is poetry?" and "Are poets born rather than made?" are problems hard to lay hold of, prob-

---

which his father and Bentham had revolted in following David Hartley. He takes pains to show that Bentham's view is essentially incomplete and discusses the limitations of one who ambitions reconstructing "all philosophy *ab initio*, without reference to the opinions of his predecessors" (*ibid.*, p. 374: see also pp. 373, 378 ff.). By his 1867 *Inaugural Address*, Mill is even more explicit in speaking of poetry as involving the "cultivation of the sentiments" (*Dissertations and Discussions*, IV, 399 ff.).

[9] His 1867 *Inaugural Address* reiterates the same views thirty-four years later (*Dissertations and Discussions*, IV, 396–407), although, as Hainds points out (Hainds, pp. 9–13), his enthusiasm for poetry was not so exuberant as it had been in the thirties.

lems which even a moderately sophisticated critic will approach very circumspectly indeed. Mill's way of swinging into them with a certain sophomoric innocence betrays that in this field he is in some sense the parvenu. Later, Mill himself apparently became aware of this fact, for, in reprinting the two poetry articles in *Dissertations and Discussions*, he puts them together under a much more modest heading, "Thoughts on Poetry and Its Varieties," which effectively takes the edge off both questions.

But these two basic problems concerning the nature of poetry and of the poet are precisely the problems which Mill must grapple with. His urge is not to instill or even to register very effectively an appreciation of poetry. His 1840 article on Coleridge, whose work Mill evidently valued highly, steers clear of the critical, appreciative approach to plunge rather into theoretical discussion.[10] It is thus not strange that the first thing of any consequence which he has to say about poetry—what he says in his 1833 essays—should engage Mill in the sort of speculative problems which occupy him. He has to get some order into this new world of poetry into which he has been precipitated, some order not so much, really, for his readers as for himself. In his two articles on poetry, Mill is groping for such order in a Hartleian universe which has suddenly run off at the edges into this strange new world. He is tidying up his mind a little—and that in public, in print.

II

Albury Castell has very convincingly examined Mill's *System of Logic* as evidence of "the impact of Newtonism on

[10] "Coleridge," *Dissertations and Discussions*, II, 5–78.

early nineteenth-century social thought." [11] The chunks of experience or "mental states" which early associationism taught Mill to deal with have at least a psychological correspondence with the quantitative parts of a universe conceived of in terms of Newtonian mechanism. Mill's *Logic*, in a sense his most definitive work, is an attempt to bring social science into line with such a universe.

Mill's view of poetry bears unmistakable, but heretofore unnoticed, evidence of the same "Newtonism." Although the close relationship between poetic theory in the late eighteenth and early nineteenth centuries and associationist teaching has been in part explored,[12] the implications of Mill's theory in terms of associationism, and in particular in terms of the associationism propounded by the University of Edinburgh professor Thomas Brown (1778-1820), has gone quite unattended to. Mill's notions of poetry find their natural setting in the kind of associationism taught by Brown in his four volumes of *Lectures on the Philosophy of the Human Mind*, which were published in 1820 and read by Mill some three years later, when he was seventeen.[13] Brown turned from

[11] Albury Castell, *Mill's Logic of the Moral Sciences: A Study of the Impact of Newtonism on Early Nineteenth-Century Social Thought* (Chicago: University of Chicago Libraries, 1936).

[12] For the celebrated connection between associationism and the views of Wordsworth, Coleridge, and Godwin, see Arthur Beatty, *William Wordsworth: His Doctrine and Art in Their Historical Relations*, 2d ed.; University of Wisconsin Studies in Language and Literature, No. 24 (Madison, Wis.: [Mayer Printing Co.], 1927); and cf. Melvin Rader, "The Transcendentalism of William Wordsworth," *Modern Philology*, XXVI (1928-29), 169-190.

[13] *Autobiography*, p. 69. See also Castell, p. 13, and cf. Oscar Alfred Kubitz, *Development of John Stuart Mill's System of Logic*, Illinois Studies in the Social Sciences, Vol. XVIII, Nos. 1-2 (Urbana: University of Illinois Press, 1932), pp. 13, 71 ff., 89 ff., and *passim*.

Hartley's physical and physiological conditioning to expound a more purely mental conditioning as basic to his view of the mind. Skirting the more recent developments of Hobbes or Hume, he returned to Aristotle to adopt the latter's enumeration of resemblance, contrast, and contiguity as the three "primary" laws of his psychology.[14]

When Mill complains of the desiccating effect which "analysis" has on feeling, he is thinking in terms of the Brownian sort of world in which chunks of psychological matter are strung together in chains.[15] Explanation or "analysis" is tantamount to disenchanting dissolution, or the resolution of these chunks into the kind of pattern they form when the mind has nothing to do with them:

The very excellence of analysis (I argued) is that it tends to weaken and undermine whatever is the result of prejudice: that it enables us mentally to separate ideas which have only casually clung together; and no associations whatever could ultimately

[14] Thomas Brown, *Lectures on the Philosophy of the Human Mind* (Edinburgh: W. and C. Tait, 1820), II, 218–269. As noted above in Chapter 9, in his work *Elements of Criticism* (Edinburgh: A. Millar, 1762) another Scot, Henry Home, Lord Kames, has as his first chapter "Perceptions and Ideas in a Train." Brown is clearly in a tradition. I have found no evidence that Mill used Kames. Cf. Castell, p. 13.

[15] Brown speaks of "[Newtonian] physics, whether of matter or of mind" (II, 270). Cf. James Mill, *Analysis of the Phenomena of the Human Mind*, ed. by Alexander Bain, Andrew Findlater, George Grote, and John Stuart Mill (London: Longmans, Green, Reader, and Dyer, 1869), I, 161, where John Stuart Mill's father reduces the function of predication, and thus the process at the heart of all human intellection, to that of marking the *order* of trains of thought, and *ibid.*, I, 175, where he expresses his resentment at the intrusion into his mechanistic world of so unmanageable a business as existence. Attaching itself by "connotation" to predication, he says here, existence "has produced the wildest confusion."

resist this dissolving force, were it not that we owe to analysis our clearest knowledge of the permanent sequences in nature; the real connexions between Things, not dependent on our will and feelings; natural laws, by virtue of which, in many cases, one thing is inseparable from another in fact. . . . Analytic habits may thus even strengthen the associations between causes and effects, means and ends, but tend altogether to weaken those which are, to speak familiarly, a *mere* matter of feeling.[16]

Mill is uneasy at his letting the associationist cat out of the bag "in many cases" in view of the fact that he must keep him in the rest of the time, but he is also uneasy because he is here very close to the crux of poetic theory (which can, of course, be formulated in ways other than that in which it presents itself to Mill): How is the artificial organization which lies at the core of any poem as precisely the thing contributed by the poet to be viewed as other than a worthless deception?

Mill's answer, hinted at in the autobiography and developed more at length in the two poetry articles, undergoes its initial formulation in terms of "feelings." Besides the "real connexion between things," there are other junctures made by the feelings (and by the will, which, however, Mill discards in his poetic theorizing). This is taken as a kind of datum. Such other junctures, as the passage just quoted makes clear, are in a way opposed to connections in the real world established by "natural laws"—so much so that the "analytic habits" sympathetic to natural laws must be hostile to poetry.

It is precisely at this point, where most analysts experience their most serious embarrassment, that Mill's associationist doctrine intervenes to effect a spectacular rescue. For, according to that doctrine, not only are there psychological connections which are other than natural connections—most people

<hr />

[16] *Autobiography*, pp. 137–138.

would admit this at least in admitting that men occasionally fall into error—but these fortuitous or whimsical connections are the very bones and sinews of associationist psychological theory. The laws of resemblance, contrast, and contiguity associate states of mind in contravention to the "permanent sequences in nature." In doing so, these laws operate in the same sort of realm as poetry itself—a realm of detachment from reality.

What would be more natural than to erect a theory of poetry which moves along the same level as the laws of association themselves? This is exactly what Mill, in the second of his 1833 essays, finds himself doing. "Whom, then, shall we call poets?" he asks. And he replies, "Those who are so constituted, that emotions are the links of association by which their ideas, both sensuous and spiritual, are connected together." [17] Mill has here in effect dredged up an additional law to set alongside the familiar three which Brown had made over from Aristotle.

There is a kind of tragedy in Mill's maneuver here, the tragedy of a man, and one with a vigorous and trained mind, who is doing the very thing which he hoped most earnestly to avoid—yielding to his own prejudices, being unable to break the links of association forged in his earlier life. Ostensibly shocked by his own interior crisis out of what he thought was the complacency of associationist attitudes, he ends here more deeply in the toils than before. His theory of poetry, seemingly conceived in rebellion, really retreats more deeply than ever into associationist territory. Mill seems never to have noticed what his behavior here hints to a later generation: that associationist doctrine was itself largely a kind of poetry, something which, charmed by the speciousness of its

[17] *Dissertations and Discussions*, I, 106.

own "analysis," pretended that it had nothing at all to do with anything but cold reason, but which at bottom was faced with the same difficulty as poetry itself. In poetry the question was: What makes the poem cohere if the items in it do not cohere in nature? In associationist doctrine the question was: What did make you keep together what came together in time or space? How, after all, *were* the associations to be accounted for?

III

If the general framework of associationist thinking forms the setting of Mill's concern with poetry, there are special aspects of this thinking, as brought out in detail in Brown's theories, which go far toward accounting for Mill's famous dictum that "Eloquence is *heard;* poetry is *over*heard." [18] Besides the three primary laws of association which he adopted from Aristotle, Brown proposes eight "secondary laws" which modify the application of the three primary laws: relative duration, liveliness, frequency, recency, previous coexistence with fewer alternative associates, constitutional differences, temporary as against constitutional variants, and habit. [19]

It is under his treatment of the law of constitutional differences that Brown gathers to his discussion the question of poetry. Poets, Brown proposes, are full of what one today would call suggestion, but what he calls "analogy." Since they write things so notably concerned with "analogies" both of "ideas" and "feelings," their being poets must be put down

[18] *Dissertations and Discussions*, I, 97.
[19] Brown, II, 270–286. Castell (p. 13), after stating that Brown lists nine secondary laws, enumerates only eight. The confusion is doubtless due to the fact that "constitutional differences" is again subdivided by Brown into laws of (a) variations in individuals and (b) greater proportional vigor.

to their having constitutions in which association trains are piled high with "suggestions of analogy," [20] and in such a way as to make them quite unlike the rest of men. With Brown, the question of feeling was not acute in the way it was to be to Mill. Brown hardly notices feeling here except in passing, and his explanation, apart from its resort to constitutional differences, proceeds in the usual way in terms of concatenations. Even so tenuous a distinction as that between a poetic genius and a mediocre poet is reduced by Brown, as noted in the preceding chapter here, to the fact that the genius' trains of association run backwards.[21]

If one regards not so much Brown's discussion of "analogy" or his neglect of feeling but rather his erection of constitutional difference into an "explanation" of poetry, Mill's discussion of poetry is evidently at least an oblique derivative from views such as Brown's. It would be too much to see in Brown's secondary laws the means of filling in all the details and emphases in Mill's scheme of poetics. But Mill is certainly akin to Brown in bringing the poet into his scheme over a route labeled "constitutional difference." For both, constitutional difference is an ultimate in explanation of the poet's activity. Labeling the poet constitutionally different is tantamount to admitting that there is simply no accounting for him in the associationist world. He is not the associationist type. If the poet is to make his way into the associationist scheme of things, Brown's or Mill's, he is going to have to do so as an anomaly.

It will be noted that at the very heart of Mill's theory, in his answer already quoted above to the question "Whom,

[20] Brown, II, 276.
[21] There is a "change of the direction of the suggesting principle" (*ibid.*, p. 274).

then, shall we call poets?" constitutional difference puts in its appearance: "Those who *are so constituted*, that emotions are the links." Mill continues:

This constitution belongs (within certain limits) to all in whom poetry is a pervading principle. In all others, poetry is something extraneous and super-induced; something out of themselves, foreign to the habitual course of their every-day lives and characters; a world to which they may make occasional visits, but where they are sojourners, not dwellers, and which, when out of it, or even when in it, they think of, peradventure, but as a phantom-world,—a place of *ignes fatui* and spectral illusions.[22]

His close association of constitutional difference with his poetic theory subtly recommends itself to Mill also in a curiously reflex sort of way. It enables him to account for the fact that the explanation he gives of poetry has waited all these years to be given. Poetry has been neglected in so far as its practitioners are exceptional instances, for metaphysicians, he goes on to say in the same place, "while they have busied themselves for two thousand years, more or less, about the few [i.e., three?] *universal* laws of human nature, have strangely neglected the analysis of its *diversities*."

Mill's concern with constitutional differences betrays itself throughout the entirety of his second essay here, for, although originally published under the title "The Two Kinds of Poetry," this essay is really a gloss on the dictum "Poeta nascitur," which forms the opening words of the essay. The matter of the two kinds of poetry, that of Wordsworth and that of Shelley, is brought into the discussion only to "assist in rendering our meaning intelligible" [23] concerning constitutional differences. Mill explains that, although poets can make

---

[22] *Dissertations and Discussions*, I, 106–107.   [23] *Ibid.*, p. 109.

themselves philosophers, philosophers cannot make themselves poets because they have not in their make-up the "poetic laws of association." [24] This kind of remark is likely to turn up whenever Mill concerns himself with poetic theory. Writing five years later of Alfred de Vigny, he maintains that the poetic nature has an "instinctive insight" and "by that faculty, *as by an additional sense,* is apprised, it cannot tell how, of things without, which escape the cognizance of the less delicately organized." [25] Mill echoes the same point toward the end of his life in the *Inaugural Address.*[26]

A doctrine recognizing the poet's case as somewhat special is by no means entirely strange in the realm of poetic theory —it goes back at least to Plato and Aristotle—but in the context in which it occurs in Mill, it is remarkably illuminating. The dictum "Eloquence is *heard;* poetry is *over*heard" occurs in direct function of Mill's notions here. Eloquence is to poetry, we are told by Mill, as social intercourse is to solitude or as Rubens' "historical" painting is to Raphael's "Virgin and Child." [27] Poetic theory is often enough more or less explicitly cognizant of the fact that art is concerned with a communication somehow different from that of a more purely conceptualized utterance, that in some fashion a work of art, whether music, painting, or poetry, is in the last analysis untranslatable into a medium other than itself. Mill's notion that eloquence is heard and poetry overheard, pointed by the examples he adduces, is plainly designed to cover in

[24] *Ibid.,* p. 117.
[25] "Writings of Alfred de Vigny" (*London and Westminster Review,* April, 1838), *Dissertations and Discussions,* I, 349; italics mine.
[26] *Dissertations and Discussions,* IV, 402.
[27] "Thoughts on Poetry and Its Varieties," *Dissertations and Discussions,* I, 97–102.

some fashion this sort of ground, to indicate the difference between poetry and more purely abstract linguistic operations.

But that is not all it is designed to do. To say that a poem cannot be perfectly rendered or grasped in any words other than those of the poem itself is not to intimate that the poet is talking to himself. Plainly, he is not, although he may tend to sound more as though he were than do other users of language. Mill's dictum comes out the way it does because he is, among other things, incorporating into it his own associationist views of poetry in all their elaborate history. Poetry is association by feelings or emotions. This kind of association operates in function of constitutional differences. It is connected not so much with "the few universal laws of human nature," about which metaphysicians "have busied themselves," as with the "diversities" of human nature. Poets are people with "constitutional differences" which amount almost to "an additional sense."

Thus, in the final analysis, Mill's poet must be overheard rather than heard because he is an anomaly. He is constitutionally different from man as associationists envision man, a misfit in a mechanistically plotted world. All chains of association are, in a sense, accidents in so far as they are not representations of "natural laws," but the poet's chains of association are much worse. They are *exceptional* accidents, ultimately inexplicable, and they pretty well exile the poet for good. Mill's poet, or Brown's, finds his natural habitat outside society because in an associationist society no one can really understand him. Mill's tracing of poetry to constitutional differences, his deciding, with Brown, that the poet is a poet in so far as he is not like other men—these "other" men being those whom the associationist psychology *could* account for

—is the readmission, under a slightly altered guise, of Mill's original difficulty, connected with his nervous breakdown, the difficulty of admitting poetry into the utilitarian associationist world. That poetry finally enters in terms of constitutional difference ultimately means that it enters where it should not and despite everything. In a world subjugated by mechanistic frames of thought, it has no place at all. In every implication of the word, it is *outré*.

This conclusion is closely borne out of Mill's way of handling, rather covertly, the problem of communication. How is it that what this pariah poet has to say can effect others at all—that the lover of poetry can make out anything whatsoever with even the most diligent eavesdropping? Mill touches on the question, significantly, in his essay on Bentham. He assures us that the communication effected by the poet is indeed quite an accident, for it is based not precisely on what the poet as poet has to say, but rather on the adeptness of the eavesdropper's own Coleridgean imagination:

The imagination, . . . that which enables us, by a voluntary effort to conceive the absent as if it were present, the imaginary as if it were real, and to clothe it in the feelings, which, if it were indeed real, it would bring along with it. This is the power by which one human being enters into the mind and circumstances of another. This power constitutes the poet, *in so far as he does any thing but melodiously utter his own actual feelings. It constitutes the dramatist entirely.* It is one of the constituents of the historian: by it we understand other times.[28]

The imagination, in terms of which one's own poetic eavesdropping is able to succeed, does not belong per se to the poet but to the dramatist. Mill is trying to be consistent here.

---

[28] "Bentham," *Dissertations and Discussions*, I, 378–379; italics mine.

Poetry must remain "feeling confessing itself to itself in moments of solitude."[29]

Mill's view evidently is not free of some confusions. Still, certain things in it come clear. That Mill's poet is in some ways the poet of the nineteenth century par excellence few will dispute. His poetic theory is at least as much at home in the intellectual climate as are Wordsworth's views on emotion recollected in tranquillity and Coleridge's views on imagination—to both of which views, as has been suggested, the theory has certain relationships.

Made a proving ground for all the issues which the preceding milieu had accumulated in the mind of his father, the consciousness of John Stuart Mill found the agonizing tensions of his own milieu all the more agonizing because it had been preconditioned to them, scarred prematurely by the earlier struggles out of which these tensions grew. Out of the agony is wrung the symbolism: the exile poet singing to the moon. Mill was unaware of the full import of the symbolism, but his preoccupation with the possibility of eavesdropping shows how unstable and unsatisfactory the solution represented by the symbolism really was. Unlike Plato, he rebelled at a poetless society, for he needed the poet precisely to save society from itself. His poet had to be both in exile, where he would be safe from a mechanistic world, and not in exile in order to redeem this world. The poet *over*heard is such a poet, inside and outside society at one and the same time.

Mill's dilemma here provides confirmatory evidence that he was never able thoroughly to abandon or to cut under the intellectual presuppositions which he says all but wrecked his

[29] "Thoughts on Poetry and Its Varieties," *Dissertations and Discussions*, I, 97.

life at the time of his nervous breakdown. Mr. Castell has found in Mill's *Logic* an example of an intellectual lag from the eighteenth century.[30] Mill's notions concerning poetry betray the same lag. In his revolt from his father's and Bentham's mechanistic utilitarianism, his criticism of the mechanistic view never really departs from mechanistic theorizing. In the case of the poet, he almost departs from it: he makes an exception to the rules of the game as the mechanist would play it. But he does not revise the rules. The poet remains a surd in Mill's philosophy, and no amount of handshaking or speeches of welcome—for the surd here is welcome—will change the fact. Here the eighteenth century is fighting its battles to a new standstill. The erection of poetry into soliloquy, although it indeed honors certain facts relevant to any poetic theory, is most of all the utilitarian associationist's monument to his own philosophy of man. It shows in subtle fashion what was happening to the concept of language when the old oral-aural world which had nourished rhetoric, basically an art of oral communication, was abandoned for a more exclusively visualist cosmology. In a cartographer's universe, communication is conveyance of a commodity called information. Since this is hardly what a romantically conceived poet does, such a poet understandably, is not "communicating" at all.

[30] Castell, p. 94.

II ⟩ Romantic Difference and the Poetics of Technology

I

It is a classic observation that the romantic movement can be defined in a great many ways, some complementary and some competing. But whatever way one defines it, one of the movement's characteristics—more or less central depending on the particular definition—is a preoccupation with otherness, with what is different, remote, mysterious, inaccessible, exotic, even bizarre. Some historians make more of this characteristic, some less. But virtually all scholarship falls back on it to set off the romantic movement from the neoclassicism which preceded it.

However, to bring out the full significance of romantic otherness neoclassicism will no longer suffice as a point of comparison. Neoclassicism was a molehill where romanticism was a mountain, and still is. Romanticism has not been a transient phenomenon. Most and perhaps even all literary and artistic, not to mention scientific, movements since the romantic movement appear to have been only further varieties of romanticism, each in its own way. Early Victorian is attenuated romantic, late Victorian is recuperated romantic, fed on Darwin, Marx, and Comte, the American frontier and the American Adam are primitivist romantic, imagism and much other modern poetry is symbolist romantic, existentialism is supercharged or all-out romantic, structuralism is formally interiorized romantic, programmatic black literature is alien-selfhood romantic, and beatnik and hippie performance is disenfranchized romantic. Insofar as it is an art form or a substitute for literature, and in other guises, too, activism is most certainly idealist romantic. The late Renato Poggioli has suggested that for the entire foreseeable future all serious developments in literature and art, and it would seem in life styles generally, will oscillate back and forth between one and another form of romantic alienation.[1] Arthur O. Lovejoy's celebrated prowess in distinguishing varieties of romanticism was probably due to the ferment of romanticism still active in all of us as much as it was due to the diversified richness of the original romantic movement itself. It takes one to catch one.

All this hints that we have not yet plumbed the depths of the otherness which romanticism was and is. One way of improving our understanding of the drive toward otherness in the movement, I should like to suggest here, is to look at its relationship to the commonplace tradition, the tradition rooted in classical antiquity, in which poetry felt itself dedicated to asserting, always in fresh form, what poetry had always asserted—"What oft was thought, but ne'er so well expressed," as Pope summed it up for his and preceding ages. The commonplace tradition was old, part of the vast rhetorical tradition which had formed many of the sinews of Western civilization since before Christianity and Judaism. Quite recently, we have

---

[1] Renato Poggioli, *The Theory of the Avant-Garde*, tr. from the Italian by Gerald Fitzgerald (Cambridge, Mass.: Belknap Press of Harvard University Press, 1968), p. 127.

become more aware of the massiveness of this rhetorical tradition and all that went with it as we have become aware of the relationship of rhetoric to the development of the communications media and of technology and thus to the total evolution of culture.

So long as it lasted, the commonplace tradition imposed itself overwhelmingly in literature and the arts, all the way from music, with its standardized opening phrases and concluding cadences,[2] and sculpture, with its epithetic owl for Athena or lily for St. Joseph, to rhetoric and politics. Some of the force of the commonplace tradition can be sensed in the line from Pope just quoted, where it imposes itself doubly. For this line not only prescribes the commonplace—"What oft was thought"—but also is itself a commonplace, a *locus communis* exploited by countless writers before Pope. Pope is saying what poetry virtually always said that poetry should say. Of course, he is saying it better. But even this is what the commonplaces about poetry and within poetry always said, too: poets, if they are real poets, always say it better.

Originality itself here commends itself within a nonoriginal frame. Obviously a tradition such as this is seriously at odds with the romantic hankering for what is different, original, strange, ineffable, inaccessible, unknown. The high contrast of the commonplace tradition with romanticism can help us understand what romanticism was and is. And this tradition is a massive one. Unlike neoclassicism, it is at least as massive as romanticism itself.

[2] Cf. the sections on "Topics of the Exordium" and "Topics of the Conclusion" in Ernst Robert Curtius, *European Literature and the Latin Middle Ages*, tr. from the German by Willard R. Trask, Bollingen Series, XXXVI (New York: Pantheon Books, 1953), pp. 85–91.

The present essay is a modest attempt to indicate certain ways in which romanticism is set off against the commonplace tradition. It is not inclusive, but I should like to think it may be germinal. To keep the attempt within feasible dimensions, after some attention to the commonplaces as such, I shall point up some contrasts by looking to Wordsworth's poetic doctrine and bits of Coleridge's comment on this doctrine as representative of romanticism in one of its important and typical manifestations. Elsewhere I have suggested very briefly a contrast between Wordsworth and the commonplace tradition by attending to the lines concluding one of his Lucy poems: "But she is in her grave, and oh, / The difference to me!" [3] In instancing these lines, my point was that the preromantic poet's public concern with things commonplace in the sense that they were known to all men ("What oft was thought") was replaced quite decisively in Wordsworth by a concern with private epiphany. In this study the focus is somewhat different. It centers on Wordsworth's attitude toward poetic diction.

"Wordsworth," Carl Woodring has stated bluntly and accurately, "was out to kill poetic diction." [4] So he was. But this suggests a paradox which had best be faced here before we go further. If Wordsworth was out to kill poetic diction, was he not thereby allying himself with commonplace, which poetic diction was calculated to transmute? Do not critics from Samuel Taylor Coleridge and Charles Lamb to Carl Woodring and Geoffrey Hartman find Wordsworth actually immersed in the commonplace rather than rebelling against it? Does not Wordsworth in fact so see himself? Coleridge and Lamb felt

[3] Walter J. Ong, *The Presence of the Word* (New Haven and London: Yale University Press, 1967), pp. 253–254.
[4] Carl Woodring, *Wordsworth* (Cambridge, Mass.: Harvard University Press, 1968), p. 145.

that Wordsworth's distinctive effects derived from his ability
to spread "the depth and height of the ideal around forms,
incidents, and situations, of which, for the common view
custom had bedimmed all the lustre." [5] Professor Woodring
finds that romanticism has two matters, "the far-away-and-
long-ago and some 'more humble' matter of today" and that
Wordsworth "proclaims in hammered verse the thrill of the
ordinary." [6] Professor Hartman observes that the poem generally known as "Tintern Abbey" begins in a mood intensely
matter-of-fact, and that Wordsworth does not make use of
lofty subject matter but, on the contrary, is "the first English
poet to consider and use personal experience as his 'sublime
argument.' " [7] All this would seem to indicate anything but the
contrast between the commonplace tradition and romanticism
which I have been suggesting, at least if Wordsworth is representative of romanticism. It would seem to indicate that
Wordsworth's typical concern was quite obviously the commonplace, so that he belongs inside the commonplace tradition, not outside it. He himself confirms the commonplaces-
ness of his themes. In his revised Preface to the second (1800)
edition of *Lyrical Ballads*, he states that he has written about
"common life" and that both his thought and his language
might well be charged with "triviality" and "meanness," [8] that
is to say, with lack of difference or distinction. He goes on to

[5] Samuel Taylor Coleridge, *Biographia Literaria*, ed. with his Aes-
thetical Essays by J. Shawcross (Oxford: Oxford University Press,
1949), I, 59 (chap. iv).

[6] Woodring, pp. 82, 214.

[7] Geoffrey H. Hartman, *The Unmediated Vision: An Interpretation of Wordsworth, Hopkins, Rilke, and Valéry* (New Haven: Yale
University Press, 1954), pp. 5, 6.

[8] [William] Wordsworth and [Samuel Taylor] Coleridge, *Lyrical
Ballads: the Text of the 1798 Edition with the Additional 1800 Poems
and the Prefaces*, ed. by R. L. Brett and A. R. Jones (New York:
Barnes and Noble, 1963), pp. 238, 240. See also Woodring, pp. 143 ff.

defend himself, of course, saying in the Preface that he uses the nondistinctive, the commonplace, with high "purpose." But is this not a *deus ex machina* or perhaps even the intentional fallacy, promulgated by author rather than critic?

II

To understand the paradox we must first note that "commonplace" can mean either of two things. Classically it has referred to the *loci communes* or commonplaces of discourse —and these Wordsworth largely rejects. "Commonplace" can also refer to the commonplace things of life—and these Wordsworth made abundant use of. The two senses are related but by no means the same. The second sense, the commonplaces of life, hardly needs explanation. But the first does, for this is the sense employed when we speak of the "commonplace tradition" which forms a central part of rhetoric from as far back as we can go in classical antiquity up to the 1800 Preface and even some time afterwards.

The commonplaces of discourse have been treated at length in Chapters 2 and 3.[9] There the twofold meaning attaching to the term has been explained. "Commonplaces" can refer to what we may label more specifically the "analytic commonplaces," that is, headings for analyzing or breaking down areas from which one can possibly draw matter of discourse—general headings such as definitions, causes, effects, related things, opposites, or more specific headings such as (in cases where a person is being discoursed on) place of origin, parentage, education, gifts of body, mind, and fortune, and so on ad infinitum. The term "commonplaces" can also refer to what

[9] For discussion of the commonplaces more detailed than in the present book, see Ong, *The Presence of the Word*, pp. 79–87, and references there.

we may label more specifically "cumulative commonplaces," which are excerpts culled from one's own compositions or from other authors on one or another standard subject and saved up in one's head (or later in writing) for subsequent use—such as purple patches on heroism, decadence, loyalty, treachery, descriptions of a dark night, of a shipwreck, etc.

The commonplaces of discourse derive directly from the old oral culture of mankind. Like this old oral culture, they did not pass out of existence immediately with the invention of writing but instead, with other oral institutions, were codified by writing to form what we have styled rhetorical culture. Rhetorical culture is basically oral culture shrouded in writing. It is an oral culture whose institutions (in the sociological sense of this term, ways of doing things, patterns of behavior) have been codified, put into manuals, made the object of reflection and of reflective training, and thus both artificially sustained and reinforced by writing—the very instrument which was ultimately to make these institutions obsolete.

Oral cultures necessarily place a premium on standardization or fixity, for in the absence of writing the major noetic effort of a society must be not to seek new knowledge but to retain what is known, not to let its scant, hard-won store disintegrate or slip into oblivion. In this economy of thought commonplaces serve an important function. They are major devices for standardizing knowledge, for keeping it in fixed forms retainable in the memory, so that it can be retrieved orally at will. Indeed, commonplaces are seen to be the most basic or central standardizing devices in the entire noetic economy once we recognize that commonplaces include epithets, which are omnipresent in an oral culture. We here understand epithets in the usual sense of expected or standard qualifiers (the *mournful* cypress, the *clinging* vine, the *sturdy* oak, *wise* Nestor, *wily* Odysseus) or standard surrogates or kennings ("tamer of horses" for an ancient Greek warrior, "The Mantuan" for Virgil, "whale-road" for the sea). Epithets are mini-commonplaces (cumulative commonplaces, of course, rather than analytic commonplaces). Like all cumulative commonplaces, they are cullings from what has been said about a subject before, though each epithet is a culling on a very small scale, often not more than a word or two.

Books in which commonplaces in the sense of cullings or excerpts were gathered could be anything from collections of moderately lengthy passages of several hundred words on set subjects, such as heroism or treachery or a description of a storm at sea, to lists of mere epithets or stock phrases, as in Erasmus' *De duplici copia verborum et rerum* or Ioannes Ravisius Textor's *Officina* or *Epitheta*, or in the *gradus ad Parnassum*, one of the last forms of commonplace books, devised especially for writing Latin poetry. In its countless editions, abridgments, and amplifications from 1709 through the nineteenth century, the *gradus ad Parnassum* ("steps to Parnassus") excerpted from Latin writers and ranged under various subject headings or "places" the myriads of standard adjectives, synonyms, paraphrases, and other *modus dicendi* which schoolboys painstakingly wove into their Latin poetry and other Latin compositions.[10]

---

[10] A *gradus ad Parnassum* is distinguished from other phrase books in that it indicates the quantities of all the syllables in all the Latin words to facilitate their use in poetry. The first Lation gradus was published in 1702 by the famous Jesuit schoolmaster Paul Aler (1656–1727) according to the *Encyclopaedia Britannica* (1968) article, "Gradus ad Parnassum," but the 1709 edition is the earliest listed in Carlos Sommervogel, *Bibliothèque de la Compagnie de Jésus*, Part I, Vol. I (Brussels: Oscar Schepens, 1890), col. 164; the continuation

The *gradus ad Parnassum* was regular equipment for schoolboys in Wordsworth's age and indeed much later and, as we shall see, explains some of what today we might style Wordsworth's "hangups." The millennia-old, chirographically sustained oral culture which we have styled rhetorical culture was here still alive, continuing to proffer commonplace material as major operational units of discourse. But the days of rhetorical culture were numbered when romanticism set in. For our purpose here it is important to note that the romantic movement began to stir exactly at the peak of the great encyclopedic age. The *Encyclopédie* of d'Alembert and Diderot appeared in the years 1751–1772. At this point the potential

---

of Sommervogel by Augustin de Backer, *Bibliothèque des écrivains de la Compagnie de Jésus,* (Liége: A. de Backer, 1899), lists no earlier edition. Earlier gradus were all in Latin, cover-to-cover, later ones in Latin-French, Latin-English, etc. They were still being published through the nineteenth century: see, for example, Paul Aler, *Gradus ad Parnassum,* new ed. (London: Company of Stationers, 1851) and François-J.-M. Noël, *Gradus ad Parnassum: Nouveau Dictionnaire poétique latin-français* (Paris: Hachette, 1875). This gradus, as others, cites both ancient Latin and Neo-Latin authors (those writing after around A.D. 1400) indiscriminately. Ausonius, Horace, and Ovid are accompanied by Theodore Beza (1519–1605), George Buchanan (1506–1582), the Jesuit professor of rhetoric and poetry Gabriel Le Jay (1657–1724), Ioannes Baptista Spanuola, known as Mantuanus (Shakespeare's "old Mantuan"—1448–1516), and so on. This spread of approved authors attests to the life of the Latin academic tradition. In the *Biographia Literaria* section from which a quotation is given below (I, 12–13—chap. i), Coleridge cites from a gradus a Latin quotation from Politian (Anglo Poliziano, 1454–1494), one of the few Western Renaissance scholars who wrote significant verse in Greek as well as Latin. There were also gradus for Greek as well as for Latin, but in far less number. Under each entry (a word) the gradus gives such material as synonyms, phrases or longer excerpts and paraphrases using the word, lines and parts of lines from the Latin poets using the word, etc.

for locating discourse in space had been maximized through alphabetic letterpress print—which is alphabetic writing locked with new firmness into space—and man's feeling for his life-world was undergoing a major change. Noetic contact with actuality was no longer maintained by repetition. What one now needed was "originality."

Romanticism and technology, as we shall be suggesting, are mirror images of each other, both being products of man's dominance over nature and of the noetic abundance which had been created by chirographic and typographic techniques of storing and retrieving knowledge and which had made this dominance over nature possible. The atrophy of the commonplaces of discourse signaled by Wordsworth's typically romantic theory and practice is simply one of the more spectacular signs of the attenuation or etiolation of the massive and venerable old oral noetic economy in the face of technological development reflected and furthered by the evolution of communications media.

III

Practice of one sort or another in the use of the commonplaces, in both senses of this term, helped form virtually all the poetry and other literature in the Western world from Homer through neoclassicism. This practice was a residue, as we have just suggested, of the oral heritage, which must place a premium on fixity.[11] The oral heritage lingered long after writing and print and persists in significant ways even into our present age of electronic verbalization, for words are basically and irreformably sounds, no matter how many ways we devise of handling them indirectly through patterns on a surface or otherwise (electronic culture dephoneticizes words,

[11] See also Chapter 2 above.

as on computers, but also maximizes their phonetic qualities as by telephone, loudspeakers, radio, and television). The degree to which the oral heritage asserts itself was and is controlled by a variety of factors. One of the principal factors used to be, until roughly the maturing or romanticism, the academic use of Latin. This use of Latin had a bearing on the development of the romantic movement, and it warrants some attention here.

The state of Latin as an academic instrument has been a major detriment in the intellectual and literary and cultural development of medieval and modern Western Europe. We have long known, of course, the influence exerted by Latin through its lexical resources and its cultural content. Generations of scholars, in our day led by such major figures as Leo Spitzer and Ernst Robert Curtius, have spelled out much of this sort of indebtedness. But we are becoming aware of a more profound and more pervasive, if more elusive, influence of Latin on the making of the modern mind. This more pervasive influence was due to the special relationship of Latin to the evolution of the media of communication.

From the past to the present, the media of verbal communication have developed in successive stages, conveniently classed as oral, chirographic, typographic (or chirographic-typographic, for in many ways these two are closely connected), and electronic. Latin relates in a special way, quite different from that of the vernaculars, to this sequence of the media, and has had a consequent influence on world views, personality structures, and culture generally quite different from that of the modern vernaculars with which it coexisted and still coexists. This influence is quite apart from the influence of the vocabulary or content of Latin literature.

Latin had remained a vernacular until sometime in the

sixth to the ninth century of the Christian era (depending on what region or social group one is speaking of). After that it split, developing along normal oral lines into the modern romance languages on the one hand, and on the other remaining basically unchanged in the academic world, where we can style it, from this point in history on, Learned Latin. In its relationship to the media, the first pertinent characteristic of Learned Latin was its chirographic nature: although it was far from a dead language (its lexicon has increased by still uncounted thousands upon thousands of words from late antiquity up to the present day), it was a language controlled in its development entirely by writing. (In this it was more or less like certain other learned languages during approximately this same period of time, some time after the invention of writing and before the spread of print: classical Greek, Sanskrit, Old Church Slavonic, etc.) It was a spoken language, to be sure. For over a millennium and a half, millions of persons spoke Learned Latin, but not a one who did not also write it. A curious state for a language. The linguistic development of Learned Latin was determined not by an oral phonemic system, not by the way people actually pronounced Latin, but exclusively by the way they wrote it. It was no longer a mother tongue, learned at the hearthside, but an exclusively extrafamilial language. In its great quantity of witty verse and prose Learned Latin manifests a wonderful spirit of play but it knows no baby talk.

Practice in the use of the commonplaces, in both senses of this term, was kept particularly vigorous by the continued academic dominance of Learned Latin. Until roughly the romantic age, all formal university-oriented academic training in expression dealt with Latin expression (with some variable and usually perfunctory attention to Greek). Inevitably, from the time when Latin had ceased to be the language one spoke at home, its users therefore depended for their materials almost exclusively upon a corpus of writings more or less isolated from the ebb and flow of their vernacular speech. Moreover, they depended on this corpus more and more as time wore on. So long as Latin remained an active oral academic vehicle as well as a written one, as it did on a large scale well into the eighteenth century at least, there could be some feed-in to Latin writing from nonwritten Latin or Latinate sources, but as the oral currency of Latin waned, its users relied less and less on flexible sources and more and more on commonplace books which were the typographic equivalent of oral conversation, purveying bits of expression (in this case culled from and in writing) such as would be digestible even by relatively unskilled users.

We must continually remind ourselves that, by and large, well through the nineteenth century the only poetry a student was allowed to study or write in preparatory school or university was normally Latin poetry (with some scraps of Greek perhaps). As late as the mid-nineteenth century we find in Newman's *Idea of a University* a section on the development of an effective Latin style.

A second pertinent characteristic of Learned Latin was that, despite this utter subjugation to chirography, it remained paradoxically a language formally dedicated to oral performance: its great classical proponents such as Cicero and Quintilian had solemnly declared, with almost all antiquity to back them, that the objective of training in Latin and indeed of all liberal education was to form the public speaker, the orator. Paradoxically, Cicero and Quintilian had put this oral view of Latin into impeccable writing, so that it was accepted by later ages dominated by writing, notably the Middle Ages and even more the Renaissance. Thus, even though Learned Latin was always a language dominated by chirography, and later by typography, it was also nominally dedicated to oratory.

Somewhat mysteriously linked to these two characteristics of Learned Latin were several others on which I have commented elsewhere [12] and which I can here only mention. For one thing, as has already been explained in Chapter 5, Latin was a sex-linked language, for almost a millennium and a half spoken and written only by males. Latin was learned in a male puberty-rite setting, normally not in the home at all but in a school frequented by males only and marked by more or less formalized violence (physical punishment) in which the basic lore of the tribe was transmitted. (For centuries Latin was the only entry to academic knowledge: even through the sixteenth century it was difficult if not virtually impossible even to state in English or the other vernaculars the contents of the physical sciences or mathematics or medicine or law or theology—there were no vernacular words current for these matters.)

To this sex-linked character of Learned Latin was allied its fourth characteristic, its polemic status, which connects again with Latin orality. Latin was identified with an academic world which from classical times had known no other way of validating or testing its teachings than through ceremonial oral polemic. The dialectical debate or rhetorical contest which until the accession of romanticism determined one's profi-

[12] See Chapter 5 above and, for more specialized recent developments, Walter J. Ong, "Communications Media and the State of Theology," in John W. Padberg, ed., *Theology in the City of Man: A Sesquicentennial Conference* [of Saint Louis University] (West Nyack, N.Y.: Cross Currents, 1970), 462–480; this same study appeared earlier in the periodical *Cross Currents*, XIX (1969), 462–480.

ciency in all academic subjects from grammar through logic, physics, medicine, law, and theology, was the academic version of the ceremonial combat which ethologists find in lower animals as well as in man to be a typically male phenomenon —females fight for immediately realistic reasons, such as the protection for their young, rather than ceremonially, for less obvious reasons ("for fun"), and consequently their fighting cannot be turned off by appropriate biological signals so commonly as can the ceremonial fighting of males.

All these characteristics are mysteriously interdetermined. Their tie-in with one another, and thus with the orality which makes the Latin tradition, appears surprisingly absolute. Always and everywhere in Western culture, so far as I can find, where one of these developments takes place the others are pretty well concurrent, and not by conscious design but by the nature of things: the opening of schools to girls, the disappearance of Latin as a functional medium of instruction and with it of the oratorical ideal, the abolition of physical punishment as an educational instrument and of a ceremonially polemic approach to learning, and the atrophy of initiatory or *rite de passage* features in the academic milieu.[13] (In our day the virtually total suppression of all *rite de passage* features of academic education is marked by our corresponding large-scale rise in identity crises.)

If we need to single out a spokesman for the new order of things, he could be John Dewey. The spokesman for the old order was all of Western culture—with, we should note, the exception of the commercial world, which, as I have undertaken to show in *The Presence of the Word*,[14] was and is not oratorical, not exclusively masculine, not violent (either in

[13] See Ong, "Communications Media and the State of Theology."
[14] Pp. 241–255.

principle or in overt drive, as Schumpeter has made clear [15]), not verbally polemic in public despite its competitiveness, and which, when it was verbal, was vernacular rather than Latin.

IV

All this is a preface to a passage in Coleridge's *Biographia Literaria* which cites Wordsworth as well as Coleridge himself in such a way as to enable us to bring some of the foregoing points to bear on an understanding of the nature of romanticism, and particularly of the romantic preoccupation with "All things counter, original, spare, strange."

Coleridge has been discussing developments in English poetic style, in particular expressing his dislike for some features of Gray's poetry and that of Gray's age.

In referring various lines in Gray to their original in Shakespeare and Milton; and in the clear perception how completely all the propriety was lost in the transfer; I was, at that early period [at the age of seventeen, when he first encountered William Lisle Bowles's sonnets], led to a conjecture, which, many years afterwards was recalled to me from the same thought having been started in conversation, but far more ably, and developed more fully, by Mr. Wordsworth; namely, that this style of poetry, which I have characterized above, as translations of prose thoughts into poetic language, had been kept up by, if it did not wholly arise from, the custom of writing Latin verses, and the great importance attached to these exercises, in our public schools. Whatever might have been the case in the fifteenth century, when the use of the Latin tongue was so general among

[15] Joseph Schumpeter, *Imperialism and Social Classes*, tr. from the German by Heinz Norden, ed. by Paul M. Sweezy (New York: Ronald Press, 1951), pp. 90–130, and the same author's *Capitalism, Socialism, and Democracy* (New York and London: Harper, 1942), p. 128.

learned men, that Erasmus is said to have forgotten his native language; yet in the present day it is not to be supposed, that a youth can *think* in Latin, or that he can have any other reliance on the force or fitness of his phrases, but the authority of the writer from whence he has adopted them. Consequently he must first prepare his thoughts, and then pick out, from Virgil, Horace, Ovid, or perhaps more compendiously from his Gradus, halves and quarters of lines, in which to embody them.[16]

In this passage Coleridge is quite aware of a special effect which Latin had on the vernacular poetry of the ages before his and of his own age. He knows that something is wrong in the reliance on prefabricated expression to which he objects and which at its worst in poetry persistently substitutes "finny tribe" for "fish" or for "birds" says "feathered friends." He identifies as best he can the root of the difficulty: the relationship between thought processes and formulary expressions in the broad sense of pre-existing, fixed phrases. He dislikes the poetic use of pre-existing, fixed phrases such as the gradus supplied. Wordsworth objects to the same sort of thing, not only in his conversation as reported here by Coleridge but also in the Preface to the *Lyrical Ballads* when he censures the performance of Gray, "who was at the head of those who by their reasonings have attempted to widen the space of separation betwixt Prose and Metrical composition, and was more than any other man curiously elaborate in the structure of his own poetic diction." [17] The special poetic language which Wordsworth scores is clearly language marked

[16] Coleridge, I, 12–13 (chap. i). I am grateful to Father John J. Welch, S.J., of the Department of Classical Languages at Saint Louis University, for having called my attention to the importance of this specific passage.

[17] Wordsworth and Coleridge, *Lyrical Ballads* (1963 ed.), pp. 246–247.

by the use of epithets, that is, of standard, expected qualifiers. We can see this from the lines he approves and disapproves in the passage from Gray which he quotes to make his point clear. In the disapproved lines epithets occur in abundance, in the approved lines at best once, perhaps not at all. Wordsworth remarks of the poem he quotes that "the only part of this Sonnet which is of any value is the lines printed in Italics." The sonnet is given here with Wordsworth's italicization.

> In vain to me the smiling mornings shine,
> And reddening Phoebus lifts his golden fire:
> The birds in vain their amorous descant join,
> Or chearful fields resume their green attire:
> These ears alas! for other notes repine;
> *A different object do these eyes require;*
> *My lonely anguish melts no heart but mine;*
> *And in my breast the imperfect joys expire;*
> Yet Morning smiles the busy race to cheer,
> And new-born pleasure brings to happier men;
> The fields to all their wonted tribute bear;
> To warm their little loves the birds complain.
> *I fruitless mourn to him that cannot hear*
> *And weep the more because I weep in vain.*

Although he does not say so explicitly but only by implication, it is clear that the false poetic language which annoyed Wordsworth is marked by standardized phrases of the sort which Coleridge also disapproves: "smiling mornings," "reddening Phoebus," "golden fire," "amorous descant," "chearful fields," "green attire," and so on. The only thing like these nouns with predictable qualifiers we find in those lines which are satisfying to Wordsworth is "lonely anguish," and this will at best doubtfully qualify as an epithet since "lonely" appears at least moderately specific or circumstantial, nonstandard (contrast the more standard "burning anguish" or "bitter anguish" or "sleepless anguish"). "Imperfect joys," the only other adjective-noun combination in the approved lines, is almost witty, an anti-epithet, a play on the standard "perfect joy."

All the lines Wordsworth likes, it will also be noted, focus explicitly on subjective feelings, the lines he dislikes more on the exterior world as processed or interpreted through standard epithets. Romantic subjectivism exists here not in contrast with an "objectivism" built on observation but rather in contrast with adherence to formulary expression.

Wordsworth's poetry is not entirely without epithetic or other standardized apparatus, but it is mostly free of such things (much freer than is Coleridge's—so that one suspects that Coleridge's reservations about the poetic doctrine Wordsworth sets forth in the Preface come not only from a valid objection to Wordsworth's way of describing his aims but also from a more receptive attitude toward what Wordsworth rejected). Typical examples of relatively nonepithetic poems of Wordsworth would be "Michael" from the *Lyrical Ballads* of 1800 or, from the same period, "The Ruined Cottage" and, from somewhat later years, "The Prelude." These could be compared with "Christabel" or "Kubla Khan." Coleridge's Alph or Alpheus in "Kubla Khan," for example, is an abundantly glossed river in Renaissance phrase books: in Ravisius Textor's *Epitheta* the first epithet given for it, *advena* (foreign, alien, strange),[18] announces the basic theme of Coleridge's reverie.

[18] Ioannis Ravisius Textor (Jean Tixier, Seignieur de Ravisi), *Epithetorum opus absolutissimum ex recognitione ipsiusmet autoris: lexicon vere poeticum ad imitationem Graecorum elaboratum, uberem, omnium et verborum copiam complectens* (Basel: Nicolaus

But is it true that the difference between good and bad poetry is that the latter avails itself of pre-existing phrases or formulas? Certainly not, and Coleridge is astute enough to avoid saying this. He situates the difference in a disjuncture between thought and formulas: as they are actually used in the tradition he resents, the formulary bits and pieces are appliqué work. Ideally, thought and expression coincide, not only eventually in the finished product but temporally in the creative act.

This is good romantic doctrine. But it does not allow enough for historical facts of which we have recently become aware. It suggests that all pre-existing formulas are at least suspect if not entirely objectionable. It does not allow sufficiently for the relationship of poetic performance to the sequence of media development and thus for the effect in fuller perspective of the Latin verse-writing on which Coleridge justly throws some blame. Coleridge's age tended, as our age still mostly tends, to think of set formulas as an additive to abstract thought, introduced as ornament or perhaps as enticement for snobs. Historically, this is not the original relationship of formulas to thought. Thought begins as formulaic, and abstract thought grows out of fixed, formulaic thought by a process of liberation made possible through writing. As the work of Eric A. Havelock in particular has shown, early thinking held its very existence in formulary patterns.[19]

Any culture knows only what it can recall. An oral culture, by and large, could recall only what was held in mne-

---

Brylingerus, 1558), p. 41—copy in Pius XII Memorial Library, Saint Louis University, which has a number of other editions of this work and of Textor's companion work, *Officina*.

[19] Eric A. Havelock, *Preface to Plato* (Cambridge, Mass.: Belknap Press of Harvard University Press, 1963).

monically serviceable formulas. In formulas thought lived and moved and had its being. This is true not only of the thought in the spectacularly formulary Homeric poems but also of the thought of the oratorically skilled leader or ordinarily articulate warrior or householder of Homeric Greece. In an oral culture, if you cannot think in formulas you cannot think effectively. Thought totally oral in implementation has specifically limited, however beautiful, configurations. A totally oral folk can think some thoughts and not others. It is impossible in an oral culture to produce, for example, the kind of thought pattern in Aristotle's *Art of Rhetoric* or in any comparable methodical treatises—and these would include, we may as well admit, Plato's dialogues, which, with their highly linear structure, can be developed only in association with writing however ostentatiously oral may be the interlocutory form in which they are cast.

The practice of Latin verse-writing, then, on which Coleridge comments to show some of the aims of romanticism (although he did not use that term) thus has a far more complex history than Coleridge sets down. I would hesitate to say that the history is more complex than he allows for. The passage quoted above is astutely worded, and one feels that Coleridge suspects that there was much more to the story than he could at the time know. In its judiciousness, his statement rings like the statement of a first-rate mind, as it is. But however that may be, the practice of Latin verse-writing on which Coleridge comments was in fact, like much else in the Latin academic tradition, a carry-over from the old oral world of antiquity. In Coleridge's day boys were still being trained to compose as they did because a continuing tradition dating back to ancient oral culture and changing only with glacial slowness favored composing the way Homer did (and Virgil perhaps thought he was doing), that is, in formulary fashion. This is not to say that in the late 1700's such a mode of composition was up-to-date. It was not, and Wordsworth and Coleridge both knew it was not. On the other hand, it probably was not without some good effects even on vernacular poetry, and if you were going to write Latin poetry, it at least put you through a stage you had to go through.

v

When we plot the romanticism of Wordsworth and Coleridge in the perspectives of the commonplaces, as we have undertaken to do sketchily here, we are plotting it in terms of noetic development, in terms of the growth of knowledge and particularly of this growth as related to the evolution of the communications media. This kind of plotting does not of course account for everything in romanticism, but it does, I believe, account for a good deal and for a good deal that is utterly central. For one thing, it provides, better than most other accounts, some insight into the vexing question as to why the romantic movement took place when it did and neither sooner nor later. Romantic features can be found in literature and culture from the beginning, in almost any sense of the term romantic. But, whatever of the various current senses—all of them interrelated, as I believe—we give the term, it is apparent that there was a massive build-up in romanticism from the late eighteenth century on. Why at this particular time? We can suggest a partial answer.

A typical manifestation of romanticism on which we have focused is interest in the remote, the mysterious, the inaccessible, the ineffable, the unknown. The romantic likes to remind us of how little we know. If we view romanticism in terms of the development of knowledge as we are beginning to understand this development, it is little wonder that as a major movement romanticism appeared so late. From man's beginnings perhaps well over 500,000 years ago until recent times, particularly until the build-up implemented by print, knowledge had been in desperately short supply. To keep up his courage, man had continually to remind himself of how much he knew, to flaunt the rational, the certain, the definite and clear and distinct. Of course, he did not do so on every occasion or in every way. Antiquity had its Pseudo-Dionysius, the Middle Ages their Walter Hilton, the Renaissance its Sir Thomas Browne. But by and large, the human intellectual enterprise had been a drive toward rationalism and explanation, toward dwelling on what was known, not what was unknown. As we have seen, the formulary mode of composition against which Wordsworth and Coleridge reacted was an integral part of this strenuously rationalist, knowledge-conserving enterprise in its early stages.

Knowledge conservation and retrieval was immeasurably helped by writing, but even more by alphabetic print. For alphabetic print made indexing—at best a feeble and not very efficient effort in a manuscript culture—into a major achievement. Chirographic man remained in many ways basically oral, for though he had knowledge on record, he had no very efficient way of retrieving it in quantity quickly. Looking up masses of material in print and in even the best manuscripts are two different things. Until print had its effect, man still necessarily carried a heavy load of detail in his mind. Memory systems flourished until typography had its full effect—until romanticism. When print locked information into exactly the same place upon the page in thousands of copies of the same book in type far more legible than almost any handwriting, knowledge came suddenly to the fingertips.

With knowledge fastened down in visually processed space, man acquired an intellectual security never known before. The enterprise of fixing knowledge in space reached a peak some three hundred years after the development of alphabetic letterpress print as, for example, in the *Encyclopédie* (1751–1772) of d'Alembert and Diderot, with its huge fold-out chart at the beginning which, in the best Ramist fashion, presented a schema of the entire "field" of knowledge (a typographically favored concept) in a spatial lay-out.

It was precisely at this point that romanticism could and did take hold.[20] For man could face into the unknown with courage or at least equanimity as never before. What if the venture into the unknown became at one or another moment too frightening? All one had to do was to glance back over one's shoulder at the *Encyclopédie* and other books on the library shelves. With all that was by now immediately accessible in print, the German proverb *Wenige Leute wissen wieviel man wissen muss umzu wissen wie wenig man weiss* rings astonishingly true. Few persons know how much you have to know in order to know how little you know. The romantic age felt it knew enough to savor its unknowing.

Because it rests on this poetic base, romantic poetic, insofar as it moved away from the formulary, was the poetic not only of Wordsworth's and Coleridge's age but also of the future. We have already noted Poggioli's studied prediction in *The Theory of the Avant-Garde* that all new movements in literature and the other arts will build on the same romantic

[20] Working largely with evidence other than that adduced here, Walter Jackson Bate has located at this same historical point the intensification of pressure on the poet to relate himself consciously to the past as past—in *The Burden of the Past and the English Poet* (Cambridge, Mass.: Belknap Press of Harvard University Press, 1970), pp. 132–134, etc.

quest of romantic difference which has marked avant-gardism from the start. The theory is well founded. The more knowledge grows, the more appealing become ventures into the unfathomable or bizarre.

Romantic poetic was the poetic of the future for another reason, too, which can be at least suggested here. Oddly enough, romantic poetic was the poetic of the technological age. In the perspectives we have just traced, romanticism and technology can be seen to grow out of the same ground, even though at first blush the two appear diametrically opposed, the one, technology, programmatically rational, the other, romanticism, concerned with transrational or arational if not irrational reality. Yet, as we have seen, in terms of the growth of knowledge and the development of knowledge storage and retrieval systems, both romanticism and modern technology appear at the same time because each grows in its own way out of a noetic abundance such as man had never known before. Technology uses the abundance for practical purposes. Romanticism uses it for assurance and as a springboard to another world.

"Wordsworth was never to find romance in steam locomotives," Carl Woodring has observed.[21] Yet romanticism manifests a feeling of control over nature based ultimately on noetic advance which allies it strangely with the technological spirit. Before romanticism, as has so often been pointed out,[22] mountains were hostile masses and other comparable manifestations of nature were regarded with similar misgivings. Upon the accession of romanticism, mountains are still awesome enough, but their alien hostility is gone. They are

[21] P. 82.
[22] See, for example, Marjorie Hope Nicolson, *Mountain Gloom and Mountain Glory* (Ithaca, N.Y.: Cornell University Press, 1959).

lovable, "ours." Even when unacknowledged, the feeling of control over nature quiets old fears of being swallowed up by nature. Wordsworth's much advertised surrender to nature is to this extent a by-product of technology. Nature was under surveillance in his world. It was no accident that, by and large, technologized societies, such as Germany and England, entered on a romantic period sooner than did the much later technologized Latin cultures, notably Italy or southern France, and most notably Spain.

The configurations in the academic structures which we earlier detailed advertise in their own way the alliance between romanticism and technology, and at the same time give a new reality to the old classic-romantic polarity. In the academic world the all-male Latinate curriculum of the British public schools prepared typically for the rhetorical and political arenas of empire. This education was ceremonially polemic and conservative, as education maximizing formal oral training virtually always is. This curriculum contracts with the nonclassical, commercial, sight-oriented and hence technologically oriented courses of the vernacular schools which grew from the late seventeenth century on and which were more receptive to girls.[23] These schools rather than the older Latin schools generated the ideal of "reading, 'riting, and 'rithmetic" which formed the basis of universal literacy programs. If only from the fact that they were devoted to the vernacular, they were also far more open to romantic literature—which even as far back as its medieval roots had depended on women readers and women patrons (Eleanor of Aquitaine and Marie de Champagne, for example) as classical Latin literature never

[23] See Ong, *The Presence of the Word*, pp. 241–252 and references there.

had. The story which this rough polarity suggests has never been told in the full. It would be well worth telling.

The vexed question of the relationship between Wordsworth's poetic ideas as expressed in the Preface to the *Lyrical Ballads* and in his poetic practice can suggest a final alignment in the perspectives this study has been attempting to work out. Obviously, we cannot hope to solve this entire question here, but only to treat it from certain limited points of view.

Coleridge had urged objections to Wordsworth's proposal in the Preface to use "a selection of the real language of men," and to "imitate and . . . adopt the very language of men" in low and rustic life.[24] Coleridge urged that rustic language was no more "real" than any other: the *lingua communis* or "ordinary" language of those using a given tongue such as English is a complex thing, varying from group to group and even from individual to individual. And, Coleridge goes on, the fact that it happens to be uttered "in a state of excitement" does not make rustic or any other ordinary language more fitting for poetry, "for the property of passion is not to *create*, but to set in increased activity."

This is no place to go into the explications or other sequels of Coleridge's critique, which proliferate into our own time. We need only note here that Coleridge had raised central issues, so that most subsequent discussion of Wordsworth's poetic, however it may qualify Wordsworth and Coleridge, does not stray too far from Coleridge's areas of concern. Keeping these areas of concern in view, I would like here merely to suggest a description of what Wordsworth's poetic style is when it appears to stay closest to his own prescriptions, as, for example, in "Michael" or "The Prelude," in the

24 Coleridge, II, 41, and 28–43 (chap. xvii).

light of what we have discussed through this study. As we have seen, what both Wordsworth and Coleridge were objecting to was a poetic style which in fact manifested in high relief the formulary features of the old oral tradition. Seen as a reaction to the residue of this old oral tradition, Wordsworth's style in the poems just mentioned, which is certainly one of his typical styles, is of course not close to the language of rustic or other "simple" folk at all. It is a highly sophisticated style, which results when language, originally oral of course, is first formalized over many generations by self-conscious rhetoric (through the long tradition coming out of classical antiquity) and then carefully diversed of many of its typically oral features. It is, in fact, the opposite of the language of the common man insofar as the language of the common man is typically rather oral in its economy, epithetic, filled effortlessly with the formulas which rhetoric elaborated by sophisticated conscious effort and which Wordsworth quite successfully avoided. The language of the common man is perhaps not commonly highly epithetic in conversation, but as it becomes more impassioned it is likely to be markedly so ("dirty scoundrel," "dastardly coward," "yellow-bellied traitor," etc.). Wordsworth's poetry is rather remarkably sparing of epithets, notably so in "Michael" and "The Prelude," as can be seen by examining it as compared, say, to *The Faerie Queene* or to Gray. The poetry of Coleridge, unlike Wordsworth's, is rather generously sprinkled with epithets, for the good reason that it is more oral in cast, allied more with the vernacular ballad tradition, for one thing. In our perspectives here, Wordsworth's typical verse style is the poetic equivalent of what is known in prose as the "plain style," the style which Bishop Thomas Sprat and the Royal Society promoted and which present-day science commonly uses. Such style may be "plain" but there is little that is prmitively simple about it at all. It is many millennia from primitive orality, the style of highly evolved and literate folk who have divested their language of some of the most typical and ornate features it had inherited from more primitive times.

12 The Literate Orality of Popular Culture Today

I

The relationship between present-day orality and the orality of preliterate man is a subject few discuss in circumstantial detail. Many are aware of the marked orality of our culture today when compared with the culture of thirty years ago, before the electronic potential first mobilized in the 1840's with the telegraph had matured and become interiorized in life styles and world views. But much talk and writing about present-day orality assumes that orality is orality and that since primitive man was highly oral and we are likewise more oral than our immediate ancestors, we are back in the state of preliterate man once more.

Orality is orality in some ultimate sense, of course. Certain points of resemblance between present-day phenomena related to our orality and primitive phenomena related to orality are startling enough. Sound always tends to socialize. The drive toward group sense and toward participatory activities, toward "happenings," which mysteriously emerges out of modern electronic technological cultures is strikingly similar to certain drives in preliterate cultures. But the startling likeness carries with it an equally startling difference.

For primitive man, happenings occurred. Today we program happenings. With all the inner-directedness we can muster, we plan unplanned events, and we label them happenings so that we will be sure to know what is going on. If I may use terms which I fondly believe I have originated, I would suggest that we speak of the orality of preliterate man as primary orality and of the orality of our electronic technologized culture as secondary orality. Secondary orality is founded on—though it departs from—the individualized introversion of the age of writing, print, and rationalism which intervened between it and primary orality and which remains as part of us. History is deposited permanently, but not inalterably, as personality structure.

The differences, as well as the likenesses, between secondary and primary orality are intricate. Discussing the differences involves the difficulty which all discussion of the communications media involves: phenomena interact in distressingly criss-cross fashion. Causes and effects are hard to distinguish. This is doubly so in studies of popular literature in its relationship to the history of the media, the subject which I propose to discuss here. I shall try to deal with the subject by centering attention rather restrictively on one phenomenon, a phenomenon complex enough in itself yet reasonably isolable. This is the use of formulary devices, which are well known to occur in great quantity at least in primary oral culture and which, as will be explained, have their place in secondary oral culture, too, and in particular in our popular arts, including literature.

Formulary materials can be examined in various ways. Here we can delimit our investigation further by examining them chiefly as knowledge storage and retrieval devices. Their serviceability as such is one of the principal factors encouraging their use in oral poetry and in early literature. Poets use formulary materials more intensively than others in primary oral cultures, as also do orators and other stylized oral performers. But, we must remind ourselves, in using these materials poets and other oral performers only carry to a special height of perfection a skill which to a greater or lesser degree all members of a primary oral culture practice. Any culture knows only what it can recollect or bring to mind. A primary oral culture may avail itself of some visual *aides-mémoire* such as notched sticks, quipus, or wampum belts, but these are extremely low-level information storage devices, not only of limited currency but also of quite minimal value because they do not represent words as such. An individual quipu may demand a special quipu-keeper to interpret what this particular quipu means. Without a true script, primary oral culture cannot "look" anywhere to find utterances, sayings, statements, or even words as such—it cannot "look up" these things. Of verbalized material it knows only what it can "call" to mind. Hence special mnemonic arrangements of words are of primary importance in such a culture. These arrangements we can speak of as formulary devices. They are often crucial even for the use of primitive visual *aides-mémorie*. The interpretation of a notched stick can well depend on the ability of the interpreter to recall in verbal formulary devices what the notches mean.

By formulary device I mean here any set or standardized verbal expression, "set" and "standardized" implying, of course, that the expression is used more than once, and generally quite often. The term "formulary device" thus understood is rather expansive.[1] It includes all manner of

[1] The problems and divergencies in present understanding of "formulary" modes of thought and expression are due to the richness of

proverbs, adages, apothegms, proverbial phrases, and the like, the "old said saws" that weave through and support virtually all early writing, keeping even learned writing close to oral performance for centuries, from antiquity through the productions of Renaissance humanists such as St. Thomas More and Erasmus, affecting the very substance of Shakespeare and Pope, and to a diminishing extent later writings, as well as the speech of residually oral folk everywhere still today. In addition to proverbs and the like, the formulary device will include programmatically mnemonic verses such as the familiar "Thirty days have September, / April, June, and November. / . . . ," alphabetic verse such as some Psalms and the Lamentations in the Old Testament, mnemonic numerological groupings such as the seven sages of Rome or the seven deadly sins or the seven wonders of the world, and so on and on. It also includes formulas of the Homeric sort, such as "rosy-fingered dawn" or "tamer of horses," which are not so obviously used for recall purposes but which occur as a regular phenomenon in verse, we now know, in great part because the culture where this verse arose reveled in the use of formulas, which it needed not only in verse but also in practical administrative activity for recall purposes. In a comparable fashion, story-telling devices such as the "seven-at-one-blow" refrain or climax are formulary devices although they are not directly mnemonic ex-

---

the material we are dealing with and the inadequacy of our present ways of conceptualizing it. This and related matters, with particular reference to the Homeric formula, are discussed with great discernment and promise by Michael N. Nagler, "Toward a Generative View of the Oral Formula," *Transactions and Proceedings of the American Philological Association*, ed. by John Arthur Hanson, XCVIII (Cleveland: Case Western Reserve University, 1967), 269–311.

pressions but rather take-offs on mnemonic formularies. Among formulary devices must be included, moreover, the commonplaces discussed in Chapters 2 and 3 in at least one of the two standard rhetorical senses of this term, that is, the "cumulative commonplaces" as we have styled them, prefabricated purple patches on some standard subject such as loyalty or mother love or dishonesty or general civic corruption or one's own incompetence ("unaccustomed as I am to public speaking"). In such commonplaces, as in other formulary expressions, some variability in the standardization does not disqualify the expression as formulary so long as the expression retains its effective identity.

Other kinds of standardization can be formulary devices in a larger but related sense, insofar as they are matrices for set or standardized verbal expressions. Such for example are standardized narrative themes which epic poets have used (the arming of the hero, the hero's shield, the message, the summoning of the council, and so on) and which, more surreptitiously, historians still use today, since without pre-existent themes to determine what of all the potentially infinite possibilities a narrator is going to attend to, there would be no way to lift up for inspection any strand at all in the unbroken web of history. A fixed theme is not of itself an expression and hence not a fixed expression, but it enters into and fixes expressions and is in turn defined by fixed expressions. A set theme generates type characters with fixed epithets: for example, wily Odysseus and bold Robin Hood.

Related to such themes and at least equally formulary are the commonplaces in their second standard sense, the "seats of arguments," as Quintilian puts it, headings, we would say, such as causes and effects, contrary matters, and so on, to which one has recourse in developing discourse about one or another subject. Rhetoricians from those of antiquity to the authors of modern newswriting manuals, use these headings to set up formulas for treating a subject: Who? When? Where? What? How?

In a different fashion, at the neurological rather than the psychological level, versification or any kind of heavy rhythmic design is a formulary device: the beat itself is a kind of abstract fixity which can lend itself to various word groupings.

It will of course be difficult to refine the definition of formulary devices so as to sort out all borderline cases. Is the expression "borderline cases" itself a formulary device? In the last analysis the reason for the difficulty may well be that, since language is grounded in repetition, there is a residue of formula in all expression, so that we should not take as our basic question, Is this formulary or not? but rather, How formulary is this expression as compared to other expressions? Nevertheless, for our present purposes we can put aside such troublesome and basic questions, since truly basic questions are often best handled rather late in any investigation, and work with the rule-of-thumb description just given: a formulary device is any set or standardized verbal expression. For our present purposes we can interpret "standardized" in the sense of markedly fixed, that is, rigidified in a way beyond what the ordinary lexical resources of the language would normally lead one to expect, restricting discussion to cases which are clearly not borderline.

In a primary oral culture, as has already been suggested here and elaborated in earlier chapters of this book, formulary devices are not merely decorative but highly functional or operational. They connect Homeric texts not only with other imaginative or creative verbalization of the time but with the day-to-day business of tribal or early civic life as well. To get a complicated message from Ithaca to Samos in pre-alphabetic Homeric Greece, the sender would have to cast it up in something like mnemonic formulas in which both messenger and receiver could retain it for recall. We must remember that there was no possibility for the originator or anyone else to write the message down so that it could be memorized verbatim. Nor, if it was fairly long, would there be any way for the sender even to compare his initial utterance of the message with any repetition of the utterance by himself or others. The message would have to be thought out in the formulary devices present, more or less, in everyone's mind.

Dependence on formulas gives not only a special kind of surface but also a special kind of content to messages sent in oral cultures. The highly analytic thought structures we take for granted in certain utterances among literates are quite simply unthinkable. The medium is not the message, for one medium will incarnate many messages. But medium and message interact. The medium in neither container nor vehicle (*pace* I. A. Richards) nor track. The message is neither content nor cargo nor projectile. Medium and message are interdependent in ways none of these carton and carrier metaphors express—indeed, in ways no metaphor can express. In the last analysis, the medium is not even a medium, something in between. Words destroy in-betweenness, put me in your consciousness and you in mine. There is no adequate analogue for verbalization. Verbalization is ultimately unique. True information is not "conveyed."

Eric Havelock has shown circumstantially and in great detail how formulary the oral culture of Homeric Greece was and indeed had to be, and how the formulary character of the *Iliad* and *Odyssey* is of a piece with this culture as a whole.[2] In their use of formulas these epics do not invent something peculiarly poetic but rather put to superb poetic use an economy of expression and thought inevitable in the world from which they derive. The world of formulas and oral techniques opened in the past few years to modern view by Havelock and, shortly preceding him, by Milman Parry, Albert Lord, and others, has appeared as something quite wonderful and new to most literary scholars. Folklorists who work with oral tradition, however, have taken it pretty much in stride and have even expressed astonishment that anyone should regard it as novel. They have always taken for granted, because it has been so overwhelmingly evident, that primary orality is radically **formulary**.

The excitement of literary scholars is understandable. Over the centuries, the *Iliad* and the *Odyssey* had been all but totally assimilated to the literary world, preserved entirely by scholars whose proper work was conceived of as attention to texts and who, almost without exception, automatically downgraded or wrote off oral performance as subacademic and *infra dignitatem*. These scholars, it is true, had acknowledged a distinction between oral or primary epics and literary or secondary epics, but on the whole they had proved themselves incapable of thinking of orality except as a variety of writing, although the converse relationship was the true one. Scholars even wrote—and indeed still write—of "oral literature," thereby revealing the fact that they could conceive of oral performance only as a variant of something which came after it and derived from it—a parallel would be to think and speak of a horse always as an automobile with-

[2] Eric A. Havelock, *Preface to Plato* (Cambridge, Mass.: Belknap Press of Harvard University Press, 1963), esp. pp. 36–114.

out wheels, and to make the parallel close we would have to make this the *only* way to think of a horse, omitting our term and concept "horse" itself, as the term and concept of (oral) performance were omitted from analyses of oral productions.[3]

We are only beginning to assess the weird results of this imposition of chirographic and typographic categories upon oral culture and on the vestiges of oral culture which linger so long after the advent of writing.[4] Chirographically distorted, epic theory grew more and more grotesque in attempting to square the account with the actuality, or it simply went its own way independently of actuality. In a paper at the December 1967 Modern Language Association meeting, Professor Robert Kellogg pointed out the hopeless discrepancies between the epic theories of the Renaissance and the actuality of Renaissance epic productions. What did *The Faerie Queene* or *Orlando Furioso* really have in common with Homer? By the eighteenth century virtually the only effective epic is mock epic: the only thing one can do with the impulse to write like Homer is to laugh at it.

A chirographic and typographic bias still governs to varying degrees all but a tiny fraction of scholarship and criticism concerned with oral productions, and this fact itself deserves serious attention in any study of popular literature. The assertiveness of the chirographic and typographic bias, which accounts in great part for the hostile reviews of *Webster III* by virtually all subscholarly writers up to and including the level of the *New Yorker*, is highly informative, for it shows

[3] There is no generic term in English and hence no readily available generic concept for artistic verbal productions which would include both oral performance and literature.

[4] See Chapter 2 above.

the massive and deep-set, subconscious defenses which writing sets up in the psyche to sustain the restructuring of personality which it brings about.[5] Or, to put it another way, it shows how the acquisition of writing brings those who acquire this skill to structure their entire world view around a feel for the written word to the positive (but not often conscious) exclusion of the oral as such. It is not easy for a literate person to enter into the realities of primary oral culture, or to feel anything but vast and sputtering hostility for any dictionary (typographic idol par excellence) which acknowledges that language usage may be determined by other than professional *writers*. These alone guard against "corruption," although even some of them are suspect. Even the best writers do talk.

Literacy thus set up subconscious defenses of great strength and depth. It is no surpise that our present breakthrough from literacy into understanding of orality has had to wait all these centuries until psychoanalysis made the depths underlying consciousness more openly accessible. Even so, abandoning the artificial securities of typography has proved too traumatic for most, particularly for the semieducated, who cling to the fixity of typographic space as a substitute for the living permanence of truth.

We do not have a complete account of the contrasts between economies of expression and thought in primary oral cultures, where they probably vary considerably from one culture to another, and the economies of expression and thought in chirographic, typographic, and electronic cultures.

[5] See Walter J. Ong, *In the Human Grain* (New York: Macmillan, 1967), pp. 52–59; and the same author's *The Presence of the Word* (New Haven and London: Yale University Press, 1967), pp. 92–161, 192–255.

But it is spectacularly evident that primary oral cultures are commonly ultraformulary. We have seen at length in Chapters 2 and 3 above how their ultraformulary economy of thought and expression established what we have styled "rhapsodic structure" and how this and other formulary features of style died out only gradually over a period of thousands of years culminating with the romantic movement. Romanticism emerges at the same time as modern technology and with a double and paradoxical relationship to the old oral world. On the one hand, romanticism appeared to favor the old primary oral world by rejecting the chirographically and typographically grounded rationalism that had been slowly destroying the oral world and by programming an academic or para-academic interest in popular literature, folk ballads, and ultimately folklore generally, where the old formulary patterns persisted. On the other hand, romanticism covertly relied on rationalism. It manifested its reliance in one way by a stylistic which writing culture and, even more, print culture had implemented.

In this stylistic, with its conscious questing for the novel,[6] romanticism was moved to reject the cliché formalism of an old knowledge storage and retrieval system (except where self-consciously simply imitating antique style, as in romantic ballad forms) and to rely instead, though not often admittedly, on chirographic and typographic storage. When truths needed no longer to be constantly reiterated orally in order to remain available, virtuosity, that is, superlative skill in manipulating well-worked material, was displaced by "creativity" as an ideal. The romantic quest for originality, the

---

[6] See Renato Poggioli, *The Theory of the Avant-Garde*, tr. from the Italian by Gerald Fitzgerald (Cambridge, Mass.: Belknap Press of Harvard University Press, 1968), p. 127, etc.

novel, the new new, reveals romanticism as a typographic phenomenon despite its avowed commitment to the past. Insofar as romanticism persists today, as it does, we are still in a typographic world. In these perspectives, it becomes possible to interpret some situations in today's media and popular culture in certain specific though not exclusive ways.

II

It is by now a commonplace that since the advent of electronic devices, from the telegraph and telephone through the radio and television and beyond, technological man has entered into a new world of sound. Our culture contrasts markedly with that of a hundred years ago, when the dominant form of verbal communication beyond direct person-to-person contact was visual, with words circulated in writing and, more extensively and controllably, in print. Today living and winged words come into our consciousnesses from across the globe—St. Louisans and Londoners simultaneously hear news reported from Hong Kong. Telephones implement personal vocal contact as letters could not, and rapid transportation multiplies person-to-person confrontation in ways unheard of until the past few decades. Over and beyond the sound of the voice itself, other related sounds, most notably instrumental music, pulse through the air from millions of radio and television and recording sets to a degree never even remotely approximated in the past. Total immersion in electronic sounds becomes an ideal and a frequent achievement in electronic music played with a volume which makes conversation at the top of one's voice inaudible and is known to cause physical damage to the organs of hearing. Experiments in noise by John Cage, Andy Warhol, and others attract thousands. The silent lifeworld into which man had moved with writing and print and, he thought, with Newtonian physics is no more.[7]

But the present age of sound, the age of secondary orality, differs from the age of primary orality as much as it resembles it. The ways in which this is so are legion. We can only suggest a few of them here in the light of our brief sketch of the development of formulary devices.

First the formulary device is no longer deeply grounded in practical living since it has now relatively limited use for knowledge storage and retrieval. It does not feed from life into either folk art or sophisticated art, or into the popular art which is concocted of the other two.[8] Very little knowledge today is really organized around proverbs or catch phrases for "*calling* to mind." Instead, knowledge is largely "structured" for visual retrieval. We write memos even to ourselves, as even highly literate medieval man had not yet learned to do. Our most sophisticated knowledge storing and retrieving tool is the computer, essentially a visual device, with a print-out. (Not all electronic machinery, we must recall, serves sound.) Where something like the catch phrase

---

[7] The silent Newtonian cosmos, incidentally, as I have elsewhere attempted to show, connects with habits of nonauditory, visual synthesis associated with writing and, more intensively, with print rather than with scientific observation. The "solemn Silence" which, in the August 23, 1712, *Spectator*, Joseph Addison predicated of the supraterrestrial universe corresponded to a psychological need, not to any findings of science: there were, and still are, ear-splitting explosions going on all over the cosmos.

[8] A useful statement of the relationship of popular art to folk art on the one hand and sophisticated art on the other is to be found in Roger D. Abrahams and George Foss, *Anglo-American Folksong Style* (Englewood Cliffs, N.J.: Prentice-Hall, 1968), pp. 5–6. What the present chapter refers to as academic would be included in the "sophisticated" classification of these authors.

or formula still functions, as perhaps it does in law, the formula is not a storage and retrieval device so much as a focal point for elaborate analytic work. Such for example might be "clear and present danger" or "last clear chance" or "Possession is eleven points in the law." In connection with set expressions such as these, the learning consists in spinning out the analysis, not in burrowing around in the rich gnomic deposit.

Discredited as an operational gambit in the practical human lifeworld, the formulary device has also become discredited in highly literate circles, academic and other. Under the encouragement of I. A. Richards in his *Practical Criticism* as well as of other New Critics, clichés have for many years now been hunted down mercilessly with a view to total extermination, although of late the hunt has somewhat cooled, possibly because it itself has turned into a cliché. Its anti-formulary program identifies the New Criticism both with print and with romanticism in the paradoxical relationship which these two exhibit, as already explained. More recent criticism attending to the societal character of literature points to an individualist bias underlying much of the New Criticism and occasionally relates this bias to capitalism,[9] al-though it could with greater plausibility and more sweep relate it to the personal isolation or inner-directness, to use David Riesman's term, which is encouraged by writing and especially by print.

Despite this discrediting of the formulary, countervailing tendencies make themselves felt. The formulary is far from dead. This is seen in one of the major manifestations of pre-

[9] See, for example, the discussion in Andrew Hawley, "Art for Man's Sake: Christopher Caudwell as Communist Aesthetician," *College English*, XXX (1968), 1-19.

sent orality, the cult of folk song: "Sing along with Mitch."
Although the modern folk song does not build by any means
entirely with blatantly formulary expressions (it is a product
of secondary orality, indisputably, if cautiously or surreptitiously, literate), it does exploit cliché themes, as any listener
can find out for himself. Such themes are common to other
popular culture productions, too, such as the western movie.
But the western does not touch the same patriotic strings as the
folk song, or does not pluck them with such feeling. Oscar
Brand points out that the principal appeal of folk music
derives from the overwhelming persuasion of its devotees
that it is of great antiquity[10] (often it is not) and connects
with their past (as they imagine this to have been, at any
rate). Evidently the oral as such here enforces some specific
orientation to the past. "Sing me a song of the days of old."
Or tell me a tale. We hardly say with comparable feeling,
"Draw me a picture of the days of old."

In the visual arts today, preoccupation with the cliché,
the formulary, the commonplace, takes a charcteristic turn.
This can be observed in Roy Lichtenstein and pop art generally. Here the visual formula, the cliché, the cartoon of Super-
man or the cornflakes box advertisement, is indeed revalidated
—in the spirit of the new orality, but with visualist caution,
for the confrontation with the cliché as cliché is made highly
reflective. The result is camp, in the sense of a presentation
of the second-rate so managed that one's responses to it can be
taken to be first-rate. This is simply supercharged romanticism:
strangeness is found even in the cliché through exaggerating
confrontation with it. Elsewhere also, as in the mixed sensory
fields of the Rowan and Martin "Laugh-In," the cliché is

[10] Oscar Brand, *The Ballad Mongers* (New York: Funk and Wag-
nalls, 1962), pp. 50–61.

self-consciously exploited and the hostilities built around it sublimated by humor: thus the already old "H'yeah come de judge." This spoofing of a fixed and in effect meaningless expression is an old in-group-type joke, but it achieves a new order of existence when it makes the entire television audience into the in-group.

At another point in the hyperactivated oral world of today the cliché appears to turn up even more assertively: this is in the slogan, the catch phrase, or the compulsive jingle with which, for example, commercially sponsored radio and television programs fill the air. But if this is a revival of the old formulary mnemonic device, it is so with a vast difference. The advertising cliché is not in fact much of a knowledge storage and retrieval device at all. It is a slogan, which is not at all the same thing. The formulary devices of a primary oral culture are conservative devices, ordered to the treasuring and use of hard-earned lore. Slogans, by contrast, are typically action-oriented, fitted to short-term goals. The slogan enters more directly into the spirit of the new orality. It is not reminiscent but programmatic, ordered to the future and thus even to something new. The graffiti on the walls in the Latin Quarter in Paris after the University rioting were generally slogans, not formulary devices in the older sense. They are more original (and by the same token more literary) than formulary devices as they are also more evanescent. A literate culture—and our secondary oral culture remains literate even in its orality—does not mobilize itself around sayings as permanently as an oral culture does.

Some of our most oral manifestations show their strong literary grounding by their studied rejection of the cliché in any but occasional ironic form. Thus the attempted total casualness of the lyric style of Simon and Garfunkel. Here the popular culture of a bookish youthful generation reproduces what it has learned from the far-gone literacy of T. S. Eliot and Jules Laforgue. Worn rhetorical clichés (relicts of the pompous old oral world, mediated by residually oratorical journalism) are put on careful display, for example, in the first two lines of the following passage only so that in the next two lines they can be subjected to heavy ironic bombardment:

> "Save the life of my child!"
> Cried the desperate mother.
> The woman from the supermarket
> Ran to call the cops.[11]

This is everyman's version of "Let us go then, you and I." The noetic abundance of the print milieu drives the Simon and Garfunkel imagination to look for something unusual, bizarre:

> Wish I was a Kellogg's Cornflake
> Floatin' in my bowl takin' movies.

Here a babyland setting—very common in the Simon and Garfunkel kind of lyric—doubly insures against the least formality.

The old orality was postpubertal, and had long been preserved in a highly polemic literature, reserved all but totally for males and taught in an all-male setting, the school, as a puberty rite to initiate the boy into extrafamilial, adult existence, as detailed in Chapter 5 above. The new orality here is of another temper, closely associated with infantilism and thus no doubt filling a real need, for there are few traces left

[11] The Simon and Garfunkel lines here and in the following quotations are all from songs recorded on *Bookends*, Columbia Stereo KCS 9529, and are used with permission of the copyright owners, Charing Cross Music, Inc.

of puberty rites or of other initiation rites in the academic
world—imagine what initiation rites must have to be in one
of the new coeducational fraternities (or would it be sorori-
ties? siblinghoods? *Geschwistergemeinschäfte?*). No wonder
the youthful concern about identity speaks to us over and over
again out of the Simon and Garfunkel repertoire:

> I'm such a dubious soul
>
> .   .   .   .   .
>
> I've just been fakin' it,
> Not really makin' it.

Indeed, we live in an age not merely of vanished puberty
but even of reversed puberty rites. First, regarding costume:
Early puberty rites often involved putting off the clothing of
childhood and assuming that of an adult. But children's clothing, as Philippe Ariès has detailed in *Centuries of Childhood*,
in all but the most recent Western societies has been modeled
on adult garb of a century earlier: late Victorian male children
wore the knee-breeches of the late eighteenth-century adult
male. Children thus were adult-oriented even before the puberty rites, learning to copy adults from the start—first, before puberty, at a respectful anachronistic distance, then, with
proper pubertal initiation ceremonies, in up-to-date fashion.
Now, however, the adults are so child-oriented that they are
copying their own children. Mom is in miniskirts, and only
eight or nine years old. Little wonder that her children are
threatened with immaturity. You have to squeeze pretty small
to be more childlike than mom.

Secondly, in the academic world much of the focus of attention regarding the relationship of the generations has reversed its field. The original puberty rites focused the attention of the immature on what lay ahead, adulthood, rather

than on themselves. Puberty rights build up adulthood expectations. Now sociology and psychology focus the entire professional attention of thousands of adults on the immature years, from infancy through adolescence. Youth is studied not only for what it will become, but for what it is, which is transitory. This massive scientific preoccupation with preadulthood was quite unknown in earlier cultures. Since the adolescent, if he is normal, is unsettled and self-conscious in the extreme to start with, little wonder that his new position as a specimen for major research projects has brought his self consciousness to an intensity he may find unbearable. Adults who weathered their own identity crisis without even knowing the name for the phenomenon are tensing today's children for this and countless other crises years before their onset. "Haven't you had your problems yet? Let me help you find them."

But at this point the connections become too complex and introverted to explicate in a brief treatment. We can only note that in the perspectives in which we are here specializing the total casualness which is also featured in this style is not merely prepubertal but also the polar opposite of the formalism of the primary oral world.

> Relaxin' awhile, livin' in style.

. . . . .

> I prefer boysenberry
> More than any ordinary jam.
> I'm a "Citizens for Boysenberry Jam" fan.
> Ah, South California.

The matching of "boysenberry" and "ordinary" is studiously informal, superrelaxed, while in the phrase "Citizens for Boysenberry" a formulary fixity is echoed only in order to be satirized. Total irony regarding the commonplaceness of everything and total casualness combine in the verbal cop-outs scattered through the repertoire: "Blah blah blah" and "Ooo ooo ooo" and "Wo wo wo" and "Hey hey hey." In its depreciation of fixed formulas, shown in its tendency to satirize those it recognizes—it is bursting with unrecognized ones—the culture which these revealing productions represent is an extremely literate one, despite its hi-fi modes of existence.

We have treated the use of formulary devices here in a highly specialized way, and skimpily at that. We have discussed simply the formulary devices themselves, not what Scholes and Kellogg call the "grammar" of formulaic composition, "the abstract patterns in accordance with which singers can produce new phrases" out of their formulary store.[12] This would be a whole subject in itself, and a very worthwhile one. So, too, would be the place of formulary material in the tall story and other American specialities discussed by Constance Rourke, or in the comic strip, the political cartoon, or detective fiction. Examined in its connection with residual orality and high literacy, the formulary device could certainly throw some new light on these and many other genres. But perhaps enough has been said here to point up certain contrasts not only between primary orality and writing-and-print culture, but also between primary orality and the secondary orality with which we are going to have to live through the foreseeable future.

[12] Robert Scholes and Robert Kellogg, *The Nature of Narrative* (New York: Oxford University Press, 1966), p. 25.

13 { Crisis and Understanding in the Humanities

"The proper study of mankind is man," Alexander Pope wrote in his *Essay on Man*, expressing in highly quotable eighteenth-century English form a thought at root so old that no one will ever be able to establish its ultimate origins. The studies collected in the present book are concerned not only with man, but often with man's study of himself as he engages in one of his central activities, verbal expression.

Man's ability to study himself, to reflect on himself and his achievements or failures, distinguishes him from the animals below him on the evolutionary scale. A subhuman animal's perceptions and responses are turned forever outward, unable to return on themselves. The brute animal is "otherized," we might say, adapting José Ortega y Gasset's Spanish term *alteración*. The subhuman animal reacts beautifully to movements, sounds, smells, to his own bodily condition, to the kindness of human beings, but not reflectively to himself as himself, unable to achieve that return of consciousness to its own center which enables a human being to say "I."

The ancient world, much more bound to the grosser inhuman side of material existence than is modern technological man, set great store by the difference between man and brute animal, making much of *humanitas* and of the *litterae humaniores* or "humane letters" which especially fostered those qualities in man that were typically human, raising man above brute beasts. From these terms and concepts, *humanitas* and *litterae humaniores* and their cognates, has come our term and concept of the humanities, those studies which have to do with the specifically human in man's activities and achievements. "Humanities" is an expansive term. The humanities can include not only the study of literature and the arts but also, in more extended but quite legitimate senses, the study of philosophy and anthropology and theology and history, which itself includes the history of science and of mathematics as well as of everything else; for all history, as Giambattista Vico well knew, is made by man, and the physical sciences, which come into being in history, are likewise made by man (though not independently of the way things actually are) and can be studied not merely for their content, which sets them off from the humanities, but also in relationship to their maker and his culture, which incorporates them into the humanities. But for all this expansive meaning the term may have, at the center of the humanities lies always the study of human expression in all its manifestations and of human culture, which grows in and around human expression as this expression grows in and around culture.

The studies collected in this book have to do with human expression and culture and the transformation of both through history, and thus can be brought to a final focus by considering the state of the humanities in our day. The state of the humanities is a large subject, but fortunately a recent occurrence has provided access to far more resources than my own. In 1968 a Conference on the Future of the Humanities was held in Brookline, Massachusetts, sponsored by *Dædalus*, the journal of the American Academy of Arts and Sciences, under the direction of Dr. Stephen R. Graubard, Editor of *Dædalus*, and Professor James S. Ackerman of Harvard University. As one of the participants in the Conference, I was asked to summarize for *Dædalus* the problems of the humanities today as these were identified by the conferees and to give my own response to the charges against the humanities which the Conference bared. The present study is an adaptation of the result to the needs of this present volume. Thus, as the concluding chapter here, whatever its intrinsic defects may be, this study has the advantage of being grounded in much more than the thought of one person; there were some forty-five participants at the Conference, representing a great variety of the central humanistic disciplines if not in fact all of them.

I. CHARGES

The humanities have a way of overlapping one another, and so do the charges brought against them. With some allowance for overlap, in the transcript of the discussions at the *Dædalus* conference on the future of the humanities one can identify some sixty-odd charges and subcharges that the humanities today must answer if their case is to be cleared. (One of the charges is that they want their case cleared.) Here I shall attempt not to itemize all the charges and subcharges, but rather to indicate areas in which these cluster before moving on to consider how it might be feasible to think of the humanities today. The charges against the humanities cannot, I think, be dealt with unless we revise our understanding of how the humanities relate to the human lifeworld, historically considered, and unless we become aware that the humanities do not yet know enough about themselves.

Predictably, the most recurrent of the charges against the humanities strike at what is most vital and precarious in the humanities—the relationship between their subject matter and the rest of life. These charges focus on the interchanges where art and actuality, literature and life, play and work, academic commitment and civic obligation come together and at the same moment diverge. The charges call up, in an endless variety of forms, the question: What do the humanities have to do and what should they have to do with the day-to-day business of living in our present world?

This question, even in this simple form, can be contracted or expanded in the restricted sense—that is, to the study of man's own creative or imaginative works. These consist of verbal creations, oral or written, as well as of art, music, the dance, and so on. Even this sense is somewhat protean, for the range of "creative" or "imaginative" varies. It would certainly include in some way architecture and many crafts, but how far the meaning is to be carried into such areas as road-building, cookery, yoga, and mountain climbing is matter for discussion. One can also take the humanities in the larger sense as the study of man in his relationship to the entire human life-world, thus including such subjects as philosophy and anthropology and history. And once history is admitted, almost everything can be got in, directly or indirectly, under one or another perfectly honest rubric.

The *Dædalus* conference in fact focused chiefly on the humanities in the restricted sense, but so as to include history, which, of course, takes the pinch out of the restriction. Even if we do not count history, however, the temper of the conference was not exclusive. Yoga and mountain climbing did make their way into the discussion. The temper of this present essay is intended to be the same as that of the conference.

The humanities and the rest of actuality. The charges under this general heading against the humanities today are urgent and wide-ranging. Many students challenge the entire humanistic enterprise, and they do so, some analysts feel, because humanistic studies are not effectively related to the extra-curricular life. The failure here can take several forms. The one which today perhaps leaps first to the eye is failure in activism: the humanities are not responsive enough to political, social, and educational crises. Some even feel that responsiveness alone does not suffice: the humanities as taught do not sufficiently exacerbate conflicts as they should and as extra-academic trends do. Desperation, and often only desperation, can lead to transfigured joy. Too many humanists protectively insulate their subjects from ongoing cultural crises. They encourage retreat from activist movements outside the classroom. Good teachers will testify that even in the classroom a public disaster, such as the assassination of Martin Luther King, far from showing up Henry James as irremediably arcane, has actually made him more urgent.

One of the advantages of having artists on campus is that artists, in contrast with many humanist scholars, do often form far-out nuclei around which students can take positions. The art world is with it. Action paintings and other happenings advertise the urgent connection between creativity and contemporary history. Humanists should be willing to put their bodies on the line.

The relationship of humanistic studies and good citizenship is often at stake here. This relationship, it is charged, has been insufficiently reflected upon, is left vague, and is rendered still more vague because we do not adequately understand what being a good citizen means or can or should mean in our age. Many feel it must mean public commitment on public issues. Students are helped if teachers are publicly committed. Movies are particularly relevant here. The reluctant and then often inept incorporation of film criticism into the curriculum goes to show how laggard the humanities can be.

But this activist critique itself does not go unchallenged, for some feel that activist over-response to political, social, and educational crises shows a failure of nerve and dedication in the humanities. They call attention to the fact that activism is often make-work and indeed is chiefly taken up by humanities students who have no clear objective yet in view: professional students with more identifiable goals have shouldered too many responsibilities to allow for this freewheeling expenditure of time. Individuals do indeed have obligations as citizens as well as obligations as humanists, but the two sets of obligations are not the same. As one critique put it, the humanist teacher as teacher should not feel responsible for the slums of Roxbury, although he should feel deeply responsible for the attitudes that his students have toward the very existence of slums. Humanists only deceive themselves if they think that their function is to make immediately recognizable physical changes.

Looking beyond the activist question, many would situate the cleavage between curricular and extracurricular life within the teacher and his way of relating to his subject. Too many teachers fail to convey any sense of the real world in which their own responses and students' responses to the material of their subject take form, although the artists or writers whose work they are dealing with do convey their own worlds, even while aesthetically transmuting them. The teacher not infrequently insulates his class and his subject from his own and his students' actual life, never daring to regard the realm of television and electronic guitars and newspaper headlines and politics and ghetto housing with the intent gaze he directs to the wit of Ben Jonson or Marvell or Rembrandt's light and shadow or Bach's fugal counterpointing. Such timidity or stupidity, of course, impoverishes the teacher's understanding of Jonson and Marvell and Rembrandt and Bach as well as of his own and his students' world. This critique implies no one-to-one correspondence between art and present workaday actuality, but it does affirm innumerable connections whereby art gives meaning to the workaday world and this world gives meaning to art. No one can mediate between the two unless he faces both honestly.

Cleavage between the humanities and the rest of actuality can also be due to a failure to relate the classroom treatment of a subject to the real subject as it exists outside the classroom. What is said in the classroom about the process of painting or writing may be miles from any painter's or writer's real activity or experience. Irrelevance of classroom treatment to the actuality of the work of art, of course, cancels out the relevance that the work of art itself has to other actuality. It also trivializes the subject. One way to remedy the situation might be for humanist scholars to attempt at least some work in the subject they teach—in this view a teacher of poetry is better off if he has had some experience writing poems, or an art historian if he has attempted some art work, and so on. The teacher need not arrive at professional expertise, although he should work toward it, but he should have experience in the medium.

Such experience would discourage the tendency to trivialize

in another way by making too much of statable rules that a writer or musician or painter or architect supposedly follows. The catalogue of such rules is limited by critics' proficiency at abstract analysis and description, and there is no guarantee that the as yet unstated rules are not more central to the performance than those everyone talks about. On the other hand, particularly in music and art departments, the presumed need to master certain technical elements, with or without laboratory work, has led to too great rigidity and paralysis of student energies.

Other forms of trivialization making the humanities unreal are mechanization of chronology, scholarly nit-picking, and toleration of research ruts, such as the stock monograph on an individual artist or writer or musician or work. It is pretty universally recognized, however, that chronology, nit-picking, and monographs on individuals can be invaluable and tremendously stimulating when vivified by imaginative interpretation.

A final cluster of charges regarding failures of humanist scholars and teachers today to relate their subject matters to other actuality centers on their handling of values. Often scholars and teachers dodge problems of value, both in art itself and in the life to which art complexly relates. Humanists often do not face honestly the conflicts in values that their subject matter raises, do not examine how values are learned, fail to incarnate in their own outlook the real extracurricular conflicts, tend to dodge conflicts in ideas, and even to avoid abstract ideas as such. For these and other reasons, they are likely to be either insufficiently pluralistic in outlook or falsely irenic and syncretist in their pluralism. Related to pacific pluralism is the tendency to find total correspondence and consequently total adjustment between all phenomena in a given culture, a tendency that is at present so strong as to constitute a kind of humanistic camp.

Parochialism, universalism, minorities, revolution, community. Another set of complaints points even more sharply at the present age than do those just reviewed. This set has to do with the relationship of subject matter and teaching to the social sense of our time, especially our feeling for a global society. One charge reads that although the humanities should educate for emotional and imaginative sharing with all men and cultures, liberating and enlarging and universalizing man's potential for sharing the entire human heritage, they too frequently fail to do so. Foreign cultures are neglected. Even domestically, humanistic subject matter is often white-oriented —black art is treated as a special phenomenon, interesting because it is a variant, just as the social sciences treat blacks not as normal people, but as special problems, handled under "race relations." Racial minorities are underrepresented in student bodies at the higher levels of the humanities as elsewhere, and they are underrepresented as teachers. Although most humanists would object to the steadfastly parochial principle that only a black can teach literature written by blacks as much as to the principle that only a white Anglo-Saxon Protestant can teach Henry James, it appears evident to all that those close to any given culture can contribute awarenesses otherwise unavailable concerning works from that culture.

Although the humanities should encourage sensitive response to all human beings, scholars and teachers in the humanities can fail to open themselves not only to racial groups other than their own but also to various other groups. Non-humanities majors are often neglected, as are students who begin or continue their education with no settled aim in view. The underprivileged generally are neglected.

Humanists should be more alert than they are to the socially unifying effects of the humanities well understood and interpreted. The arts can create community, enabling people to define themselves as communities or cultures. The humanities should make more of intersubjectivity and of participation. It is distinctive of our times that the audience often wants to get into the act. Energies really at work in our society have been neglected by overly academic minds. In particular, the humanities have failed to incorporate sufficiently within themselves the new perspectives in various revolutionary movements—for example, in ethnic and other subcultures. Too little is made of the virtuosity of Negro American culture in welding together the most diverse traditions, as Dorson has pointed out,[1] or of the Beatles' achievement in making art out of Liverpudlian nonart. A good number of artists and musicians and writers are responsive to the new revolutionary energies, but the academy and the practitioners are insufficiently in contact, although it would be prejudicial to lay the blame for this entirely on the academy.

Humanist scholars and teachers, moreover, often fail to respond to current movements dealing with the body and space. Not only have they neglected the liberating effects of dance, yoga, swimming, mountain climbing, and the like, but they have also done little or nothing to interest man in the power he should have over the spaces created in society by architecture and engineering, and in the human relationships that could or do result.

On the other hand, overvaluation of participatory activities

[1] Richard M. Dorson, *American Folklore* (Chicago: University of Chicago Press, 1959), pp. 184–198.

and of the related workshop approach to learning, often fostered by stress on the social and experiental, can scorn purely verbal involvement, which, by and large, is perhaps the most liberating of all humanistic procedures for the largest number of persons.

Criticism as an object for study. Apart from the charges against the humanities relating to current social crises, those that are most distinctive of all for our present age have to do with the place of criticism—art criticism, music criticism, literary criticism, and so on—in humane studies. With reference to criticism, however, one meets less with charges against the humanities than with concern and puzzlement over what the place of criticism should be.

Insofar there is a complaint, it is that criticism itself is being studied to the neglect of works of literature, art, music, architecture, and other fields. This is bad because criticism is secondhand art. But such an observation elicits the immediate rejoinder that criticism, if authentic, is not secondhand anything, but as real as art itself and as worthy of direct study. If teachers and students turn to criticism rather than to works of art themselves, this attests a weakness in the art in question: criticism is often where the action is today.

Both sides to the dispute seem to agree that art, music, literary, and other criticism has today attained a degree of complexity and richness hitherto unrealized. It is for the first time competing with its own object as a focus of attention. Moreover, criticism catches interest because of the art explosion, the by now well-known diversity of styles in all creative fields, and the need to interrelate them and to discriminate. Interrelating the diverse styles calls for interrelating whole cultures, so that critics of necessity extend themselves into every imaginable subject, with more or less proficiency or amateurishness, trailed by hordes of avid followers. With no formal philosophy requirements in most curricula, criticism masquerades as a universal wisdom, substituting for moral and political philosophy as well as for metaphysics and theology. Charlatanry is rampant, sometimes draped in scatology, less often in wit.

On the other hand, criticism is necessary because art implies something beyond itself and because the aesthetic distance demanded by an art object is tolerable only for a time: art must be reintroduced by reflection into contact with extrafictional actuality or translated into its equivalents or analogues in such actuality. Moreover, criticism as complex as that today is itself a rich manifestation of culture, and the history of criticism can be a highly sophisticated and enriching pursuit. Besides, the students love it.

Teachers, physical conditions. By implication at least, most critiques of the humanities thus far rehearsed here include problems concerned with teaching personnel and physical conditions. But a few charges push directly into these specific areas.

Many other problems in the humanities would be solved if so many teachers were not dull persons, one critique insists. More effort should be made simply to create and recruit more interesting people. Even interesting people, however, can be dulled by a creeping conservatism fostered by an undue concern over one's personal career which masquerades as professionalism. Many humanists, interesting ones as well as dull ones, fail to speak out on issues related to their profession because they want to get ahead.

Groups underprivileged in other ways are likely to be underprivileged in their physical educational surroundings as well as in their teaching personnel. Reform is urgent here. But the charges noted above that humanists have not awakened to the possibilities of architectural and engineering planning for educational environments refer across the board, *mutatis mutandis*, to privileged as well as underprivileged settings.

Aims, curriculum, and achievement. If the aims and achievement of the humanities were easy to formulate, charges relating to them should come first in an essay such as this one. But since the aims and achievement are in fact very difficult to formulate because they are complex and embedded in the density of existence, the charges relating to them can climax this summary here, together with the related charges regarding curriculum.

Charges regarding aims, curriculum, and achievement tend to be sweeping rather than particularized. Humanistic research and teaching are insufficiently reflective and philosophical about their aims. Man's sense of history is changing; he is consciously making it today as much as studying it; and the humanist's sense of identity does not build sufficiently on this fact. Many things are going on today that could be, but are not, brought into the scope of humanistic studies. The curriculum is outmoded and, as noted above, resists even such accessible innovations as study of the film. The work of social and revolutionary forces has not been looked at for what it is worth. Has SNCC done some of the most important sociology of our times?

Paradoxically, on the other hand, the quite real achievement of the humanities is one of the things that holds them captive. They are content to proceed with the same business as before. On the other hand, it can be countered that the humanities have produced not merely self-satisfaction but also, quite deliberately, intense self-suspicion—the sense of desperate urgency and the drive to self-criticism that mark our times. Since Nietzsche, the humanistic enterprise has been publicly advertising that it itself should not be taken too seriously. A reason for this could be that the humanities are degraded by their own material, for the subject matter of the humanities—literature, art, music, architecture, and the rest —at one point or another always merges with that of the mass-produced stereotyped culture that the humanities should actually counter.

Humanist aims, in a further charge, are commonly distorted by maladjustment to actuality. Technology and machines are treated all too often as alien to the humanities, as bad dreams best forgot, rather than as part of human life to be mastered, improved, and used.

Finally, the achievement in the humanities, insofar as it is not merely a liability to itself but an asset to mankind, is not made known to many who for various reasons should know about it—such as the general public, congressmen, and the culturally underprivileged whether in the ghettos or in the achievement-powered suburbs, where the very word humanities may suggest some quaint import from an overprogrammed and stuffy past.

II. NOETIC STRUCTURES AND HUMANISTIC RELEVANCE

It would be ingenuous to respond to this array of charges by undertaking to "answer" them if by this we mean to exonerate the humanities. Most of the charges are to some degree true. They are, moreover, truer of some times and some places than of others. Merely to determine exactly where and how far they are valid would take more man-hours than we can afford. Why try to exonerate the humanities or humanists anyhow? The humanities and humanists are always going to work with uneven success, and we need all our energies to make them work better, not to prove how good they are. I propose to respond to the charges by simply accepting them as more or less valid and by examining some of the things that have happened in the humanities to give rise to the present state of affairs. By improving our understanding of where we are and how we got here, such examination should facilitate truly effective improvement. It should also, I hope, help resolve some of the conflicts in the charges regarding the nature of the humanities. Some of these conflicts, though not all, arise from factitious suppositions about the humanities' history.

Charges against the humanities are likely to assume that the humanities are inherently more stable than they have been and that they must be prodded into changing as soon as possible because society has already changed before them. We need to be more generally aware than we are that the humanities have been far from stable and have themselves changed a great deal, both in the distant past and in recent years. They have perhaps altered as greatly as society itself has, although the changes may have been obscure or guarded and not always advantageous. We have often failed to discern the changes because we have not discerned adequately the varied ways in which study of humanities and man's other responses to actuality have interlocked. The interlocking is dismayingly complex. I shall consider here only how it has shifted in terms of some larger neotic structures which have held together man's life world and which have been the concern of studies in this present book.

The Knowledge Explosion

Growth in knowledge has probably been the principal agent working changes in the humanities. This growth has affected our very notion of the humanities by polarizing the notion in a different way. Today we tend to set off the humane against the mechanical. It was not always thus. Originally, as suggested earlier in this chapter, the cluster of concepts that has yielded our present "humanities"—*humanitas, litterae humaniores,* humane letters, and so on—was set off against the bestial. The humanities lifted man above the lower, unreflective, gross, animal, menial, slavish world of physical toil and made him a full reflective human being as distinct from brute beasts or from the large numbers of men whom pretechnological cultures assumed were condemned to a menial, animal-like existence. The humanities fostered the work of the intellect and thus raised man out of ignorance. Today the humanities are still assumed to make man more human, but they locate the threat that they must counter not in the animal world but elsewhere, in the world of machines. The humanities are commonly set off against science and its mechanistic offshoot, technology. The inhuman other is no longer a population of brutes to which man's lower, nonintellectual nature threatened always to hold him in bondage, but a population of nonliving things that he has made.

The growth in knowledge of the physical world that has created this state of affairs is often referred to as the "knowledge explosion." But it is not the only growth of knowledge affecting the humanities. If the humanities have been repolarized against machines by the enlargement of the physical sciences, they have also been restructured internally by a comparable explosion of humanistic knowledge itself. In Chapter 1 above I have touched upon this knowledge explosion in the humanities, where the sheer "quantity" of knowledge available in language, literature, history (of art, music, technology, science, philosophy, religion, theology, and so on, as well as cultural and psychological history), anthropology, the performing arts, and other subjects qualifying as humanistic, has become so vast as to be beyond description and is still accelerating its growth. By far the greater part of this knowledge is hardly a century old. In many cases, an entire field of knowledge is even younger than that. Such is the case, for example, in art history, in modern literary analysis, and in the study of all but a handful of the hundreds of languages academically accessible today.

Growth of knowledge in the humanities has, of course, had profound effects on the relationship of the humanities to the entire human life world. Many of these effects lie back of our present crises. Attending to themes that have played through this book, I should like to single out four that have especially altered noetic and psychological structures. These are the atrophy of puberty rites in humanistic education, the romantic cultivation of the unknown as an effect of the evolution of the communications media, the development of our sense of a synchronic present, and the anthropologizing of knowledge.

The Atrophy of Puberty Rites

Through the Renaissance and thereafter decreasingly up to the romantic period, humanistic studies, in the strict sense of the study of *litterae humaniores*, served the function of a puberty rite, at least in the schools of the West as explained in Chapter 5 above. The feeling for humanistic studies as an initiation rite enforced by the old Latin-grounded education can no longer be engendered in our less polemic, more permissive, coeducational world. Today's teachers seldom declare the once conventional war on the students, for if they do, they lose face with their peers. And thus, since war there must be—because you cannot simply skip initiation if you want to become a self-respecting adult—the students have to declare conventional war on the teachers. Since females, among human beings as among lower animals, are not commonly ritual fighters as males are, given to ceremonial duelling, but fight typically for realistic and often desperate ends (such as the life of their young), the coeducational setting which marks today's education helps the conventional war shade into real war.

Attention to the initiatory quality of earlier humanistic education shows several things. First, it shows the complexity of "relevance." Initiation rites achieve supreme relevance often by prescribing precisely actions that have no apparent relevance at the time they are performed: the assumption is that their relevance can be seen only when one has crossed the border that the initiation rites mark—at which point the relevance then is commonly found to be largely symbolic. In primitive puberty rites a youth may, for example, be forbidden to touch any part of his body with his hands or may be made to fast or abstain from certain foods or to undergo severe beating. In the Renaissance, schoolboys were made to master a language that for most would have no practical use. In both cases what was achieved was largely a sense of belonging. You went through what others had gone through for no other reason than that they had gone through it and wanted you to do the same. Irrelevance proved that your eye and heart were not on the object but on the "gang."

Secondly, the initiatory quality of earlier humanistic education reveals the nature of some problems concerning subject matter for the humanities. Much more is at stake—or may be at stake—than finding subject matter that is immediately satisfying or useful. We have to start with what is accessible, of course, for learning means proceeding from the known to the unknown. But the ultimate aim is to place the student in an area that he knows was previously inaccessible to him—and, in a way, the more inaccessible the better. This area must also be one where the lore of the culture is centered.

The content problem of the humanities today lies in finding a place where the lore of culture—and now even the lore of all human cultures, as these merge into one—can be centered. If the humanities function as an initiation rite, an induction, an entrance into some area, what area are we to choose? There is no longer one narrow door through which all must pass to gain access to the spot where all the earlier initiates stand. There are hundreds of doors, all opening into corridors that connect with one another only through a maze. A recent poem by the British poet Adrian Henri well illustrates the state of affairs. It is made up of twelve stanzas, mixed quatrains and triplets. Here is a four-line sample:

> Belà Bartók Henri Rousseau
> Rauschenberg and Jasper Johns
> Lukas Cranach Shostakovich
> Kropotkin Ringo George and John

The entire poem is of this same substance, nothing but a list of name after name after name, with only an occasional "and" interspersed, running in crossdisciplinary zigzags from St. John of the Cross and Matthias Grünewald to Andy Warhol. The poem is quite conspicuously an initiation poem, a quest for identity, a do-it-yourself puberty rite. It is straightforwardly entitled "Me" and carries the subtitle "If you weren't you, who would you like to be?" After banging through the litany of names in response to this patently identity-crisis question, the poem whimpers to an end with the plaintive "and / last of all / me." Henri's piece is utterly representative of a widespread and now even conventionalized state of mind. It appears in a paperback got up with standard psychedelic jacket design under the title *The Mersey Sound.*[2]

Some of Adrian Henri's cultural heroes present perhaps only questionable credentials, but he whips up with his mixture a good deal of real excitement. The mixture suggests that the world into which students step today is full of humanistic material and even humanistic fireworks, but that it is a terrifyingly vast world. Our poet, in fact, could have shown it to be far more vast than he does, for he has in fact taken a pretty provincial view, since his litany of persons to identify with includes only Western figures and is limited to those currently in fashion in undergraduate courses or in the mass media.

The Romantic Cultivation of the Unknown

Ours is not the first age troubled by an overcharge of knowledge. Other ages have had the problem and have found some kind of relief. From one point of view, romanticism for example can be seen as a defensive response to such an overcharge, perhaps the major defensive response in past ages. As explained above in Chapter 11, the overcharge to which romanticism responded had been built up by print and had cul-

[2] Adrian Henri, Roger McGough, Brian Patten [no indication of editor], *The Mersey Sound*, Penguin Modern Poets, Vol. 10 (Hammondsworth, Middlesex, 1967). Henri's poem "Me" is on pp. 27–28. Copyright by Penguin Books, Inc.; quoted with permission.

minated in the Enlightenment, a heady experience of noetic power typified by the more than Olympian self-assurance of the *"Discours préliminaire"* to the Diderot and d'Alembert *Encyclopédie*, with its propensity to think of knowledge control reassuringly and not unpatronizingly in terms of "assembling" knowledge in a mental map (*mappemonde*) or geometrical layout (*système figuré*).[3] This kind of ambition, of course, stirred up strong subconscious fears and countermovements. It is noteworthy that the age immediately preceding the Enlightenment saw the first stirrings of hostility toward the printed book as such. The Renaissance had frequently attacked printers and publishers, chiefly for malfunction in reducing expression to satisfactory typographical condition. Indeed, dispraise of the printer or publisher was a standard topos of Renaissance rhetoric as this art manifested itself in prefaces. But the age of Swift and Sterne was the first to spoof the structure of the book as a physical object, its footnotes, indexes, and storage-and-retrieval apparatus generally —the very features that recommended the printed book to the visualist *encyclopédiste* mind.

In the light of these preceding developments, romanticism has been examined in Chapter 11 above as the consequence of an overload of organized knowledge. Romanticism withdrew to a world of mystery, darkness, remoteness, and inaccessibility, the polar opposite of the world of the Enlightenment. But the Enlightenment had made this withdrawal possible. Ro-

---

[3] See "Discours préliminaire des éditeurs," *Encyclopédie ou Dictionnaire raisonné des sciences, des arts et des métiers*, ed. by [Denis] Diderot and [Jean] d'Alembert, 3d ed. (Geneva: Jean-Léonard Pellet, 1778 ff.), I, ii, xxvii–xxviii, xxxiii, xxxv, and after p. xciv the folding chart (*système figuré*) of knowledge, reminiscent of the similar foldouts "displaying" knowledge which earlier are found most often, although not exclusively, in Ramist works.

manticism would have been intolerable in earlier ages, which labored under a noetic economy of scarcity. Today's avant-gardism—which in its cultivation of the different and strange is a specialized and seemingly permanent extension of romanticism, as the late Renato Poggioli has shown [4]—manifests the same relative insouciance of those who have as a matter of fact a great deal of knowledge under ready control. At the same time, this romantic avant-gardism affords relief from the pressure inevitably built up by having to deal with vast stores of knowledge. It provides a vacation from oppressive rationalism.

But the vacation time has now played itself out, and that because of the very success of the romantic movement. The remote, the seemingly formless, the purportedly unstructured, the dark, the vaguely limned areas of human consciousness have themselves become part of our noetic store, as reflectively organized in their own way as the rational, structured areas were before. Consciousmess of the unconscious is a permanent part of our thinking. In comparable fashion, the once remote past has become part of our present conscious life, and the geographically remote feeds into our local airports and enters our living rooms. The Tokyo or Mexico City Olympics come on live television broadcasts into St. Louis, Boston, and Seattle. The cult of the exotic has proved self-defeating. All remoteness is naturalized now, including the four-billion-year-old rocks of today's recently visited moon. In the Adrian Henri poem just cited and in hundreds of similar outpourings, running from the high sophistication of Pound and T. S. Eliot down to echoic babbling in beatnik

[4] Renato Poggioli, *The Theory of the Avant-Garde*, tr. from the Italian by Gerald Fitzgerald (Cambridge, Mass: Belknap Press of Harvard University Press, 1968), *passim*.

and now hippie poetry, many of the figures out of the once remote past still circulate, but they do so arm-in-arm with other figures out of yesterday's newspapers. The past lives no longer in its own time or in indeterminate, somewhat legendary "days of old." It lives in the synchronic present where everything lives today.

The Synchronic Present

Paradoxically, absorption of the past is what has given the present its own identity. Because we have access to so much history, this is the first age that has undertaken a concerted academic study of itself. Earlier, one always lectured on general, scientific principles or on something far removed in time —Plato, or Aristotle, not one's own contemporaries. But once knowledge of the past achieved a certain size or density, the present began to take on a distinctive face of its own. It now is seen to differ from the past, but does so in terms out of the past. In earlier ages when there was not enough historical background to project the present against, the present appeared largely inevitable, not something positively determined to be this rather than that, but rather a neutral circumambient medium, nonisolable, like water for a fish. Now, however, with knowledge of the past brought to the state it has reached in recent generations, the present can be examined for Renaissance, medieval, classical, preclassical, Christian, Hebrew, and countless other elements, or for its lack of one or another of these. Knowledge of the past thus bears in on us to define the here and now, where all ages meet. The resulting sense of a synchronic present is further intensified and deepened by our knowledge of the histories of a great many other cultures and how they accord with and differ from our own. The complexities of American culture today, seen in terms of its past history, can thus be contrasted with the complexities of Japanese or Indonesian or Kwakiutl culture, seen in terms of their own past histories, too, as feeding into or not feeding into the present.

Not only in the world of learning and of art, but also in political life, technology, and clothing styles a distinctive sense of the present as present rides through contemporary culture. This sense has shifted the structure of the humanities, which now focus intently on ourselves in our own life-world. Probably no other culture has ever had to bear the burden of so acute a self-consciousness as does the present United States. Coached into identity crises by psychiatrists whose livelihood depends on their studying adolescents, students stampede from personal self-scrutiny into the collective self-scrutiny of "American studies." Self-consciousness among us has reached such a pitch as to produce in the United States that unprecedented phenomenon, the autobiographical drama, seen in O'Neill, Tennessee Williams, Arthur Miller, and others. Nearly all literature today that takes itself seriously—and too much does—becomes confessional literature.

Our stampede into the present may prove self-annihilating if it crowds out firsthand knowledge of the past by neglecting the linguistic and other tools that make such knowledge possible. Humanistic studies can by default shrink the present once more to a flat and toneless surface. One of our problems today, not talked of much at *Dædalus* conferences, is to provide a continuous supply of persons with the hard-nosed skills needed to study humanistic subjects other than oneself.

Meanwhile, in any event, our feel for a synchronic present fosters the demand for relevance in all studies today. Antiquarianism in history is dead. Its equivalents in other subjects are also dead. There is no longer any point in studying history or anything else as a whimsical assortment of anecdotes or a merely numerically sequential chronology, for everybody knows that the past has all had its effect on the present somehow, and to fail to expose the effect or effects is simply to condone or foster ignorance. History can indeed no longer afford to be merely retrospective. The questions that we put to the past to generate history for ourselves are our own questions—late-twentieth-century questions, not those of earlier ages. We might as well admit the self-interest that necessarily and quite honorably calls into being the history and the other studies this age needs at the same time as we acknowledge that the questions are fruitless unless we can do the grubbing research needed to understand other cultures on their own grounds.

Our historically based self-study (different from the rather atemporal self-study of earlier philosophy, as in the Stoic *gnōthi seauton*, "know thyself," and related rather to the self-study of Hebrew-Christian asceticism) has been responsible, certainly in part, for the hypertrophy of criticism reviewed earlier. The arts—literary, graphic, musical, and other —open into the archetypal consciousness, and depth psychology has taught us to bring the archetypes to focus on ourselves—if indeed it is even possible to attend to them without such a focus. A surprisingly large amount of the most insistent criticism today bears on contemporary works. Indeed, in literature at least, the development of a New Criticism and the emergence of the sense of a synchronic present have come about simultaneously.

The mass media, together with radically improved storage and retrieval systems, have intensified the sense of an active synchronic present by keeping afloat in the public consciousness vast stores of information concerning past, present, and future previously available only to the few and after years of study. The media and the storage and retrieval systems have, of course, not supplied the reflection that converts mere information into understanding and makes it truly serviceable. But the instant science and neatly topical philosophy of the slicks combine with round-the-clock newscasts and stereotyped radio and television analyses to make the present weighty and momentous, and, at the peak of its own quite normal self-consciousness, youth understandably finds itself either intoxicated or terrified or both.

If the demand for relevance in the humanities obviously connects with the feeling that we live in a synchronic present, the demand does no more than insist that the humanities live up to their best potential. Yet the demand to realize this potential is essentially dismaying. How can everything that is known in humanistic studies possibly be brought to bear on the here and now? And especially when one is thinking not merely of the resources of one individual culture—Western European, or Eastern Mediterranean, or United States culture, or Latin American culture in its manifold phases, or Chinese culture, or Polynesian culture, or the vast spectrum of African cultures—but of the resources of all these cultures simultaneously bearing in on us here and now in the United States or, for that matter, on man anywhere else in the world?

Whatever the answer, let us not pretend that the assignment to do something like this does not in fact bear down on us. The hippie imagination at its zaniest speaks for mankind here. Tom Wolfe's account of the Merry Pranksters recently at large in California, *The Electric Kool-Aid Acid Test*, notes among the equipment of the Ken Kesey place: "Day-Glo-paint . . . Scandinavian-style blond. . . . sculpture of a hanged man [redolent of Tarot cards, though this sculptured hanged man was seemingly right-side-up or at least not up-side-down]. . . . Thunderbird, a great Thor-and-Wotan beaked monster . . . a Kama Sutra sculpture. . . ." [5] with of course Indian beads and English boots. Against this is contrasted a "conventional" establishment, "a house or apartment decorated with objects of the honest sort, Turkoman tapestries, Greek goatskin rugs, Cost Plus blue jugs, soft lights—not Japanese paper globe lights, however, but untasselated Chinese textile shades . . . with Mozart's *Requiem* issuing with liturgical solemnity from the hi-fi." [6] Other than hippie in-group-out-group faddism, there is really little to distinguish the two cultural armories. Each is one more manifestation of modern technological man's quest for identity—in the case reported on by Wolfe complicated by an infantile quest for total security and acceptance despite one's own refusal of all responsibility. (The Pranksters recorded their own outrageously infantile behavior on tape and film so as to be able to replay themselves and prove over and over again to themselves that they are not rejected—that is, not by one another).

It is of course fatuous to think that all human awarenesses and learning can become simultaneously present to every mind at all times, or even that most of it can be accessible to every mind at all times. Moreover, the demand for relevance frequently warps understanding of the past beyond all recognition: the Renaissance humanist is interpreted as a proto-revolutionary of the early-nineteenth-century sort or Cotton Mather as a Jeffersonian democrat. If the totality of humanistic awareness is to be accessible and relevant, it is going to have to be on some more general grounds. What can such

[5] Tom Wolfe, *The Electric Kool-Aid Acid Test* (New York: Farrar, Straus, and Giroux, 1968), p. 138.

[6] *Ibid.*, p. 245.

grounds be? We cannot survey them all but we can note some that have been well worked already and others where possibilities lie.

The Anthropologizing of Knowledge

One way in which cultures across the globe grow together in humanistic studies is simply through the natural development of concepts suitable for relating diverse cultures to one another. We are likely to forget the achievement we live with here. Terms such as civilization, culture, world view, form criticism, literary history, and many hundreds more have no equivalent in classical or medieval or, for the most part, Renaissance languages, for the concepts which they represent and which are the basis for intercultural awareness simply had not yet been developed. As these generalizing concepts have emerged, the humanities and all knowledge have become increasingly anthropologized or centered on man rather than on schemata. To a degree, even the humanities can be studied nonanthropologically in relatively abstract or schematic form, as, for example, rhetoric was studied from classical antiquity through the Renaissance. When, as discussed in earlier chapters of this book, classical or "grammatical" rhetoric was supplanted in the eighteenth century with a "psychological" rhetoric of the feelings and this in turn from the early 1930's on with a "sociological" rhetoric, the fullness of the human life-world was increasingly honored.[7] In comparable fashion, psychology has been increasingly anthropologized as it moved from faculty psychology to depth analysis; philosophy has been anthropologized as it worked out from an epistemological starting point in Kant through to personalist existentialism

[7] See Douglas Ehninger, "On Systems of Rhetoric," *Philosophy and Rhetoric*, I (1968), 131-144.

today; and so have literature, history, and even grammar itself in its psychological connections. Pierre Teilhard de Chardin's synthetic work undertakes further anthropologizing of all actuality by interpreting the entire globe (and eventually perhaps the entire cosmos) as "hominized" through the phenomenon of man. Teilhard's work shows, incidentally, that anthopologizing our interpretation of actuality is not the same as secularizing, but can proceed with an intensification of religious insight at least as well as on purely secular grounds.

Anthropology itself as a new science has been making itself more immediately relevant. Starting, as science always does with what was more remote from the here-and-now human life-world of the scientist himself, striking off into the South Sea Islands, Alaska, and Australia to report on the primitives, anthropology has lately been making its way home and forcing itself to do anthropological studies of the contemporary culture in which the science of anthropology itself originated. This is itself an impressive instance of movement into a synchronic present in a particular field.

Developments in many other fields of humanistic study similarly relate the remote and the proximate, or the past and present, in ways clearly encouraging a sense of relevance. Only a few such developments can be singled out as samples here. In the study of literature the message in T. S. Eliot's magistral essay on "Tradition and the Individual Talent" has by now pretty well been driven home. Critics and historians alike are aware that the new works of art both derive from a tradition and by their own presence alter the position of previous works in the tradition. The business of a contemporary writer is to take hold of the maximum in the tradition and transform it as completely as possible. Eliot did not invent this message, for it comes home of itself to anyone who understands the historical dimension of actuality, but he gave it its classic formulation. Artists and musicians today, as well as writers, are intimately aware of it and of course are tortured by it. It implies that, since they come later in history, they must know more about more predecessors than earlier artists ever did, and in an increasing diversity of cultures, and that they must nevertheless somehow bring off a new incarnation never before realized. Poggioli believed that through the entire foreseeable future this situation will perpetuate the avant-garde pursuit of the *outré*.

Here perhaps lies the theoretical basis, if one is needed—as perhaps it is not—for the participatory study of humanistic subjects that is plainly going to play some part in our educational programming for the present and future. If the present is the front of the past, engagement in creative or imaginative work in the present gives access to the past, which, when turned to for its own achievements and values, can further interact with and enrich the present.

Readers can supply further instances of anthropologizing, but we cannot close this list without mentioning one of the most central of all—the interiorizing or hominizing of history as it grows from a study of external events to an account of the reorganizations of the human psyche.[8] Early history had tended to be political and military in content and more or less oratorical or ceremonial in manner. Human beings were indeed portrayed, but their relations to one another were stylized often in standard terms of virtues and vices. With Giambattista Vico we arrive at the explicit awareness that civil society has been made by men and that is somehow correlates with the structures of the human mind. This inward

[8] See Walter J. Ong, *The Presence of the Word* (New Haven and London: Yale University Press, 1967), pp. 176–179, 111–138.

turn has been further pursued by Freud, Jung, and other psychologists, in historical studies such as Stuart Hughes's *Consciousness and Society*, and, most sweepingly, by Pierre Teilhard de Chardin. It entails anthropologizing because it centers history not in the movement of materiel and the redefining of political boundaries, but in the human consciousness and in the patterned shifts in personality structures which in great part determine the externals of cultures and of history and at the same time are determined by these externals.

The current study of the communications media belongs with history of this sort, for the media in their succession from oral to chirographic and typographic and electronic restructure the personality, as has been shown not only by present-day psychological studies of oral cultures [9] as compared to literate cultures, but also by the study of epic singers undertaken by Albert B. Lord, following on Milman Parry, and by Havelock's study of the roots of Plato's philosophy in the slow shift from an oral to a functionally literate culture in ancient Greece.[10]

Permanent Crisis

The relevance achieved in the humanities in the instances rehearsed here does not of course automatically become relevance in a particular classroom situation. These achievements can only lay the ground for particular classroom relevance,

[9] For example, J. C. Carothers, "Culture, Psychiatry, and the Written Word," *Psychiatry*, XXII (1959), 307–320; and Marvin K. Opler, *Culture, Psychiatry, and Human Values* (Springfield, Ill.: Charles C. Thomas, 1956); and other studies cited in these two works.

[10] Albert B. Lord, *The Singer of Tales*, Harvard Studies in Comparative Literature, Vol. 24 (Cambridge, Mass.: Harvard University Press, 1960); Eric A. Havelock, *Preface to Plato* (Cambridge, Mass.: Belknap Press of Harvard University Press, 1963).

which is always going to have to be achieved by the individual teacher. But the teacher familiar with what has been going on in his field is in a very favorable position today. He has something to work with. Certainly matters do not stand where they did in the days of Quintilian or John of Salisbury or Erasmus or even Matthew Arnold. The humanities, we must keep ourselves aware, have changed greatly in the past and will change still more in the future.

They do not change whimsically or without pattern. In the last analysis, it is the business of the humanities to discern and try to understand the patterns whereby they have altered themselves or been altered. The foundation for the relevance of the humanities today must be a humanistic understanding of the development of the humanities themselves in relation to past cultures and to today's cultures. If we can reach some understanding here, we can hope to achieve the insight into the present situation that will enable us to enlarge the humanities more—for example, by growing new areas of humanistic studies out af the resources of minority cultures in the ways so badly needed today.

If we can achieve a better historical understanding of where the humanities have stood and now stand today, we may grasp better the urgency of our present crisis. Some of the charges summarized earlier were urged in a spirit of desperation. I believe myself that the situation is indeed desperate. But I do not feel thereby obliged to agree that it is more desperate than it used to be. I know of no time or place in the past where, if we attend to all human beings in a given culture and not to the small privileged class, the situation of the humanities was even remotely satisfactory. It is not more desperate today than before, but desperate in newly desperate ways. Moreover, acknowledgment of the desperate situation or even keen insight into its causes does not necessarily mean that we can ever solve the problems intrinsic to the study of the humanities. If the humanities are worth anything, desperation belongs to their texture, for they give us knowledge and experience of the tragicomedy of the human situation. They themselves do not stand outside this situation. They are part of it. It is by living on the brink of desperation that we are most truly alive.

Index

RHETORIC, ROMANCE,
AND TECHNOLOGY

Designed by R. E. Rosenbaum.
Composed by Vail-Ballou Press, Inc.,
in 11 point linotype Janson, 3 points leaded,
with display lines in monotype Janson.
Printed letterpress from type by Vail-Ballou
on Warren's 1854 text, 60 pound basis,
with the Cornell University Press watermark.
Bound by Vail-Ballou Press
in Columbia Bayside Linen
and stamped in All Purpose foil.

Twayne's United States Authors Series

EDITOR OF THIS VOLUME

David J. Nordloh
Indiana University

Andrew Carnegie

TUSAS 355

ANDREW CARNEGIE

By GEORGE SWETNAM

Cumberland County College
Library
P.O. Box 517
Vineland, NJ 08360

TWAYNE PUBLISHERS
A DIVISION OF G. K. HALL & CO., BOSTON

HD
9515.5
C 37
S 83

80-165

Copyright © 1980 by G. K. Hall & Co.

Published in 1980 by Twayne Publishers,
A Division of G. K. Hall & Co.
All Rights Reserved

Printed on permanent/durable acid-free paper and bound
in the United States of America

First Printing

Photo Credit: Carnegie Library, Pittsburgh

Library of Congress Cataloging in Publication Data

Swetnam, George.
Andrew Carnegie.

(Twayne's United States authors series; TUSAS 355)

Bibliography: pp. 173–78
Includes index.
1. Carnegie, Andrew, 1835-1919.
2. Steel industry and trade—United States—Biography.
3. Authors, American—19th century—Biography.
HD9515.5.C37S83   338.7′67′20924 [B]   79-17472
ISBN 0-8057-7239-1

Contents

About the Author

Recently retired after a distinguished career in journalism, George Swetnam has led an interesting life as a clergyman, commercial photographer, college teacher, hobo, newspaperman, historian, dramatist, and folklorist.

Born in Ohio of Kentucky mountaineer parents, he grew up in the deep South. He began his writing career early, and was listed in *Reader's Guide to Periodical Literature* while still in high school. After attending the University of South Carolina and University of Alabama, he graduated at the University of Mississippi, majoring in English. Entering Columbia Theological Seminary, Swetnam won the degree of Bachelor of Divinity with the highest grades in Hebrew in the school's 100-year history and was ordained a minister in the Presbyterian Church.

After holding pastorates in Alabama and Mississippi, he received a graduate fellowship from Auburn Theological Seminary (now merged with Union), where he received the degree of Master of Theology, majoring in Semitic Languages. He then attended Hartford Seminary Foundation, where he won his Ph.D. degree in Assyriology, his dissertation being a translation of 100 previously undeciphered Sumerian tablets from the Third Dynasty of Ur (about 2000 B.C.).

Following a second period in the pastorate, he taught English at the University of Alabama while taking two years of post-doctoral work in English and modern languages, founded a photographic firm, gave it away, and was a hobo for two years. Returning to respectable life, he edited a weekly newspaper in Tennessee, was managing editor of a daily in Pennsylvania, and became a columnist and feature writer for the *Pittsburgh Press*.

He is the author of a dozen books—including a three-volume history of Pittsburgh and the only history of transportation in Pennsylvania—and coauthor of a half a dozen more, as well as three produced historical dramas. He edited the *Keystone Folk-*

lore Quarterly for six years, founded the Institute of Pennsylvania Rural Life and Culture (now in its twenty-third year), and is a former member of the Council of the Pennsylvania Historical Association. Currently he is directing a comprehensive historical preservation survey under the auspices of the Pennsylvania Historical and Museum Commission. Outside the newspaper field his work has been widely published, principally in historical and literary journals.

Preface

Andrew Carnegie presents today's scholar with the paradox of an internationally known author whose writings have been almost totally eclipsed by his deeds. Ninety years ago, when he was just one among many American millionaire industrialists, his articles were welcomed by the best literary magazines on both sides of the Atlantic, his books were best sellers, and his writings were translated into many languages. His ideas influenced statesmen and scholars all over America and Europe. Today every school child recognizes his name as a steel-maker and philanthropist, but knows nothing he wrote except the *Autobiography*.

This situation has resulted in part from controversy over his character, controversy which has raged since the peak of his career. He has been held up as an angel or a devil, but seldom as the brilliant, ambitious, tender-hearted, puckish, and somewhat vain human being he was, with much the same virtues and vices as many of his contemporaries. And it has resulted from the rise of ghost writing, a development which has caused some historians to assume that his works were produced in that fashion.

The purpose of this study is to bring to American attention an outstanding author whose works have been neglected and even forgotten; to demonstrate that these are actually his own, the ideas and style following a consistent pattern from early youth to old age; and to point out the relation of his background and experiences to the ideas and attitudes evidenced in his writings.

To accomplish these purposes, the study will include only enough biographical material to throw a clear light on Carnegie's reasons and purposes in writing, and to demonstrate the more significant influences on his life, and how they affected his works. It will present a broad spectrum of his production in this respect, including a discussion of all his books, a wide selection of his articles and published speeches, and a few of his letters, against the background of their origin.

No attempt will be made to offer moral judgments on Carnegie's character or actions, or to excuse or explain away his ambivalence as a greedy businessman and an open-handed philanthropist and social reformer. Neither will the study enter into details regarding his economic and political beliefs. These are the concern of biography and technical treatises, not literary study.

Because of the diverse nature of Carnegie's writings, they will be taken up here in categorical and topical, rather than strictly chronological, order, although following their time sequence as nearly as is convenient. To attempt any other approach would invite utter confusion. Insofar as a general time sequence appears logical, it is indicated at the head of each chapter.

Most of Carnegie's writings are available to the reader in his books, published miscellaneous works, magazine files, or—principally the speeches—in pamphlet form in larger libraries. Because of the immense volume of his letters and the fact that few have been published, attention to them is here limited to representative examples based on texts reprinted in biographical works.

In regard to the personal origin of his writings, it has not seemed necessary to resort to verbal analyses, since both style and external evidence are clear and convincing. Carnegie wrote to convince, to spread ideas, to influence others. By his own statement he considered his writings more important than even his most outstanding accomplishments in other fields. Perhaps, despite their neglect in the past half century, he was right. His pride would never have allowed him to present the work of a ghost writer as his own. The day of the ghost writer had not yet arrived. As the first great writing industrialist, it was he, more than any other, who created a place for this faceless phantom.

Yet the assumption that Carnegie employed such aides is not surprising; rather it is in keeping with the career of a man whose life was a tissue of contradictions: the boy impoverished by industrialization who used the machine to become one of the world's richest men; the apologist for labor who condoned and perhaps approved one of the greatest crimes against unionization; the flag-waving American whose heart still held tightly to his native Scotland; the man who "must push inordinately" the pursuit of wealth, then used the same energy to give it away. Thus it seems natural, if not poetically just, that the monster he helped to create should have been used to siphon off the one thing in life he valued most: credit for his own ideas.

The author is particularly grateful for the assistance given him by Joseph F. Wall, who provided him with materials from the United States Steel Corporation's Carnegie files, which are no longer open to the public; to Kitty Sullivan Lowder for endless hours of search in Pittsburgh newspaper files; and to the reference staff of Carnegie Library of Pittsburgh, chiefly Marie Zinni and her aides in its Pennsylvania Room. Special thanks are owed to Helene Snyder Smith, the author's coworker, for research, advice, encouragement, ideas, and other assistance without which the book would probably never have been completed.

GEORGE SWETNAM

Chronology

1835   Born at Dunfermline, Scotland, November 25; first of
        two sons of William Carnegie, unemployed weaver, and
        his wife, Margaret Morrison.

1843   Enters the Rolland school, near his home. Brother, Tom,
        born.

1848   Comes to America with parents and brother, settling in
        Allegheny, now a part of Pittsburgh. First job as a bobbin
        boy in a cotton mill.

1850   Becomes telegraph messenger.

1851   Becomes telegraph operator. Takes press wire for news-
        papers.

1853   Operator for Thomas A. Scott, Pennsylvania Railroad divi-
        sion superintendent. Soon his assistant. First published
        work, letters to the *Pittsburgh Dispatch*.

1855   William Carnegie dies.

1856   Becomes assistant division superintendent. First printed
        book (probable), *Rules for the Government of The Penn-
        sylvania R. R. Company's Telegraph.*
        First investment: $500 for ten shares of Adams Express
        Co. stock.

1859   Becomes division superintendent of railroad.

1861   Called to Washington to organize telegraph service.

1862   First trip back to Scotland. Silent partner in bridge firm.

1863   Income over $47,000, only $2,400 from railroad.

1865   Leaves railroad. Organizes bridge and iron firms. Goes on
        walking tour of Europe.

1866   Expands and consolidates industrial holdings.

1867   Moves to New York and comes in contact with cultural
        influences.

1868   Writes memo planning early retirement and life of culture
        instead of money-making.

1869 Begins bond sales overseas, totaling $30 million in four years.

1873 Builds Edgar Thomson steel mill largely with money from bond sales.

1878 Trip around the world. *Round the World* printed for private distribution.

1880 Organizes newspaper syndicate in Britain.

1881 Coaching trip in England and Scotland. Gives library to Dunfermline—first of over 2,800 to towns all over the world. Begins organizing chain of liberal newspapers in Britain.

1882 First article published in a major journal, *Fortnightly Review*. *Our Coaching Trip* printed for private distribution.

1883 *Our Coaching Trip* revised and published as *An American Four-in-Hand in Britain*.

1884 *Round the World* revised and published.

1885 Abandons and sells newspaper chain.

1886 *Triumphant Democracy* published. Mother and only brother die within a few days.

1887 Marries Louise Whitfield in New York.

1889 *Wealth* published in *North American Review*.

1891 Writes first draft of *Autobiography*. Carnegie Hall built.

1892 Homestead steel strike. Carnegie bitterly attacked by press.

1893 *Triumphant Democracy* revised and republished.

1896 Gives Carnegie Institute to Pittsburgh.

1897 Daughter Margaret born.

1898 Buys Skibo Castle, in Scotland.

1900 *The Gospel of Wealth* published. Founds Carnegie Trade Schools, later Carnegie Institute of Technology, now Carnegie–Mellon University.

1901 Sells out steel holdings and gives up industry. Founds Carnegie Relief Fund for retiring workers. Makes first Scottish Education grant.

1902 *The Empire of Business* published. Becomes Lord Rector of St. Andrews University. Builds residence at Fifth Avenue and 91st Street, New York. Founds Carnegie Institution in Washington.

Chronology

1903   President of the Iron and Steel Institute. Builds Hague Peace Palace.

1904   Establishes first Hero Fund.

1905   *James Watt* published. Establishes Foundation for the Advancement of Teaching.

1906   Resumes writing and begins revising *Autobiography*.

1907   Visit to Emperor Wilhelm II of Germany.

1908   *Problems of Today* published.

1910   Sets up Endowment for International Peace.

1911   Sets up first Carnegie Corporation for the advancement and diffusion of knowledge.

1912   Last published address.

1914   Leaves Scotland for last time. Abandons *Autobiography*.

1915   Health weakened by respiratory attack.

1916   Last magazine article published, in *Woman's Home Companion*.

1919   Daughter marries Roswell Miller. Dies August 11, at summer home in Massachusetts.

CHAPTER 1

White-Haired Scotch Devil: 1835-1865

EVEN in his teens Andrew Carnegie was a writer, and writing was his greatest outlet. Late in life he recorded that, "Indeed, my early letters to friends in Scotland, when fourteen, are said to be better than those of this date."[1] At eighteen he used his pen for the same purpose as in later life: to spread his ideas and to accomplish his ends.

Writing, like speaking, came to the young immigrant naturally. He was born into a voluble family of Scottish liberals, and wrote just as he talked. As his ability at speaking improved, so did his literary style. Early experiences even determined much of his subject matter. Poverty and wealth, politics, industry, and patriotism were all concerns of his childhood surroundings.

Andrew Carnegie was born on November 25, 1835, in Dunfermline, Scotland, first of the two sons of William and Margaret Morrison Carnegie. A sister died in infancy, and the younger son, Thomas, was his junior by eight years. As an only child Andrew must have been rather spoiled, since he was not required to go to school until he was past eight. Then he attended a one-teacher institution near his home.[2] Perhaps finances may have had something to do with the delay. His father was a weaver—one of many in Scotland's ancient capital who had been thrown out of work by the introduction of the power loom.

Faced with absolute want, the family borrowed money and came to America in 1848, settling near relatives in a poverty-stricken area of Allegheny, Pa., now part of the city of Pittsburgh. Things were little better in the new land. William Carnegie, unable to endure work in a cotton mill for long, picked up a little money weaving and peddling linens. Mrs. Carnegie, obviously the strong member of the family, largely supported them by binding shoes, a trade she had learned from her father when a girl. Andrew had to give up any hope of further schooling and go to work, first at $1.25 a week as a bobbin boy in the cotton mill, shortly afterward at $2 a week tending a small boiler and oil-treating the product in a bobbin factory owned by a friend of the family.

J. F. Wall notes that although Carnegie was aware of what indigence meant, he never fully experienced grinding poverty, as many of his playmates did. His parents, in particular his mother, shielded him from the full impact, and as a result he was "more an observer than a participant." With this early protective environment it became easy to extol the virtues of poverty —one of his favorite themes throughout life.[3]

Immigrants from the British Isles were little if any more popular in Pittsburgh in that "Know Nothing" era than were their successors from Central Europe a generation later. Andrew and a few other young "Scotchies" had to fight their way out of some difficulties, which may have contributed to his later pugnacity and determination to win every contest. It also encouraged his Scottish patriotism, but before long he was becoming strongly American in his sympathies.[4]

I *Would-Be Reporter*

Trusted with the task of doing his employer's small bit of accounting and billing, young Carnegie took night classes in Pittsburgh to learn double-entry bookkeeping. But what he always considered the most important break of his life came early in 1850 when his uncle recommended him for a job as an extra messenger boy for the recently opened telegraph office in Pittsburgh. His small stature made him look even younger than his fourteen years, and he wisely insisted that his father stay outside while he went into the office to apply for the work. He admitted later that he had also thought of the effect of his father's broad Scottish dialect, preferring to trust to his own more Americanized tongue.[5]

The new job paid more money and had many perquisites, including free admission to the town's theaters, where the youth acquired a lifetime devotion to Shakespeare. He worked hard, memorizing the locations of all the city's business houses and learning to recognize the faces of its prominent men, many of whom came to know him. Since Pittsburgh then had a population of somewhat over 46,000 and Allegheny almost half that number,[6] this represented considerable effort. He also gave strict attention to the telegraph instruments, and in a little more than a year was promoted to a position as operator.

One of the sixteen-year-old operator's duties was to take the press wire and make copies for the reporters of the five newspapers which subscribed to the service. The contact gave him a feeling of importance and a desire to become a newspaperman— a yearning that was to persist for more than thirty years, until bitter experience convinced him that there were some battles even he could not win.

Meanwhile, as he labored with key, sounder, and pad, he was offered a position as personal operator to Thomas A. Scott, Pittsburgh Division superintendent of the Pennsylvania Railroad, which had been opened for through service between Philadelphia and Pittsburgh just a year earlier. Regular train dispatching by telegraph was still in the future when the appointment was made on February 1, 1853, but the railroad had enough business to need its own line. Though a youth of barely seventeen may seem a strange choice for the position, Scott was a good judge of men. And besides, Andy was reputed to be only the third operator in America who had learned to take messages by sound, discarding the clumsy tape rolls designed by Morse to record messages until the dots and dashes could be translated into letters.

Always a voracious reader, young Carnegie had been using a library of some 400 books which Col. James Anderson of Allegheny had set up, and later enlarged.[7] Anderson, a retired manufacturer, had allowed working boys to use the library free, taking out one book a week. But after it was turned over to the city, a new librarian interpreted this as applying only to bound apprentices, and demanded that Carnegie pay a fee of $2 a year. Small though the fee might be, it was still an important amount to Carnegie and many of his friends who were affected by the ruling. On May 9, 1853, he took up the cudgels in a letter to the *Pittsburgh Dispatch,* probably published through the good offices of a reporter friend. "Mr. Editor," he wrote. "Believing that you take a deep interest in whatever tends to elevate, instruct and improve the youth of this country, I am induced to call your attention to the following."[8] After citing the principal facts of the situation, he continued: "Every working boy has been freely admitted only requiring his parents or guardian to become surety. But its means of doing good have been greatly circumscribed by new directors who refuse to allow any boy *who is not learning a trade* and *bound* for a stated time to become a member. I rather think that the new directors have misunderstood the generous donor's *intentions*. It can hardly be thought that he meant to exclude boys employed in stores merely because they are not bound."

Young Carnegie signed the communication "A Working Boy though not bound," and the reference to store workers may have been intended to conceal his identity. After all, a dispute over $2 a year might be thought beneath a telegraph operator making $35 a month.

The letter was printed four days later, and on May 16 a reply appeared, signed X.Y.Z., criticizing it, and stating that for a time others than apprentices had been admitted free, "but for reasons it became absolutely necessary to admit none but those for whose benefit the donation was made." Next day Carnegie replied with a letter which argued; "The question is, was the donation intended for use of apprentices only in strict meaning of the word, viz. persons learning a trade and *bound,* or whether it was designed for working boys whether bound or not? If the former be correct then the managers have certainly misunderstood the generous donor's intentions—[*sic*]."

Three days later appeared a notice in the *Dispatch*: "A 'Working Boy' will confer a favor by calling at our office." There the librarian quietly agreed to make the concessions demanded. The success so elated the young railroader that he was certain he wanted to become a reporter. For years afterward, in speeches to news groups and in conversation with newspapermen, he would often confide: "The only reason I am not one of you is because no Pittsburgh paper in those days would give me a job."

II *Writing Railroader*

Lacking such an offer—if he really had not already made up his mind in regard to a change—young Carnegie continued as operator and soon as assistant to Scott in the railroad office. But his work never interfered with his continuing to write for publication. At first there were more letters to the *Dispatch,* then to the *Pittsburgh Commercial Journal,* and on May 12, 1854, to Horace Greeley's *New York Tribune*—printed under Carnegie's name.

Writing letters to newspapers was a custom he did not give up until his health broke when he was almost eighty. During the last twenty years of his active life, 177 letters from Andrew Carnegie were printed in such journals as the *New York Times, Tribune,* and *World*; London *Times*; *Manufacturer's Record*; and *Iron Trade Review.*[9]

Carnegie was never a devotee of formal religion, but at this period he was attending the Church of the New Jerusalem (Swedenborgian) in Allegheny with his family, taking part in its Sunday School and singing in the choir. A year after the "working boy" incident he was writing to a church periodical, *Dewdrop,* commending its stand against all war in terms that anticipate his peace efforts half a century later. His letters to his relatives in Scotland at this period foreshadow the super-patriotism that was to ripen in thirty more years into *Triumphant Democracy.* In an 1853 letter to his cousin George Lauder, Jr., he states the main thesis of that work pretty clearly: "But you may reply, Government has little or nothing to do with the state of affairs. Why then, I would ask, the contrast between the U. States and the Canadas? They were settled by the same people, at the same time, under the same Government—and look at the difference! Where are her Railroads, Telegraphs and Canals? her . . . potent Press? We have given to the world a Washington . . . —Ah, Dod,[10] what has Canada ever produced?"[11]

It was about this period that young Carnegie again demonstrated the courage and the executive ability that were to characterize his later life in such a high degree. Only a short time after he had begun working for the railroad, he arrived at the office one morning to discover that there had been a

serious accident which had halted all traffic, except that the
eastbound express was creeping along from station to station,
a flagman signaling its progress at every curve. Knowing that
nothing else could move without direct orders from the superintendent, who might not arrive for some time, the youthful
operator ascertained the position of each train, issued orders
under Scott's name, and soon had everything in order.

When Scott arrived at the office some two hours later, the
youth explained the situation and handed him the stack of
telegrams, not sure whether his temerity might bring instant
discharge. The superintendent didn't say a word, but a few
days later Carnegie learned that his employer had confided to
an associate:

"Do you know what that little white-haired Scotch devil of mine
did today? . . . I'm damned if he didn't run every train on the division
in my name, without the slightest authority."
"And did he do it all right?"
"Oh, yes, all right!"[12]

Carnegie's father died in 1855, and a year later Scott was
transferred to Altoona, Pa., his young assistant and the family
going along. It was at this time that the railroad began operating
all trains on telegraphic orders, and there can be little doubt
that Carnegie wrote the rule book for telegraphers, of which
an 1863 issue was in the files of the railroad at Pittsburgh.[13]

Certain phrases not ordinarily used in Pennsylvania at the
time, such as "tea" for the evening meal,[14] make it almost certain
that one of the Scottish contingent Carnegie had assembled for
the railroad is the author. And despite the printed statement
that it was Robert Pitcairn—then in the General Superintendent's
office—who inaugurated telegraphic train operation for the road,[15]
no one familiar with Carnegie's style could doubt that he was
the author. Certainly it could not have been Pitcairn, the only
other Scottish operator then in a position to compile such a rule
book. No one who has read any of his writings—full of confusion and with interminable sentences of piled-up, loosely
linked phrases—could possibly believe he could have produced
its terse, forceful style.[16] Another Scot, David McCargo, was in

position to have written the book by 1859,[17] but it seems incredible that the railroad would have operated for three years without such a set of rules. If the objection be raised that Carnegie fails to mention it in his *Autobiography*—the same thing is true of several of his more important works.[18]

Early in 1859 Scott was again promoted, and the twenty-three-year-old Carnegie was made superintendent of the Pittsburgh Division, returning to the city to make his home. No longer did the family live in the flats of Allegheny, however. Carnegie and his mother secured a house in Homewood, an eastern suburb of the city, sparsely settled up to that time, but near the railroad. The choice was a fortunate one, for the house they chose was in a small pocket of education and culture. As always, Andy's charm and wit gained him a ready entrance into the group, where he was able to gain a better appreciation of music, and expand his knowledge, which had been largely limited to the English and Scottish classics. In this atmosphere, also, he found occasion to polish his formerly awkward manners, and improve his grammar and speaking ability.

While still in Altoona he had borrowed money to invest in the Woodruff sleeping car, which was an immediate success, giving him his first substantial income from stocks. Soon after Drake's discovery of oil in the Allegheny valley, Carnegie and some friends leased a farm there, which brought them a fantastic return not long afterwards.

With the beginning of the Civil War, his former chief called him to Washington to organize the telegraph service, Scott himself being head of railroad transportation. Carnegie was only there for a few months, however, and soon was able to spend most of his time back in Pittsburgh again, investing and making money. He was present at the first battle of Bull Run, managing a telegraph office, and became so much interested in sending a press story on how the Union army was winning, that he barely escaped being captured by Confederate forces. Never able to tolerate high temperatures well, he suffered a heat stroke soon afterward from which he was long in recovering. In an effort to regain his health he and his mother took a trip to Scotland in the summer of 1862.

On his return he plunged again into business with such

success that a memo he made of his 1863 income discloses that it totaled nearly $48,000, of which his railroad salary provided only $2,400. For the remainder of the war period, it appears, the young industrialist gave virtually all his time to investment, with great success.

CHAPTER 2

Young Man of Affairs: 1865-1886

FOR a man who amassed one of the great fortunes of his era, and whose principal reputation is based on industry, Andrew Carnegie was surprisingly casual about most business matters.

May of 1865 would have seemed a most unlikely time for him to take an extended vacation. George Pullman was beginning to threaten Central Transportation Co., the sleeping car firm Carnegie had formed with Woodruff.[1] He had just enlarged and reorganized Piper and Shiffler Co. into his Keystone Bridge Works a month earlier.[2] And in March he had merged his five-month-old Cyclops Iron Co.—his first venture into basic metal production—with a rival firm, as the Union Iron Mills.[3] Yet without the slightest apparent hesitation he left these and his numerous other business affairs in the hands of his twenty-two-year-old brother, Tom, and sailed off with two companions for a nine-month tour of Europe. The obvious explanation appears to be that Carnegie had already decided something he would put on paper three years later: that his most important business was not to make money, but to acquire culture and be of influence in the world.

I The Quest for Culture

Since his return from Scotland in 1862—to some extent even earlier[4]—Andrew had studied with tutors, making up for the education he had missed because of having to work hard in childhood. Just as he had studied hard at double-entry bookkeeping in night classes at thirteen, he labored at French and cultural subjects after returning from Altoona to Pittsburgh as a railroad divisional superintendent at twenty-four. He had been inspired by Rebecca Stewart, Scott's sister-in-law, who came to Altoona to care for his children after his wife's death.

But a stronger force was the society in which he found himself after returning to Pittsburgh.

There after a few months he bought his mother a house in the small, new, residential suburb of Homewood. It had taken its name from the mansion of William Wilkins, a former judge, cabinet member, and ambassador to Russia, an elder statesman and dean of the Pittsburgh bar. Eager young Andy found himself welcomed at the Wilkins home and that of William Coleman, a well-known businessman, later Tom Carnegie's father-in-law. But he gave principal credit to another neighbor, Leila Addison. Miss Addison was the daughter of a physician and granddaughter of a judge, both deceased, and lived in Homewood with her widowed mother. Mrs. Addison was a native of Edinburgh, and had reportedly been tutored by Thomas Carlyle.

Up to that time Andy had been a brash, well-read but intentionally rugged and very successful young man. Meeting with the Addisons he suddenly glimpsed the gulf between him and the world of real culture. He felt that it was "the wee drap o' Scotch bluid atween us" which aroused their interest in his welfare.

"Miss Addison became an ideal friend because she undertook to improve the rough diamond, if it were indeed a diamond at all," Carnegie later wrote.[5] "She was my friend, because my severest critic. I began to pay strict attention to my language, and to the English classics . . . also to note how much better it was to be gentle in tone and manner, polite and courteous to all—in short, better behaved."

By the time of the 1865 trip to Europe this training had borne fruit. Although still far from the polished, urbane man he would become, Carnegie was able to converse in understandable—if somewhat gapped—French, and to make the most of the cultural opportunities which the tour afforded. If he failed in full appreciation of scenic wonders,[6] it was not so much from ignorance as from his interests, which always ran much more to people than to things.

Carnegie sailed from New York in mid-May, along with two boyhood friends, Henry Phipps (a partner in his iron venture) and John Vandevort. They stayed in Britain until September, and even after they had gone to the Continent, Carnegie passed up the trip to Switzerland in order to improve his French in Paris, and return briefly to London to secure the rights on a new process for making rails.

II *Correspondent for the* Commercial

Young Carnegie had been greatly impressed by Bayard Taylor's *Views A-Foot,* which came out in 1847 and had been read and discussed by the group at Homewood.[7] Taylor, just twenty-one years earlier, had gone overseas as an almost penniless young apprentice printer, financing two years in Europe on $500 made by writing for newspapers.[8] Carnegie, though far from penniless, and estimating the cost of his jaunt at $3,000,[9] planned to emulate his idol by writing letters to the Pittsburgh newspapers. A careful check of all available newspaper files for his period abroad, however, discloses only two published letters, both in the *Commercial.* It is likely that not more than three were written, as he explained in letters to his family and friends that he would like to write more, but found himself too busy making the most of his opportunities to see, learn, and enjoy the trip.[10] The third letter may have been declined by the newspaper's editor, C. D. Brigham, because of its controversial content. On September 29 Carnegie wrote from Paris to the Pittsburgh printer and news-dealer W. S. Haven that he had received the paper "and have occupied myself giving my views on Protection, a subject Mr. Brigham is stirring up thoroughly and well.... Mr. Brigham, I suppose, will give my protection letter his care. It isn't the Pittsburgh doctrine, I know, but it will do good in stirring up the subject. I have had it hot and heavy with the English, but that Carisbrooke Castle illustration I have always found effective."[11] Carnegie's reference makes it appear probable that he wrote and sent the article, but apparently its sentiments were so far from the "Pittsburgh doctrine" that Brigham failed to print it.

Whatever may have been Carnegie's sentiments in regard to Bayard Taylor's work, none of the three subjects chosen for his letters show any indication of copying Taylor's style or subject matter.

The first—dated at Inverness, Scotland, August 21, and printed in the *Commercial* on September 21—deals with fish breeding, and opens with a disclaimer of any intention to write the usual type of travel letter:

Descriptive letters from abroad are now-a-days so readily prepared from the omnipresent guide books, whch give not only dates and particulars, but pages, written in excellent style, upon the memories and associations which a visit calls forth, that one feels a laudable disinclination to seek such publicity as your columns would afford, for what the guide would mostly be entitled to the credit of. But what we saw yesterday is so utterly foreign to the domain of the professional *litterateur* that we are prompted to write you in regard to it, in the hope that the subject may receive some attention at home.

Carnegie then proceeds to give a careful and accurate report of a salmon hatchery near Perth, including the method of gathering and fertilizing the eggs, dimensions of the ponds, costs of labor and materials, and the results. The most impressive of these results (and italicized by the ever profit-conscious writer) was that the fishing privileges on the Tay returned *"thirty-five thousand dollars* in gold every season, over and above former rentals." The investment was "not exceeding $2,500, and maintained for not exceeding $1,000 per annum," to produce 300,000 salmon every other year weighing seven pounds each in four to six months.

Proving that he would have become a good newspaperman if he had gone into that field as a young man, Carnegie bore in mind that a regional paper would be interested in the story's local angles: "There are certainly many streams, throughout the United States, formerly noted for a bountiful supply of various kinds of fish which have become scarce, and which could readily be restored, at a trifling expense. Who will be the first to move in this matter, and thus add to the resources of our sacred land?"

Using the names of various Pittsburgh friends who shared his passion for fishing, he continued:

For the sake of John Hampton, Perry Knox, Tom Scott, David Book, Geo. Findley, and sundry other friends, ardent disciples of old Izaak's, and not without a slight degree of personal interest in the premises, we made particular inquiries as to whether the coveted brook trout could be multiplied in similar manner, and was assured that there was not the shadow of a doubt about it. Oh! Shades of Tub Mill, Big Ben creek, Kittanning Point and Bell's Mills! how you grow upon me. We will have committees formed at once, subscriptions raised, and fill you full of "bonnie trout." There will be no such thing as empty baskets any more, even for Judge ———. No longer will we condescend to use the worm, but the speckled beauties so numerous, so large, so gay, will continually, "rise to the fly. . . ."

[We] set ourselves thinking . . . what he should be called who caused twenty salmon to spring up where one existed, if he who made two blades of grass grow where one flourished before were justly called a benefactor.[12]

Perhaps because of the apparent rebuff on his tariff article, Carnegie sent nothing more to the *Commercial* on political and economic matters, although we know from his later writings that these were among his principal interests. Always the pragmatist, and with little time for such correspondence, he would limit his effort to something which gave more promise of results.

His third and apparently final letter to the *Commercial* is dated from Amsterdam, November 15, although he was in Berlin before that time, and had planned to leave Holland around November 7.[13] It appeared in the *Commercial* on December 13, like its predecessor under the heading "Our Foreign Correspondence," taking up the same space, a full column. Like the earlier letter it was signed only with the initials, "A.C."

Once more Carnegie begins his letter by demonstrating that its contents are newsworthy:

Ask twenty Americans who have made the "Grand Tour," and a majority of them will confess that they omitted to visit Holland, while nineteen out of the twenty will never have seen the northern portion of it. It is even doubtful whether a Pittsburgher ever trod the dykes at the Helder unless James Park Jr. extended his travels thus far when "doing" Europe twenty years ago. And yet there is no country under the sun where a traveler meets with such strange sights as in the region indicated. Nor is there one which will more amply repay tourists for the time necessary for an excursion through it.

The article makes scarcely any mention of Amsterdam itself, and none at all of the city's foreign trade, cultural institutions, and botanical and zoological gardens—subjects which certainly must have drawn Carnegie's interest. Instead, he devotes the whole article to the dykes and reclamation of land from the sea. He launches his discourse in most interesting fashion: "To all other places the fates have at least vouchsafed dry land ready prepared for occupation, and all that was requisite was to locate and begin the cultivation of the soil in the preparation of which they had taken no part, but to the unfortunate Dutchman nature seems to have distributed no favors. The alluvium from the Rhine and other rivers flowing into the North Sea, mixed with such mud as that ocean might throw upon the shoals when agitated by storms, was all that he received."

The description of the dykes is more pictorial than the technical style that might be expected of Carnegie, He lays out in steps the work that was necessary to reclaim the land—dyke-building, then pumping out the water:

After the Hollander had completed the dykes, he was called upon to begin the second part of the work, which consisted in pumping by windmills, out of the inclosure made, the water which had accumulated from the last raids of the ocean, and for this purpose he put nine thousand windmills to work, many of them being of immense size, the arms frequently exceeding one hundred feet in length. The monotony of the scenery is greatly relieved by these mills, which abound everywhere. From the top of one in Rotterdam we counted fifty-two, all busily at work.

Evidently Carnegie had by this time discovered Cervantes, even if he had not perfected his own syntax: "Coming up the Maas by boat in a beautiful moonlit night, they seemed to us like a squadron of giants combatting imaginary foes, as they cast their weird-like arms about, and displayed so much activity while all around slumbered."

After a concise but comparatively complete discussion of the dangers—"At one time an inundation . . . drowned eighty thousand persons. . . . At another period it is stated that one hundred thousand persons were cut off. . . . These were ocean's victims.

The losses by internal floods have been equally severe."—
Carnegie uses the courage of the Dutch as a text in chiding
faint-hearted Americans:

It may tend to reassure some of our timid citizens at home, who
shudder at the expense of our national debt, to know that competent
authority has estimated the cost of the Hydraulic Works of the
Dutch, all of which are unnecessary in our favored land, at one and
a half thousand million dollars (gold). If to this can be added the
annual cost of maintenance, we have a total not far from equal
to our entire indebtedness, all sunk and settled for by three and a
half millions of Dutchmen. What cannot America do with its ever-
increasing population and its manifold resources?

Although not much given to efforts at humor in writing—
although fond of jokes in real life—Carnegie does become playful
in discussing the problem of building on unstable land, where
piles had to be sunk more than a hundred feet to support foundations for the houses. Then turning serious, he discusses the
courageous fight of the Dutch to maintain their independence,
showing some feeling in the account of how twice in history
they had opened the dykes to repel invaders. He adds, half
serious, half tongue-in-cheek: "No country has so secure a
defense as Holland. Other nations may continue to spend millions
on fortifications, and throw away, before they are used, the im-
plements which have cost so much, only to replace them with
others which the march of events, or the progress of military
science, shall in turn render useless. He aspires not to be iron-
clad, but why should he, when nature has made him water-
proof!"

The letters are not great literature, but they are interesting
as examples of Carnegie's early style and thought. It is regrettable
that his letter on the tariff question was wasted. It would be
interesting to compare it, for style as well as ideas, with his
later writings on the subject. Slight as they are, the letters
give some opportunity to observe changes in his style and thought
before the beginning of his truly serious writing, more than a
decade later. Except in depth of treatment, they do not greatly
differ from similar descriptive matter in his family letters. But
these, of course, were largely taken up with personal references

and discussions of business affairs. One passage from a letter written while still on board ship, outward bound, reveals his intent and suggests how far ahead his planning sometimes ran: "An ocean voyage is good fun, and I think we might do as the Duncans do (They are on board)—have a summer estate in Scotland and go over every summer as they do to enjoy it. How does the idea strike the two members of the Carnegie firm at Homewood? Will Pittencrief, or Pittrearie, answer? . . . It would suit me exactly."[14]

Although Carnegie made no more walking tours, there was seldom a summer in his life from that time on which he failed to spend in Scotland, England, or on the Continent. He sometimes took his mother along, although they had a summer home at Cresson, in the mountains of Pennsylvania. And following his marriage in 1887, he and his wife spent every summer in Scotland until the war clouds of 1914 made it impossible. In 1902 he did buy Pittencrief, making it a park for his home town.

III   A Bachelor in New York

Carnegie returned from his walking trip in the early spring of 1868, and threw himself into business affairs with renewed vigor. But as his world had expanded, Pittsburgh had become too small for either his interests or his desires. Late in the autumn of 1867[15] he opened an office in New York. His mother went with him, and they lived in the city's finest hotel, the St. Nicholas, leaving the house at Homewood to Tom, who had recently married the daughter of a Carnegie partner.

The move was a fortunate one. New York was not only the center of American business, but of culture as well. Here Carnegie was able to attend theaters, art galleries, opera, and other important events, and to engage the best of tutors for studies he felt he still needed. He studied literature, history, philosophy, and economics, but found languages difficult.

Money was still rolling in. Not only were his steel works and other investments successful, but he had come to be trusted by the leading European bankers, and made a handsome commission in selling over $30 million in bonds to them during the next five years.[16]

Both in the mountains and in the city, one of his favorite pastimes was riding horseback. In New York he was—as a young, well-spoken millionaire—one of the most eligible bachelors. His riding companions in Central Park were among the cream of society, and his mother often invited groups of them as guests in summer, providing rooms at the Mountain House, Cresson's best hotel.[17] He may have caused many hearts to beat fast, but there was no serious romance in his life until he was almost fifty.

IV *Proto "Gospel of Wealth"*

Not much more than a year after moving to New York to live, Carnegie sat down at year's end to take stock. The result is a singular memorandum not known until it was found among his papers more than half a century later, after his death, attached to a brief listing of "Income for 1868." Headed "dec[r] 68, St. Nicholas Hotel, New York," it reads:

Thirty three and an income of 50,000$ per annum.

By this time two years I can arrange all my business as to secure at least 50,000$ per annum—Beyond this never earn—make no effort to increase fortune but spend the surplus each year for benevolent purposes. Cast aside business forever, except for others.

Settle in Oxford & get a thorough education making the acquaintance of literary men. This will take three years active work—pay especial attention to speaking in public. Settle then in London & purchase a controlling interest in some newspaper or live review & give the general management of it attention, taking a part in public matters, especially those connected with education & improvement of the poorer classes—

Man must have an idol—the amassing of wealth is one of the worst species of idolatry—no idol more debasing than the worship of money—Whatever I engage in I must push inordinately, therefore should I be careful to choose that life which will be the most elevating in its character—To continue much longer overwhelmed by business cares and with most of my thoughts wholly upon the way to make money in the shortest time must degrade me beyond hope of permanent recovery.

I will resign business at thirty five, but during the ensuing two

years I wish to spend the afternoons in receiving instruction, and in
reading systematically.[18]

Much of that plan did work out, but not so soon nor so
easily as Carnegie had anticipated. The end of two years found
him still rich, but locked in a financial and industrial fight
to the death. Like the hunter who was sent up the tree to catch
a bear, he had to learn in the hard school of experience how to
turn loose of the bear of business. Thirty years later he thought
he had reached his objective, only to have the sale of his
interests break down. It was almost exactly thirty-two years
after the writing of the memorandum that Carnegie gained his
retirement objective.

For all that, he was but partly turned aside from his purpose.
He made the most of every opportunity for education and influence, until he became one of the best-known men of his era.
The brash young steel-maker and bond salesman seemed to
gravitate towards important and interesting people, and they
toward him. Just as he had quickly come to know and be known
by the principal men of Pittsburgh when he was a messenger
boy and then a railroader, he somehow found ways of getting
to know the people in New York—not just the rich, but the
intelligent and cultured. Unlike most Pittsburgh millionaires
and other new rich, Carnegie made no effort to court the favor
of Mrs. Astor or become one of the Four Hundred,[19] and had no
hesitation in expressing his disdain for the plutocrat.[20]

V *Pursuit of Culture*

His most fortunate contact after moving to the big city was
becoming acquainted with Courtlandt Palmer through a mutual
interest in technical—as opposed to classical—education.[21] Carnegie, who often commented on the importance of chance happenings, noted in his *Autobiography*[22] that for a while, until
new friendships were made, he and his mother were emotionally
and by mutual interest largely dependent on Pittsburgh.

"The literary life of New York was a sort of Dutch-English
family party as late as the beginning of the nineties,"[23] and
Palmer, the son of a rich merchant, was prominent in intellectual

circles. Shortly before Carnegie's arrival he had organized the Nineteenth Century Club, which met in his home in Gramercy Park until its growth required larger quarters. At first a discussion group, it later included the reading of assigned papers, in which Carnegie took part. There Carnegie met a stimulating assemblage of people of widely varying opinions, for although Palmer inclined to Positivism his guests included those of all attitudes.

At Palmer's club meetings and less formal gatherings, Carnegie could discuss economics, politics, religion, and literature with many of America's leading thinkers of the day. Among them he met such writers as Moncure D. Conway, George Washington Cable, and Thomas Wentworth Higginson; educators such as Thomas Davidson and Horace Porter; public figures like Abram S. Hewitt, son-in-law of Peter Cooper and later Mayor of New York; statesmen such as Theodore Roosevelt and journalists like John Swinton, chief of the *New York Times* editorial staff. Here were Jewish rabbis, Catholic priests and bishops, Unitarian clergymen, Protestant ministers, and freethinkers such as the great Robert G. Ingersoll. In such company he gained a much wider view of the world than had been available in Pittsburgh. Here he came into contact with the work of Comte, Darwin, and Herbert Spencer,[24] with the latter of whom he formed a warm and enduring friendship. Through acquaintance thus made, Carnegie also came in contact with many of the principal minds of Britain, among them John Morley, who influenced him and was in turn influenced.

One of the most important benefits of Carnegie's friendship with Palmer was being introduced in 1871 to Anne C. L. Botta, for many years the unquestioned queen bee of the New York intelligentsia. Born in 1815 as Anne Lynch, Mrs. Botta was in her mid-fifties at the time they met, already white-haired, always dressed in black, a poet, "author, sculptor, critic, and not least the charming woman of the world."[25] At the time of her death twenty years later Carnegie recalled how they met: "One of her chief characteristics was that of recognizing unknown men and women, and giving them opportunities to benefit, not only from her own stores of wisdom, and from her charming manners and conversation, but from the remarkable

class she drew around her.... The position of the Bottas
enabled them to bring together not only the best people of this
country, but to a degree greater than any, so far as I know,
the most distinguished visitors from abroad, beyond the ranks
of mere title or fashion."[26]

He had been brought to her home by Palmer, and years
afterward she told him that he had been invited to return
because "some words I had spoken the first night struck her
as a genuine note, though unusual."[27] Carnegie rightly considered her home the nearest "in the modern era" to a real
salon. "Millionaires and fashionables," he wrote, "are poor substitutes for a cultivated society. Madame Botta's lions could all
roar, more or less; they were not compelled to chatter, or be
dumb."[28]

As a young poet in her early thirties, Anne Lynch had been
praised by Edgar Allan Poe, who was a regular visitor at her
Saturday evening soirees, as were many of his friends. "All the
lights of contemporary letters brightened these gatherings,"
notes Hendrick. They included N. P. Willis, Lydia Sigourney,
Fitz-Greene Halleck, Daniel Webster, and Henry Clay on their
New York visits, and many others. Following Anne's marriage
to Vincenzo Botta, professor of Italian literature at New York
University, the salons were continued in their home on Murray
Hill. During his years of attendance at such functions Carnegie
could have met there William Cullen Bryant, Ralph Waldo
Emerson, George Ripley, Charles A. Dana, Bronson Alcott, John
Bigelow, Henry Ward Beecher, E. C. Stedman, Andrew D.
White, Richard Watson Gilder, Julia Ward Howe, Charles
Dudley Warner, and scores of other personages of the day,
including his early idol, Bayard Taylor. Many of them he cer-
tainly did meet, and with some, such as White and Gilder, he
formed close friendships.[29]

Distinguished foreigners were also among the visitors at the
Botta home, including Charles Kingsley, Justin McCarthy, James
A. Froude, and Matthew Arnold. In bringing Arnold to the
Botta home in 1883, Carnegie was repaying part of his debt.
He had met the poet in England at a party given by Mrs. Henry
Yates Thompson, who was one of the great hostesses of Britain,
and whose husband owned the *Pall Mall Gazette*, edited by

Morley. When Arnold came to America that October for a lecture tour of 100 appearances, he was Carnegie's guest. His brother, Edwin, also became a close friend of the industrialist.

Now a polished man of the world, Carnegie had no reason to kowtow to anyone, however important. Nor is there any evidence that even in his salad days he had ever met anyone except on a frank and equal basis. Not even his mentor, Herbert Spencer, found himself treated by Carnegie as anything other than an equal. An incident mentioned by both Carnegie and Spencer in their autobiographies is detailed by Hendrick:

At the end of a particular dinner the waiter proffered cheese. Spencer gazed at it unsympathetically, and angrily pushed it aside. "Cheddar! Cheddar!" he shouted in tones so loud that they carried over the dining room. "I said Cheddar, not Cheshire! Bring me Cheddar"—and his fist came down upon the table.

The three companions then began talking about their meetings with well known men. Did they appear as great in the flesh as in their books and public careers? Carnegie insisted that such meetings were usually disappointing; distinguished characters turned out to be so different from the previously formed conception. Mr. Spencer evidently sniffed a personal reference.

"In my case, for example," he asked, "was that so?"

"You more than anybody," replied Carnegie. "I had imagined you, the great philosopher brooding over all things. Never did I dream you could become so excited over the question of Cheddar and Cheshire cheese."[30]

The incident did nothing to interfere with their friendship, and Carnegie commented on it that "Spencer liked stories and was a good laugher."[31] Nor did the incident appear to lessen the philosopher's influence on Carnegie's thinking—at least for some time.

CHAPTER 3

Travel Writer: 1878-1884

DESPITE the fact that Bayard Taylor's *Views A-Foot* had been one of Carnegie's early inspirations,[1] the commercial publication of his own two books of travel came to the steelmaker almost as an accident. Reaction to the newspaper articles written during his ramble over Europe had apparently been minimal, except for the pats on the back that friends give anyone whose work unexpectedly appears in print. Carnegie and his friends largely gave up their original idea of a walking trip, using public transportation most of the time. Because of this, the chatty approach to his subject appeared only in a brief portion of the first printed letter. As a result, Carnegie stumbled into what was to become the more modern style in travel writing, although there is no evidence that he ever realized what he had done.

Both of Carnegie's travel books, in their original form, were written and published for private distribution to friends, partly to serve as a "thank you" for their loyal service in staying at work while he played.[2] Eventually this pattern of devotion would become such a part of his normal course of life that he no longer paid it any heed. But up to the spring of his forty-seventh year, the feeling of gratitude remained strong.[3]

The works which first brought Carnegie to public notice as a commercially published writer of books were—like the letters to the *Pittsburgh Commercial* in 1865—accounts of extended tours, one around the world in 1878, the other a coaching trip for almost the entire length of England and Scotland three years later. Internal evidence would indicate that *Round the World* was written (or largely written) as the trip progressed, while Carnegie's *Autobiography* specifically states that *Our Coaching Trip* resulted from brief notes, later amplified.[4] Both were privately printed soon after their composition, and apparently both by Charles Scribner's Sons.[5] A copy of the later volume having come to the notice of Charles Scribner himself, he suggested it might be of interest to the public. With considerable revision, it was published commercially in 1883, and met with enough success that the earlier work was also revised and brought out the following year by the same firm.

In writing the two books, Carnegie harked back for method to Taylor, whom he had probably come to know well during the decade and more since the transfer from Pittsburgh to New York. They moved in the same cultural and literary circles, and could hardly have failed to become friends.[6] The day-by-day diary style of *Views A-Foot* had been for a century the common way of writing a personal account of travel, and even as a boy, young Carnegie must often have noted it in his voluminous reading. He also followed Taylor in using the first person, although to a notably lesser degree, and with far less intrusion of his own tastes and preferences. The principal differences lie in their choice of materials. Taylor spends much more time in describing scenes, especially landscapes, while Carnegie's principal interests are in people and their doings. Taylor views the present scene for itself alone, while Carnegie continually digs into reasons and background, especially history, which Taylor almost completely ignores. Taylor interlards his pages with original poems, while Carnegie constantly inserts quotations—mostly poetry—from his favorite authors, including Shakespeare, Milton, Burns, Scott, and Sir Edwin Arnold, a close friend whose work he much admired. Taylor evidences little interest in economics, politics, and religion, deep concerns which Carnegie constantly introduces into his travel writing.

I Round the World

In its original form, *Round the World* is a purely personal account of a trip taken for pleasure and education. Carnegie wrote it as he went, day by day, with the evident intention from the beginning of publishing it immediately on his return for private distribution to his many friends, too numerous for individual correspondence. Reaching New York on June 24, 1879, he wrote the two final paragraphs at Cresson the following day.[7] The dedication to Tom Carnegie, dated "Braemar Cottage, Cresson, July, 1879," reads: "To my brother, and trusty associates, who toiled at home that I might spend abroad, these notes are affectionately inscribed by the grateful author." Young as he was—not yet forty-three—Carnegie already felt a parental relationship with his partners and employees, which apparently squared with their own feelings toward him.

Long devoted to newspapers, Carnegie had learned to copy their slam-bang style, and leaped head-first into his story:

> New York, Saturday, Oct. 12, 1878.
> Bang! click! the desk closes, the key turns, and good-by for a year[8] to my wards—that goodly cluster over which I have watched with parental solicitude for many a day; their several cribs, full of records and labelled Union Iron Mills, Lucy Furnaces . . . but for the present I bid them farewell; I'm off for a holiday, and the rise and fall of iron and steel "affecteth me not."

He immediately recalls his former long vacation, the walking trip through Europe, and how at its end he and two companions had pledged themselves to take another which would extend all around the globe: "Years ago, Vandy, Harry, and I, standing in the crater of Mount Vesuvius . . . resolved that some day, instead of turning back as we had then to do, we would make a tour round the Ball."[9]

For the "long weary hours" of the Pacific voyage he turned to literature. His mother had sent him a set of Shakespeare "in thirteen handy volumes," and that would do. But immediately he must explain to Robert Burns that the Scot is not being neglected: "No, no Robin, no need of taking you in my trunk; I have you in my heart from 'Tam O'Shanter' to the 'Daisy.'"

Five of the next six sections (each headed by the date) open in the explosive newspaper style of the day: "What is this? a telegram!" "All aboard for 'Frisco!'" "Desolation! In the great desert!" "A palace, truly!" "At last!" But from this point the technique is used only at the approach to land, and very rarely in the remainder of the book. For the most part, after the first dozen pages the sections open smoothly—most often with indications of progress in the trip, but about a third of the time with discussions of places or events.

Perhaps partly in relation to his time for writing, Carnegie alloted his space on a fairly even basis of time spent, with certain exceptions. The number of pages (229) is nearly the same as the number of days (243). But the cutting down sharply on the coverage of the approximately 100 days spent on long rail trips or sea voyages, he was able to give from two to two-and-one-half pages a day to the time spent in visiting Oriental lands. In making these reductions, he allowed less than one page for each day which elapsed from his leaving the office to landing in Japan. To the six weeks from his leaving Alexandria to arrival at Cresson, he gave only twenty-two pages. This was apparently due not to failure in keeping his diary, but rather to concern for keeping the book's length within bounds. In his *Autobiography* Carnegie quotes two "illustrations from our 'Round the World' trip," one of about 100, the other of 150 words. The first is from the time spent at Singapore, and the second details a visit to a camp of Laplanders "on the way to the North Cape" of Norway, within the Arctic Circle.[10] Singularly, neither appears in the published versions of the book.

There is a noteworthy progression in outlook as Carnegie proceeds with the journey and the writing. His style is always intensely personal (although avoiding the use of "I" and the intrusion of individual tastes so common in Taylor's *Views A-Foot*), but the personal view becomes increasingly public in its concerns. Where he reports arguments with persons he meets in his travels, Carnegie usually offers a fair assessment from his point of view (a characteristic of all his writing), even when the discussion doesn't go his way. For example, he presents a colloquy with a Parsee over the disposal of the dead:

Parsees cannot burn the dead, because fire should not be prostituted to so vile a use. They cannot bury, because the earth should not be desecrated with the dead, neither should the sea; and therefore God has provided vultures, which cannot be defiled, to absorb the flesh of the dead. I said to him that the mere thought of violence offered to our dead caused us to shudder. "Then what do you think of worms?" he asked. This was certainly an effective estoppel. "It comes to this," he continued, "a question of birds or worms." "You are right," (I had to admit it) I said; "after all, it's not worth disputing about."[11]

Carnegie very seldom attempts humor in any of his writing. One of the evidences that he did not contemplate commercial publication when he first wrote is his occasionally becoming playful in the original version of *Round the World*. Most of these instances occur near the beginning, largely when he is on a train or the ship, with little else to write about, it would seem. He tries, not too successfully, with a jest on passing the International Date Line: "Gone, November 5th, 1878, a *dies non*, which never was born. Lost, strayed, or stolen—a rare diadem, composed of twenty-four precious gems—some diamond bright, some rubies rare, some jet as black as night. It was to have been displayed at midnight, but when looked for it was nowhere to be found."[12] He comes out somewhat better soon afterwards following a passage in which he grows enthusiastic about the beauties of a night at sea. It ends, "One does feel in such moments, when beauty and sublimity are overpoweringly displayed, that there are worlds and life beyond our ken." The next day's entry opens with: "I know I went to bed sometime early this morning, but after reading last night's effusion in the cold, sober light of day, it strikes me I must have been rather enthusiastic."[13]

In his predictions, which are few, Carnegie comes out right about half the time in *Round the World*. For instance, he appears to have been correct that "America will continue to lead in fast horses,"[14] and in asserting that "by and by" Japan would adopt a representative government (49). He didn't do so well in predicting that America would not soon be exporting live cattle (5); that it would within a short time and for the foreseeable future have no federal taxes but on luxuries (9); or that there would not be more than a few hundred miles of railway in Japan "for centuries" (70).

In general, Carnegie's travel writing, except for promoting his ideas, is straightforward and factual. But there are moments

of emotion, humor, or—as Winkler describes them—"pure ec-
stasy."[15] In his description of the burning ghat, for instance,
Carnegie writes: "My heart bled for a poor widow whose husband had just been taken to the pile. She was of very low
caste, but her grief was heartrending; not loud, but I thought
I could taste the saltness of her tears, they seemed so bitter"
(210). On the other hand, he can cap a thrilling tale with a
bit of his inward prejudice that makes it almost ridiculous:
perhaps—like the knocking on the gates in *Macbeth*—to relieve
tension. After telling the story of Jessie, the heroine of Lucknow,
he adds: "I have been hesitating whether the next paragraph
in my notebook should go down here or be omitted. Probably
it would be in better taste if quietly ignored, but then it would
be so finely natural to put in. Well, I shall be natural or nothing,
and recount that I could not help rejoicing that Jessie was
Scotch, and that Scotchmen first broke the rebels' lines and
reached the fort, and that the bagpipes led the way. That's all.
I feel better now that this is also set down" (229). With that
he passes on to a discussion of Lucknow's lack of fine buildings.

Perhaps the most highly praised passage in all his writings
was created after he had reluctantly gone to see the Taj Mahal
by moonlight, fearing he would be disappointed. He opens the
next diary entry:

We have seen it, but I am without the slightest desire to burst
into rapturous adjectives. Do not expect me to attempt a description
of it, or to try to express my feelings. There are some subjects too
sacred for analysis, or even for words, and I now know that there
is a human structure so exquisitely fine, or unearthly, as to lift it into
this holy domain. Let me say little about it; only tell you that,
lingering until the sun went down, we turned in the noble gateway
which forms a frame through which you see the Taj in the distance,
with only the blue sky in the background, around and above it, and
there took our last fond sad farewell, as the shades of night were
wrapping the lovely jewel in their embrace, as if it were a charge too
sweetly precious not to be safely enveloped in night's black mantle,
till it could again shine forth at the dawn in all its beauty to adorn
the earth. . . . But till the day I die, amid mountain streams or moonlight strolls in the forest wherever and whenever the mood comes,
when all that is most sacred, most elevated, and most pure recur

to shed their radiance upon the tranquil mind; there will be found
among my treasures the memory of that lovely charm—the Taj.
(252–53)

One very important thing the trip around the world did for
Carnegie, which was to have a lasting effect on his life: it gave
him an entirely new attitude toward religion. "A new horizon
was opened up to me by this voyage," he wrote later in his
Autobiography. "It quite changed my intellectual outlook. . . .
The result of my journey was to bring a certain mental peace.
Where there had been chaos there was now order. . . . All the
remnants of theology in which I had been born and bred . . .
now ceased to influence my thoughts. I found that no nation
had all the truth in the revelation it regards as divine, and
no tribe is so low as to be left without some truth."[16]

II   Our Coaching Trip

Carnegie's second book of travel, *Our Coaching Trip,* was,
like *Round the World,* written for the amusement of his close
friends, and discloses the fact even more thoroughly than the
earlier work, by playfulness and numerous personal references.
Its dedication "to my brother and trusty associates" is almost
the same as that of its predecessor. It is a happy, simple, and
straightforward account of a trip from "Brighton to Inverness,"[17]
driving four-in-hand in a coach with ten companions, including
his mother. The trip, measuring eight hundred thirty-one miles,
took forty-eight days elapsed time. But this figure includes sixteen on which no traveling was done, leaving thirty-two days for
coaching—an average of just under twenty-six miles a day, with
time for regular luncheon stops and much sight-seeing.[18] Included in the party were "Lady Dowager, Mother, Head of the
Clan (no Salic Law in our family); Miss Jeannie Johns (Prima
Donna); Miss Alice French (Stewardess); Mr. and Mrs. Mc-
Cargo (Dainty Davie); Mr. and Mrs. King (Paisley Trouba-
dours, Aleck good for fun and Angie good for everything);
Benjamin F. Vandevort (Benjie); Henry Phipps, Jr. (H. P., our
Pard); G. F. McCandless (General Manager); ten in all,
making together with the scribe the All-coaching Eleven."[19]

McCargo was one of the "Scotties" of his Pittsburgh boyhood, and later his superintendent of telegraphy in railroad days. King and Phipps were associates in steelmaking, as was Gardiner McCandless. The two girls were friends (Miss Johns a singer and Miss French in charge of luncheons) and Vandevort was the younger brother of John, who had gone with Carnegie on the tour of Europe and the trip around the world. Carnegie had also invited Louise Whitfield, later to become his bride. She wished to accept, but was prevented by the frown of Margaret Carnegie, who recognized her as the first real threat to the mother's complete domination of her son.[20]

Our Coaching Trip is almost exactly the same length as Carnegie's earlier book, not counting a fifty-three-page quotation from the Dunfermline papers concerning his stop there near the end of July for his mother to lay the cornerstone of a free library he was giving to the city.[21] It displays almost all the typical Carnegie themes, including—besides those of *Round the World*—admiration for Great Britain, maternal devotion, ascendency of Anglo-Saxon blood, interest in economics, unflinching faith in democracy, hatred of imperialism and speculation, and love of peace.

The book opens in almost exactly the same way as its predecessor: "Bang! click! once more the desk closes and the key turns! Not 'Round the World' again, but 'Ho for England, for England!' is the cry, and 'Scotland's hills and Scotland's dales and Scotland's vales for me.'" (2). Again, as if feeling the need for a precedent, Carnegie cites the time in 1865 when he, his cousin "Dod" (George Lauder, Jr.), Henry Phipps, and John Vandevort had "walked through Southern England with knapsacks on our backs" (1). Andrew had then announced that some day "when my 'ships come home,' I should drive a party of my dearest friends from Brighton to Inverness." Somewhat defensively he points out that he made this statement before the appearance of *Adventures of a Phaeton*, by William Black, whom he would take on a later coaching trip.

Immediately, however, Carnegie turns to a long philosophical digression on the realization of "air-castles"—a charming though somewhat overextended passage of almost 2,000 words which passes from whimsey into allegory. It begins with—and was undoubtedly inspired by—the absence of John Vandevort, who on the walking trip had exclaimed how if he ever achieved an income of $1,500 a year he would give up work forever: " 'Catch me working any more like a slave, as you and Harry do!' Well, well, Vandy's air-castle was fifteen hundred dollars a year, yet see him now when thousands roll in upon him every month. Hard at it still—and see the goddess Fortune laughing in her sleeve at the good joke on Vandy. He has his air-castle, but doesn't recognize the structure" (2–3).

It is uncertain, and speculation on the point would be futile, whether Carnegie had definite friends in mind as he goes on to speak of *Miss Fashion*, the speculator, and the society woman, although the definite details in the first case might indicate she was a real person, the others less probably so (3–8). At the end of the passage, as with his description of a night at sea in *Round the World*,[22] he brings the reader back to the story by tweaking his own nose: "I am as bad as Sterne in his 'Sentimental Journey,' and will never get on at this rate" (8).

The start from Brighton was made on June 17 after a brief visit to London. Instead of eleven passengers as planned, the group totaled fifteen, in addition to Perry, the coachman, and Joe, the footman. The Kings were to meet the coach later, after taking their children to visit grandparents at Paisley, and their places were temporarily taken by Carnegie's cousin, Maggie Lauder, and Emma Franks of Liverpool, sister of his companion on the 1865 trip through Europe. After coaxing the host a bit, the latter was added as a permanent member of the party. In addition, a Londoner, Theodore Beck and his daughter, and Mr. and Mrs. Thomas Graham from Wolverhampton began the journey with the "Charioteers." Beck left the coach at Windsor, the second day, but was replaced by his son for five days. The Becks stopped at Banbury, and the Grahams at Wolverhampton, where the visitors were entertained in homes instead of at an inn. The Kings had rejoined the party at Banbury, and at Wolverhampton the McCargos, Miss Johns and young Vandevort left for two weeks for a trip to Paris, to return at Carlisle. There were other occasional brief changes in the party (25, 27, 107–108, 149–50).

Carnegie expanded the first day's journey to about 7,500

words with a rather complete account of the arrangements, interlarded so neatly with literary quotations and jokes and incidents that it is as readable as the rest. Instead of—or in addition to—advance reading for the trip, he had armed himself with guidebooks on some of the towns (54, 90).

The narrative runs well, with description, incident, serious discussion, and humor (which is very rare in Carnegie's other writing) well proportioned and distributed. There are some choice tidbits, such as one Carnegie inserts in his discussion of Coventry, home of Godiva and George Eliot:

A friend told me that a lady friend of hers, who was staying at the hotel in Florence where George Eliot was, made her acquaintance casually without knowing her name. Something, she knew not what, attracted her to her, and after a few days she began sending flowers to the strange woman. Completely fascinated, she went almost daily for hours to sit with her. This continued for many days, the lady using the utmost freedom, and not without feeling that the attention was pleasing to the queer, plain, and unpretending Englishwoman. One day she discovered by chance who her companion really was. Never before, as she said, had she felt such mortification. She went timidly to George Eliot's room and took her hand in hers, but shrank back unable to speak, while the tears rolled down her cheeks. "What is wrong?" was asked, and then the explanation came, "I didn't know who you were. I never suspected *it was you*." Then came George Eliot's turn to be embarrassed. "You did not know that I was George Eliot, but you were drawn to plain me all for my own self, a woman? I am so happy." She kissed the American lady tenderly, and the true friendship thus formed knew no end, but ripened to the close. (89)

A few pages later appears one of the characteristic expressions of the philosophy which was more and more beginning to shape Carnegie into the man the world remembers:

In this world we must learn not to lay up our treasures, but to enjoy them day by day as we travel the path we never return to. If we fail in this we shall find when we do come to the days of leisure that we have lost the taste for and the capacity to enjoy them. There are so many unfortunates cursed with plenty to retire upon, but with nothing to retire to! Sound wisdom that the schoolboy displayed who did not "believe in putting away for tomorrow the cake he could eat today." It might not be fresh on the morrow, or the cat might steal it. The cat steals many a choice bit from Americans intended for the morrow. Among the saddest of all spectacles to me is that of an elderly man occupying his last years grasping for more dollars. (95)

The party entered Scotland July 16, and from this point onward there are frequent references to the various Scottish heroes and their doings, to replace the previous comment on the currently great men of England. At one halt, the seventy-two-year-old Margaret Morrison Carnegie and Mrs. King had waded barefoot in the brook, and "kilted their petticoats and danced a highland reel on the greensward, in sight of the company, but at some distance from us" (122). Surprisingly little is said of Mrs. Carnegie after they crossed the border. Perhaps she was overcome by emotion. But there Carnegie records his unforgettable sentiment: "It's a God's mercy I was born a Scotchman, for I do not see how I could ever have been contented to be anything else" (152). For the moment his adulation of America was silenced. But not for long.

No other work of Carnegie—not even the *Autobiography*—gives such a complete look at the man himself as *Our Coaching Trip*, nor such a glimpse at his broad literary reading. During its first fifty pages he refers to or inserts quotations from Henry Scott Riddell, Shakespeare, Burns, Wordsworth, Coleridge, William Black, Elizabeth Barrett Browning, Thomas Carlyle, William Cullen Bryant, the Bible, John G. Saxe, Lawrence Sterne, Pope, Homer, Aesop, Thomas Campbell, William Dean Howells, Thomas H. Huxley, Oliver Goldsmith, Tennyson, Izaak Walton, "Josh Billings" (Henry W. Shaw), Walter Scott, Matthew and Edwin Arnold, William Winter, Artemus Ward, John S. Kennedy, W. Robertson Smith, Robert Ingersoll, William Clark, Milton, *Harper's Magazine*, *Fortnightly Review*, *Nineteenth Century*, and the *Encyclopaedia Britannica*. The fact that a good many of the quotations show slight variation from the original might indicate that they were written from memory, without checking.

Unlike *Round the World, Our Coaching Trip* was not written during the trip, although in accordance with the day's style in travel books it is set down and arranged in diary style. Carnegie described its production:

All the notes I made of the coaching trip were a few lines a day in twopenny pass-books bought before we started. As with "Round the World", I thought that I might some day write a magazine article, or give some account of my excursion for those who accompanied me: but one wintry day I decided that it was scarcely worth while to go down to the New York office, three miles distant, and the question was how I should occupy the spare time. I thought of the coaching trip, and decided to write a few lines just to see how I should get on. The narrative flowed freely, and before the day was over I had written between three and four thousand words. I took up the pleasing task every stormy day when it was unnecessary for me to visit the office, and in exactly twenty sittings I had finished a book.[23]

III   An American Four-in-Hand in Britain

In some respects, Carnegie's adaptations of *Round the World* and *Our Coaching Trip* from private publications for friends to trade books are as revealing and of as great interest as the original works themselves. He tells how it all began: "I handed the notes [manuscript of *Our Coaching Trip*] to Scribners' people and asked them to print a few hundred copies for private circulation. The volume pleased my friends, as 'Round the World' had done. Mr. [John D.] Champlin one day told me that Mr. [Charles] Scribner had read the book and would very much like to publish it for general circulation upon his own account, subject to a royalty. The vain author is easily persuaded that what he has done is meritorious, and I consented."[24] Apparently he noted later, as it is bracketed in the published *Autobiography*: "Every year this still nets me a small sum in royalties. And thirty years have gone by, 1912."[25]

But the change from a private to a public book was not so light a matter as Carnegie would make it sound, or perhaps as he recalled it many years afterward. A comparison of *Our Coaching Trip* with the trade version, retitled *An American*

Four-in-Hand in Britain, shows extensive changes. Somewhat oversimplifying, he explains in the preface: "The original intent of the book must be the excuse for the highly personal nature of the narrative, which could scarcely be changed without an entire remodelling, a task for which the writer had neither time nor inclination; so with the exception of a few suppressions and some additions which seemed necessary under its new conditions, its character has not been materially altered."[26]

Careful collation of *Our Coaching Trip* with *An American Four-in-Hand* reveals scores of changes, but a majority of these are relatively minor; initials instead of names; variation in punctuation; change of tense; switch in word order, or other such details. Ignoring these, there remain sixty-seven principal differences, including fifty-eight insertions of more than a few words, three changes (two to correct minor errors, the other on the ground of taste), and six omissions. All the omissions are of personal material, varying from a few lines to the fifty-three-page quotation of stories on the stop at Dunfermline, referred to above.

Insertions are principally of two kinds: twenty-two are of what might be referred to as "guidebook material"—not taken directly from such sources, but descriptive and historical matter of the kind usually found in guidebooks; sixteen are inserted to give the author a chance to present his ideas on religion, economics, political science, and similar matters; eleven are casual material (mostly from a few lines to a paragraph) on weather, travel conditions and contrasts between Britain and America; four are transitional, to smooth over omitted or changed passages; three are humorous incidents and two poetical quotations.

Many of the "guidebook" and "idea" inserts are of considerable length, so that the text (originally about 66,000 words without the Dunfermline quotations) is increased by about a third, to some 87,000 words. In general the changes are wisely made, and tend to improve the style and interest of the book, and its appeal to the public. The first edition, 2,000 copies, was immediately sold, and within the next few years it was reprinted eighteen times, sales reaching almost 15,000—a considerable figure for that day.[27]

IV Round the World (*Trade Book*)

In revising *Our Coaching Trip* for the public, Carnegie—never one to underestimate his own prospects—may have expected a public edition of *Round the World* to follow, since his first omission was the slap-dash opening which was almost identical with that of his earlier work.[28] He retained the opening in his revision of the Scribner's edition of *Round the World,* which proved considerably more extensive than that of its mate.

Disregarding minor variations, this revision totaled 123 instances, including eighty-eight additions, twenty-six changes, and nine omissions. Six of the latter were deletions of personal material, two were to avoid repetition, and one to facilitate a transition.

Twelve of the changes were to correct errors, ten were to improve style, one to soften a comment on religion and one to update the previous text. The other two are puzzling, and similar: the first, in his apostrophe to Burns, substitutes *A Man's a Man for a' That* and *My Nannie's Awa'* for *Tam O'Shanter* and *To a Daisy;*[29] the other, in commenting on differing tastes in music, puts *Lohengrin* in place of *My Nannie's Awa'.*[30] The additions include thirty-one of "guidebook material," twenty-two on politics and economics, fifteen on religion, nine for miscellaneous comments, four each for humor and to update earlier references, and three for clarity. Even the choice of guidebook material is significant, including principally data suited to illustrate or prepare the way for his principles and ideas. With this revision, Carnegie's writing had come of age, using books to promote his ideas, as he had formerly used letters to newspapers.

Carnegie evidently planned the expanded version with care, increasing the space for each part of the trip almost always in proportion to the number of pages in the first edition. Thus, the transcontinental journey goes from eleven pages to thirteen, the voyage from seventeen to twenty-two, Japan from nineteen to twenty-five, the coasting trip from eleven to fifteen, China from forty-nine to seventy-one, Malaysia from eleven to seventeen, the Ceylon trip from eight to twenty, India from sixty-three to 100, the Suez journey from six to nineteen, Egypt from ten to twenty, Italy from six to fourteen, and the conclusion is enlarged only two pages, from fifteen to seventeen. Since it would have been simple and easy to expand the volume by restoring the Scandinavian trip and other parts omitted from the early printing, it is evident that Carnegie's intent was to promote his ideas and beliefs, not merely to increase the number of pages.

The expanded *Round the World* was published in 1884 in New York and London.[31] As a trade book, it did not prove so successful as *An American Four-in-Hand.* It was reprinted eight times, at least until 1902, sales totaling about 5,000 copies.[32]

While it cannot be said that Carnegie added a new dimension to travel writing—such books had been used for many years to promote ideas and policies—he did push this phase farther than most of his contemporaries, and more successfully in the sense of using ideas without detracting from the readability of the work.

CHAPTER 4

Millionaire Journalist: 1882-1916

ANDREW Carnegie's career as a journalist, like some of his other successful ventures, apparently had its origin in a chance event. Yet it is also probable that he would have eventually come to magazine articles as an outgrowth of his lifelong habit of writing letters to newspapers. Certainly one who could say "Dynamite is a child's toy compared to the press," and "The weapon of Republicanism is not the sword, but the pen,"[1] could hardly have ignored the rising magazine field for long.

His delay until he was past forty-five was partly due to the fact that the modern type of magazine, for which Carnegie eventually wrote so many of his articles, is a recent development, dating back little more than a century. Those few which were founded before the close of the Civil War were usually short lived and of minor influence. Before 1870 most of the ones which did prove successful were devoted principally to belles lettres, a field for which he was poorly prepared, and sometimes biography and travel, in which his interests were limited.

From childhood days Carnegie had been an avid reader of newspapers, and after reaching manhood a constant writer of letters to their editors. In 1868 his dream of retirement from business had included, "purchase a controlling interest in some newspaper or live review and give the general management of it attention, taking a part in public matters...."[2]

At the height of his interest in British politics, from 1881 to 1885, Carnegie joined Samuel Storey and Passmore Edwards in setting up a syndicate of radical papers to support William E. Gladstone and fight to abolish the monarchy and House of Lords. But he did little if any writing for the seven daily and ten weekly journals in which he and his partners held controlling interest. At last, in the backlash which followed the passage of the Reform Bill of 1884, he saw the futility of his effort and disposed of all but a few of his holdings in the field, dropping any serious involvement, though not all his equity was liquidated until 1902. In later years he referred to the venture as part of sowing his wild oats.[3]

It was soon after the beginning of this period of newspaper publishing that Carnegie entered what was to prove his most fruitful field of literature—contributing to the magazines.

I First Essay at Journalism

As with so many of his activities, the opportunity came by chance—almost by accident. In his years of becoming a lion in London, Carnegie had formed a lifelong friendship with John Morley, editor of *Fortnightly Review*, for many years one of Britain's most prestigious liberal journals. The coaching trip from Brighton to Inverness was creating attention, and someone suggested it might serve for an article. It was natural that Carnegie would offer the finished piece to his friend, even if Morley was not in the company when the suggestion was made. Carnegie, as usual seeking to justify his entering a new field, began the article by reporting the matter: " 'Why don't you give us, in one of our reviews, some account of your coaching trip . . . and tell us what your dozen of American guests thought of us? I'm sure it would be interesting.' Upon this hint I write." The question may possibly have been asked and the article arranged with the idea that the account would be a recital of his guests' reactions to the scenery, hospitality, and occasional incidents of the journey. Carnegie was beginning his drive to reform Britain and overthrow the monarchy. And as was his custom, he took the suggestion in the way he wanted it, if not in the sense in which it had been made. But he was careful to hedge by pointing out a ground for his interpretation: "The speaker was a noted politician—one busy with affairs of state— and, therefore, in this article I shall confine myself to the impression which political questions made upon the minds of my Republican friends."[4]

He opens softly, with a charming and complimentary paragraph on the trip—perhaps designed as a peace offering, but more likely (for Carnegie seldom wrote without weighing his words and their effect)—to provide maximum shock when his criticisms began: "Indeed, it would be impossible in one article to do more than consider one of the many interesting subjects the journey suggests; nor could any article tell how delightful, beyond all anticipation, our excursion through your quietly beautiful island proved to be, while the happiness, the joyousness of the party from beginning to end is not to be described by words. Suffice it to say that the experiment has left us all unable to think of any mode of spending our coming summers which is not tame and insipid in comparison with coaching through Britain."[5]

Stage set, Carnegie immediately plunges into his criticism of the political situation in Britain:

1. Whereas in the new land, fundamental principles of government were all settled, in "this so-called old and settled land there was nothing settled whatever, and the people were in a ferment, satisfied with nothing." (156–58)

2. Americans were surprised by discussion of the Land question, and to learn of entail and primogeniture, and that a man renting land is not allowed to clear it of game, if he desires.

3. Members of the party were shocked at seeing an established church clergyman in prison at Lancaster "because he conscientiously thought it wrong, in his ministry, to wear something or not to wear it. . . ." (158–59)

4. They were surprised to learn that even the House of Commons was the subject of severe criticism because of unfair and unequal election laws. (159–60)

5. They were surprised at problems with the courts, and that English legal institutions represented, not "the people" but "His Majesty." (160–61)

6. The party was much disturbed over the Irish question, felt that it should be settled in the way that of Scotland had been, by allowing home rule within the United Kingdom. (161–62)

7. Then he took up the question of tariff and free trade, arguing it at some length. (162–64)

8. And at last, the question of the monarchy and hereditary privilege. (164–65)

Sketching out the uncertainty of political matters in England, Carnegie describes how, during the Land Bill excitement which was going on at the time of the trip, many people were calling for the abolition of the House of Lords. He continues:

This seemed to [my friend] revolutionary. Imagine a proposition in America to abolish the Senate. While total abolition would be deprecated by the company, still the more moderate opinion would seem to be that a radical change in the constitution of the House of Lords was bound to take place ere long. And the subject would be dismissed with the remark that "if the Lords set themselves up against the opinion of the country," or, as it was gently put by one speaker, did not "behave themselves (*i.e.* register the decrees of the Commons), they would be swept away." . . . Of course, the American . . . had heard that an Englishman "dearly loved a lord," and now he doubts it. "Why don't they elect their Second Chamber somewhat as we do, and then it would have some real power, as springing from the people like the Commons?" . . . That was what one of my friends wanted to know, but I could not very well answer his question. (157–58)

Carnegie seldom resorted to irony or ridicule. But he is hitting pretty hard, and in regard to the jailed clergyman, he writes—

I did what I could to explain to my indignant friends how heinous Mr. Greene's offense had been, inasmuch as he had made a "bargain" with the State to worship God as the State directed, but the word "bargain" only created more disgust, and my friends left the prison saying, "And this is England! Shame!" This incident was not easily effaced from the minds of the Americans, and Church and State presented a frequent topic of conversation. (159)

Only on the final point did Carnegie release the pressure:

I have purposely left the only remaining political edifice till the last. The *Throne itself*. Surely here is something so high as to float in serene calm above the storms. . . . Such, however, was not the impression received by my American friends. (164)

He goes on to point out that while there was considerable

agitation against the whole system of hereditary privilege, and even the monarchy, the great majority of the people who had expressed an opinion were loyal to Queen Victoria herself. "This thought pleased my American friends, and allowed them to claim that she was "just the same as if she had been elected," and therefore "a good-enough Republican" (164).

Then came the whirlwind finish:

In conclusion, if the constitution of the Second Chamber is in danger, if even the House of Commons is on the eve of decided changes, if the tenure of the very soil of the realm is unsatisfactory, if the system of law is to be recast, if the sacred Church itself is a bone of contention . . . and if the throne itself be dependent upon the personal character of one man; in short, if England is not pleased with any of her political institutions, was it any wonder that the sympathies of my American friends were deeply touched by the sad spectacle of a dissatisfied, divided, wrangling people, irritated by the pressure of old forms. . . . (165)

The solution for Carnegie, of course, is to change Britain to a republican system of government. With this hope for Britain, "a happy issue out of all her troubles," the Americans had sailed for America with "feelings warmed and quickened into fond affection for the old home." While Carnegie had not yet solved the problem of how to win friends and influence people in such situations, he was evidently seeking.

II *Early Rebuff*

Having tasted success, Carnegie lost no time in trying again. Back in Britain to manage his newspaper syndicate and push his liberal ideas, he fired off another article to the *Fortnightly*.[6] But Morley had resigned to go into politics almost immediately after the publication of "As Others See Us," being succeeded by T. H. S. Escott. The article was apparently rejected, since it was never published. Perhaps Carnegie's hurt was so great that he never tried the magazine again. At least nothing else by him ever appeared in its pages.

Two more years passed before he succeeded in getting into print again in a leading journal. This time it was *Macmillan's*

Magazine. The story, "The Oil and Gas Wells of Pennsylvania," was completely noncontroversial—largely an account of his participation in the early oil strikes at the Storey Farm, and of the discovery of gas around Pittsburgh some fifteen years later. He described the great Haymaker gas well at Murrysville, and other strikes forming almost a circle around the city.

Despite the work of completing *Triumphant Democracy,* (1886) Carnegie succeeded in turning out a story on a related theme for the first issue of the *North American Review* in 1886, and two more on the labor question for the newly founded *Forum* in April and August, before suffering his worst illness and the loss of his mother and brother that autumn. Recovery and his subsequent marriage so took up his time that only an interview in the *New York Times* and a published speech appear to have come to wide public notice. But in 1889 Carnegie attained his full stride again, with three articles in *North American Review,* a report on the Pan American Conference, to which he was a delegate, and a published address on *Pennsylvania's Industries and Railroad Policy,* given March 18 before the Franklin Institute.[7]

III   *Sudden Popularity*

The second of these articles Carnegie titled simply "Wealth," and with it his journalistic labors hit their high-water mark. Reprinted in England, it was given the caption, "The Gospel of Wealth," and quickly became famous, catapulting its author into worldwide prominence. Reprinted with other essays and one speech as a book under the new title in 1900, it became his best known and most successful work.[8]

No longer did Carnegie have to seek outlets for his opinions. They were besought on every hand, and from that time forth he contributed a stream of articles to leading (and occasionally to little known) periodicals in America and Britain. Some were reprinted in later books or editions or in pamphlet form. His output was not large—never more than five in one year—but it was steady. During the twenty-three years from 1889 to 1911 inclusive, he published fifty-seven magazine articles, not counting interviews, letters to editors, or published speeches. In 1887 Carnegie's income had been $1,800,000,[9] so obviously any remuneration received from his writing was relatively unimportant. Yet there is nothing to indicate that he was reluctant to accept the money, which to him was a part of the assurance that his writing was considered of value, not merely used in order to capitalize on his name. In his *Autobiography*[10] he wrote with evident satisfaction of the royalties received from books. And long after retiring as one of the world's richest men to give away his fortune, he indicated he was still being paid for his literary work. Writing to his cousin, Dod, who had taken up painting, he commented: "Yes, you are at last an artist. As an artist I too am in demand. My painting is word painting and I'm all 'ordered' and 'sold'. Folk maun do something for their bread."[11]

Carnegie's total magazine record, counted as above, shows sixty-eight articles, forty-six in American and twenty-two in British publications.[12] Twenty-nine of these dealt in one way or another with political matters, thirty-three with economics. Prior to 1890 his interests appear to be chiefly in finance, with only two out of eight articles concerned with politics. During the troubled years from 1890 through 1905 there was a sudden upsurge of writing on British, American, and international political affairs, totaling twenty, with only fifteen on economics. During the final eleven years of his writing Carnegie had only eight on politics, and eleven on economics.

Of his total, twenty-five—a little more than one-third—were later printed in book form in his lifetime, thirteen in *The Gospel of Wealth*, eleven in *The Empire of Business*, and one in *Problems of Today*. All but six of these articles were on economics. A number of others on both disciplines were circulated in pamphlet form.

IV   *Four Related Articles*

Four articles on closely related matters provide an interesting insight into the manner in which Carnegie clung to his convictions and pursued his ends over a period of more than two decades. These are: "Do Americans Hate England?" (*North American Review*, June, 1890); "Does America Hate England?" (*Contemporary Review*, November, 1897); "Britain and Her

Offspring" (*Nineteenth Century,* May, 1911) and "Arbitration" (*Contemporary Review,* August, 1911.)

The first of these formed the second part of a symposium by seven principal Americans[13] in reply to "The Hatred of England," a rather intemperate blast in the May, 1890, issue of *North American Review,* by Goldwin Smith. Carnegie went into the question rather more at length than most of the others, beginning logically with a distinction between the attitudes of adult Americans and schoolboys fired up by history lessons about the Revolution, and between hatred and dislike. He then took up the question of rivalry, pointing out that the masses of British people liked America, but that the privileged classes resented its danger to their position. He dismissed as mere equality what Smith had declared an affront—that James G. Blaine dropped the traditional "America ventures to hope . . ." in favor of the Briton's curt, "Her Majesty expects that. . . ." In conclusion he cited one of his traditional themes, asserting that despite rivalries and other differences, England and America are father and son, both feeling keenly that "blood is thicker than water."

Seven years later the same question was being discussed, and through a British journal Carnegie approached it in much the same fashion as before; first the matter of competition, then switching immediately to the matter of *race.* Pointing out that the United States had recently challenged Britain's refusal to arbitrate a territorial dispute with Venezuela, he quoted the statement of a member of the Senate: "This is our own race . . .; of course, we have difficulty of our own with her, and we do not intend to let even our Motherland light the torch of war upon our continent . . . but this is a little family matter between ourselves. It does not mean that German, Russian, Frenchmen or any foreigners may combine to attack our race to its destruction, without counting us in. No, sir-ee."

He then took up the two principal bones of contention—the Venezuelan question and the killing of seals—on their merits, asserting that America was in the right in both instances, concluding with an assurance of friendliness, based on common race.

The third article, in another American periodical, without directly raising the former question, pointed out how Canada, Australia, and other British dominions were patterning their institutions more upon the United States than the British system, predicting that "From this time forth the dear old Mother and her children are to draw closer together ... until our entire English-speaking race enjoy the blessings flowing from government founded upon the equality of the citizen, one man's privilege every man's right." At the end of the essay, adding a section which he dated as the time of reading the proof, he burst into a paean of happiness at the announcement that the two nations had agreed to proclaim that all international disputes should be submitted to arbitration. He concluded with the mother-wife theme that would be so common in his biographies:[14] "Should the writer be spared to see his native land and adopted lands—Motherland and Wifeland—united hand in hand ... life will possess for him a charm unknown before ... almost leading him to murmur with bowed head, 'Now let thy servant depart in peace.'"[15]

Carnegie opened the fourth article—evidently inspired by the same event as the peroration of the third, directly with his pet theme of the importance of race: "History confirms the claim that the abolition of war between civilized nations by arbitration of all disputes is emphatically the mission of the English-speaking race." Briefly citing previous difficulties between Britain and America, Carnegie urged the frequent practice of what now bears the name of *summitry*, noting, "That we only hate those we do not know is good as a general rule." Then, referring to the exception in most previous arbitration treaties for "questions affecting 'honour' or vital 'interest,'" he argued: "The first subject reserved *i.e.*, that of Honour, is the most dishonoured word in the language. No country ever did, or ever could dishonour another. No man ever did, or ever could, dishonour another. It is impossible. All Honour's wounds are self-inflicted."

Urging the enormous cost of war—"Ruinous though this may be, it is as dust in the balance compared with its sin"—he went into a discussion of the decline of slavery and the duel. He cited the experience of his own Hero Fund to prove that doing away with war would not end heroism. Then, characteristically, he related an argument with Gladstone over the first

Irish Home rule Bill, quoted and paraphrased Burns, and concluded: "Just as it was with slavery and with duelling, when our race banishes war within its wide boundaries, as it is on the eve of doing, it sounds its death knell. Long may my native and adopted lands (Motherland and Wifeland) hand in hand, lead the world to its upward and onward march to higher and higher stages of civilisation [sic], tending to make earth a heaven, which is the mission of our race. Let us falter not!"

Had he been a younger man, and had the onset of World War I been delayed a few years, there is little doubt that Carnegie would have expanded this series into a book on the attainment of peace. It could have included his brief article in *Outlook*[16] proposing a league of nations; *A League of Peace*, his Rectorial Address at St. Andrews University;[17] "The Cry of 'Wolf' ";[18] "Peace Versus War: The President's Solution";[19] and "The Decadence of Militarism."[20]

V *Creative Efforts*

Carnegie seldom attempted anything but straight, formal prose. The few instances in which he tried verse were all brief and on the ludicrous side, except for a revision of the old *Ballad of Sir Patrick Spens*, for his little daughter, to give it a happy ending.[21] The result indicates that if he had tried poesy seriously, he might have done reasonably well. He is also said to have amused his daughter for hours at a time with antic tales, most of which he invented.[22]

One most interesting departure from his usual style is "Britain's Appeal to the Gods,"[23] a friendly but not-too-gentle satire on England's imperialism and high tariff, while seeking to get lower rates from America and other nations. At the head of the article the magazine editor published an extract of the author's letter explaining his purpose in writing on the subject, but saying nothing as to why he used his novel approach and style: "My aim has been to show your countrymen how absurdly grasping they are, how *unreasonable*. Never has the world seen such a nation, and there is much excuse for the feeling that Britain is entitled to continue to inherit the earth. She still wants more, when what surprises everyone conversant with her position is how she ever succeeded in getting and doing so much. I am impressed every time I look into the figures."[24]

The whole article is cast as an appeal by Britain, taking up, point by point and addressed to the various appropriate Olympians, the instances and statistics of the nation's greedy actions and attitudes. For each she is rebuked and answered by *Chorus*—obviously the author—until near the end, when Britain cries out, evidently in despair, to know the future. This time the answer is reassuring. But when more definite information is sought, no answer is given, and Carnegie—*in propria persona*—closes with a message of hope and confidence.

The article opens:

Hear us! hear us! mighty Jove, and ye dread gods who dwell upon Olympus.

Mark ye, our Foreign Commerce is only 903,363,000£. per year.

Chorus.—Ungrateful favourite of the gods! It never was so great either in Imports or Exports. No nation ever approached it in amount. *Per capita* it is 21£. 10s. France has only 8£. 11s. 9d.; Germany 8£. 6s. 8d.; United States, 6£. 3s.

Neptune, great god of the Sea, and thou, Triton the Trumpeter!

Mark ye, Guardians of Britannia's rule over the waves, our Shipping is only 16,600,000 tons.

Chorus.—Insatiate greed! It never was so great and is constantly increasing. All the other nations combined have not as much. Beware lest thine ingratitude offend the gods. . . . Thou wert first; now others build ships and must share with thee.

There follow appeals to "Midas, great king of Gold," for greater national wealth; Vulcan for an increase in iron, steel and coal production; Deucalion, god of Increase, for greater population; Minerva, goddess of the loom, for a greater share of the spinning industry; "Jupiter, and all the gods together," asking why rivals increase faster (the answer ending with, "Cease to be children crying for the moon"); Ceres for greater food production; and again to Jupiter, seeking to know destiny, and asking to be preserved from becoming "a fifth-rate power." The answer: "No such destiny for thy race impends. . . . Be of

good cheer and of stout heart. Let this suffice; trust the gods.
Farewell!"

The piece ends in a double conclusion:

Stay! Stay! Let us know more! How? When? What shall we do?
Muta, goddess of silence, floated above. No audible response
came, but the babbling air seemed to spread abroad in whispering
sounds—"Seek to know no more: how all is to be wrought lieth upon
the lap of the gods; to the one mortal who has presumed to forecast
their plans we waft this message: 'Thy lips are sealed.' "   .

So keeps the mortal his own sweet reveries, happy in the knowledge
that for his native land all is well, since all is to be better than yet has
been, which is saying much, and for his race—the English-speaking
race (language makes race)—its future is far to surpass its past.
To it the gods have decreed the leadership of the world for the
good of the world. The day of its power is not afar. There be many
who read these lines who shall behold its dawn.[25]

Considering the friendly acceptance by the editor, and
Carnegie's success in the novel presentation of his theme, it
is surprising that he seems never to have tried another such
radical departure from his usual journalistic style.

VI   *Final Contributions*

Rarely even in his final decade did Carnegie bother to write
on anything but serious subjects. His principal concerns during
this period were peace, labor, the tariff, international affairs,
and hereditary transmission of property. But on one occasion he
unbent enough to write "Doctor Golf" for the *Independent*.[26]
Like several of his other essays during those years, he used
his beloved simplified spelling, as in: "We are under the sky,
worshipers of the 'God of the Open Air.' Every breth seems to
drive away weakness and diseas, securing for us longer terms
of happy days here on earth, even bringing something of heven
here to us." It is a charming bit, opening with some history on
the introduction of golf into America (from Scotland, and by
Dunfermline Scots, of course). Then he recites how it was
originally a game for the wealthy, and talks of its pleasures,
introducing some amusing incidents, emphasizing that golf en-

courages friendly rivalries and requires concentration and is good exercise.

Carnegie's last publication, three years before his death, returns to the one great purpose of his life, and is titled "Principles of Giving."[27] It differs little from what he had written and said on the same subject in his early retirement years.

VII *Summary*

Looking over Carnegie's production as a journalist, we become aware that it shows only minor changes in the more than three decades of his activity in this field, or even in a much longer period of his life. Even his language and style vary little from things he wrote in his youth. In "Wealth" in 1889 he once uses "bequested" for "bequeathed" as he had in his letter about library use in 1853. In his "Gospel of Wealth—II," in 1906 he opens with a newspaper-style lead as he had his *Round the World* of a quarter century earlier.

Carnegie always wrote of things that had caught or would catch the public eye, and on which he sought to influence his readers.

If we seek a reason for this lack of change, perhaps it is to be found in the time-sequence involved. By the time Carnegie entered the field of magazine writing he had reached maturity, his ideas and habits largely formed. What few changes we find are easily traceable to changes in his own ideas and in the world around him. Anyone arguing against the industrialist's sole responsibility for the ideas and language of the books, articles and other works bearing his name must be hard-put to find anything which could rebut the complete homogeneity of his writings.

CHAPTER 5

Political Pragmatist: 1881-1911

ANDREW Carnegie plunged into political writing as he plunged into everything else—headlong. His first magazine article,[1] instead of being the interested or amused view of British ways that Morley may well have expected, was a slashing attack on the political and social system of the Mother Country. Carnegie was ever a strenuously political man. The basis of his opinions and attitudes may be readily seen in his antecedents and early life.

His family, especially the Morrisons on his mother's side, were nonconformists in religion and Chartist in politics. His home town, Dunfermline, although the ancient seat of Scottish kings, had been one of the principal centers of Chartism. And although the movement's near-revolution was coming to an end at the time the Carnegies removed to America, his formative years had been filled with news and talk of its planning, campaigning, marches, strikes, and even threats of outright revolt.

The Chartist movement, except for the brief insurrection of the "physical force men" at Newport in 1839, was not excessively radical by today's standards, though far out for Britain of its day. The name "Chartist" came from a six-point program, the "People's Charter," which was formulated by the London Workingmen's Association late in 1837. Its aims were: 1. equal electoral areas; 2. universal suffrage; 3. payment of members of Parliament; 4. abolition of property qualifications for voting; 5. voting by ballot; and 6. annual parliaments. Most of these were eventually adopted, several of them by the Reform Bill whose passage was secured by Gladstone nearly half a century afterward, at least partly through Carnegie's agitation during his newspaper period.

66

In America Carnegie found what to him—individualist that he was—proved the key that opened all doors: the right of every man to full citizenship and the ballot. Even before he was old enough to vote he was bombarding newspaper editors with letters on the abolition of slavery, and soon afterward, opposing secession.

With the move to New York in 1868 and entry into the city's intellectual life, Carnegie discovered evolution and came under the sway of Herbert Spencer, whom he never fully understood and sometimes misquoted. But although he made it his motto that, "All is well, since all grows better," and proudly announced his belief in socialism, it is extremely doubtful whether he at any time really agreed with its tenets. Rather, as Wall ably demonstrates,[2] Carnegie developed a philosophy of his own. It was made up of his early religious and political training, rugged individualism, desire for mastery and achievement, greed, generosity, and a conviction that the world—and especially those close to him—needed his ideas and guidance. No small element was his struggle of conscience over having indulged in what in 1868 he had alluded to as the "worship of the golden calf."

By 1880 his political credo might be described as a benevolent capitalistic individualism in a republican setting. Barring special privilege and inherited wealth, opportunity was within reach of everyone, and affluence could be attained by those with the energy and intelligence to seek it. Poverty was useful in developing manhood ("Adversity makes men, prosperity monsters") but severe suffering from the same source should be alleviated by legislation and both public and private charity.

I Triumphant Democracy

Carnegie's great statement of this belief came in 1886 in *Triumphant Democracy, or Fifty Years' March of the Republic.*[3] It was the best planned and most widely read of his works, except for the magazine article on "Wealth," which appeared in the *North American Review* three years later and was reprinted all over Europe in many languages.[4]

In his *Autobiography*, Carnegie relates the genesis of the work:

My third literary venture, "Triumphant Democracy," had its origin in realizing how little the best-informed foreigner, or even Briton, knew of America, and how distorted that little was. It was prodigious what these eminent Englishmen did not then know about the Republic. My first talk with Mr. Gladstone in 1882 can never be forgotten. When I had occasion to say that the majority of the English-speaking race was now republican, and it was now a minority of monarchists who were upon the defensive, he said:

"Why, how is that?"

"Well, Mr. Gladstone," I said, "the Republic holds sway over a larger number of English-speaking people than the population of Great Britain and all her colonies, even if the English-speaking colonies were numbered twice over."

"Ah! how is that? What is your population?"

"Sixty-six millions, and yours is not much more than half."

"Ah, yes, surprising!"

With regard to the wealth of the nations, it was equally surprising for him to learn that the census of 1880 proved the hundred-year-old Republic could purchase Great Britain and Ireland and all their realized capital and investments and then pay off Britain's debt, and yet not exhaust her fortune. But the most startling statement of all was that which I was able to make when the question of Free Trade was touched upon. I pointed out that America was now the greatest manufacturing nation in the world. . . . I quoted Mulhall's[5] figures: British manufactures in 1880, eight hundred and sixteen millions, sterling; American manufactures eleven hundred and twenty-six millions, sterling. His one word was:

"Incredible!"

Other startling statements followed, and he asked:

"Why does not some writer take up this subject and present the facts in a simple and direct form to the world?"

I was then, as a matter of fact, gathering material for "Triumphant Democracy," in which I intended to perform the very service which he indicated, as I informed him.[6]

Perhaps it was somewhat later that a second reason for the book occurred to Carnegie. In his preface, after setting out "the lamentable ignorance concerning the new land which I have found even in the highest political circles of the old," he adds:

I believed, also, that my attempt would give Americans a better idea of the great work their country has done and is doing in the world. . . . During its progress I have been deeply interested in it, and it may truly be regarded as a labor of love—the tribute of a dutiful and grateful son of an adopted country which has removed the stigma of inferiority which his native land saw proper to impress upon him at birth, and has made him, in the estimation of its great laws as well as in his own estimation (much the more important consideration), the peer of any human being who draws the breath of life, be he pope, kaiser, priest or king . . . a citizen.

II *Theme of the Work*

Having disposed of such preliminaries (probably as an after-thought following the completion of the book) Carnegie opens with an appropriate quotation from Milton, and then roars on—still in his early style learned from the newspapers: "The old nations of the earth creep on at a snail's pace; the Republic thunders past with the rush of an express. The United States, the growth of a single century, has already reached the fore-most rank among nations, and is destined to outdistance all others in the race. In population, in wealth, in annual savings, and in public credit; in freedom from debt, in agriculture, and in manufactures, America already leads the civilized world."

The first chapter of the book is titled "The Republic," and in it Carnegie does not delay in supporting his theory that democracy is one of the important reasons for the nation's startling growth. He does not, as he has sometimes been accused of doing, aver that the political system has been the sole reason for such rapid growth. He sets up the three "most important factors in this problem." The first, in keeping with his constant emphasis on the vital nature of heredity, is "the ethnic character of the people";[7] next, the environment, the "topographical and climatic conditions"; and third, "the influence of political institutions founded upon the equality of the citizen."

"The Republic" had faced two great dangers, human slavery, and "the millions of foreigners who came from all lands to the hospitable shores of the nation, many of them ignorant of the English language, and all unaccustomed to the exercise of

political duties." The first had been voided by emancipation and
giving full citizenship rights to the former slaves. The second
had been changed into a benefit by:

> The generosity . . . with which the Republic has dealt with these
> [immigrant] people. They are won to her side by being offered for
> their *subject*ship the boon of citizenship. For denial of equal privileges
> at home, the new land meets them with perfect equality, saying,
> be not only with us, but be of us. They reach the shores of the
> Republic *subjects* (insulting word), and she makes them citizens;
> serfs and she makes them men, and their children she takes gently
> by the hand and leads to the public schools which she has founded
> for her own children, and gives them, without money and without
> price, a good primary education as the most precious gift which
> she has, even in her bountiful hand, to bestow upon human beings.
> This is Democracy's "gift of welcome" to the new comer. The poor
> immigrant can not help growing up passionately fond of his new
> home . . . and thus the threatened danger is averted—the homoge-
> neity of the people secured.[8]

This, in Carnegie's view, was the key to America's greatness
and rapid growth. Other nations of similar blood and equally
helpful environment might fail of achievement because of
internal strife, ignorance, and lack of individual opportunity.
"The Republic" was going forward as the Social Darwinian
credo envisioned it should. All was well, and all growing better.

III *Structure of the Book*

Having set up his thesis, Carnegie proceeds to elaborate it
to some extent, but principally in regard to its results. He
evidently considers his point proved by his logic, but even
more certainly by his statistics. Perhaps in the back of his mind
were the religious teachings of his youth,[9] such as "By their
fruits ye shall know them,"[10] or "Show me your faith without
your works, and I will show you my faith by my works."[11]

The main body of *Triumphant Democracy* includes four un-
designated sections: The American people and their lives,
Chapters II–VIII; the nation's material progress, Chapters IX–
XIII; its cultural advancement, Chapters XIV–XV; and its gov-

ernment, Chapters XVI–XIX. The final chapter serves as a conclusion.

Opening the section on the American people, Carnegie expands and elaborates on the subject as introduced in the first chapter, using numerous statistics. The white American (he conveniently omits the black element until late in the chapter) is still, as of 1880, four-fifths British, the remainder largely German and French. This admixture has proved a helpful one, providing more imagination and softening the grim harshness of "those island mastiffs": "Toleration in the Briton is truly admirable; the leading Radical and the leading Tory-Democrat are found dining together.... Well, the American is even more tolerant.... Once in four years he warms up and takes sides, opposing hosts confront each other and a stranger would naturally think that only violence could result whichever side won. The morning after the election his arm is upon his opponent's shoulder and they are chaffing each other" (29–30).

Far from being opposed to immigration, Carnegie considers it one of the most helpful factors in expanding the country. It is usually the best who are dissatisfied in old lands. The new arrivals are a minority; with all that have come, America's population is still seven-eighths native-born. And he shrewdly predicts that "At these rates of advance the 'Wild West' is rapidly becoming a thing of the past, and in a few years it will be a thickly settled land" (39). Even the apprehensions that idleness and trouble-making would develop among the slaves freed by the Civil War have proved groundless, he finds.

Turning to the growth of cities and towns, Carnegie keeps the reader's head swimming with facts and figures which show how municipalities have mushroomed almost overnight, without perceptible diminution of the agrarian population. But America has no reason to fear such growth, since it is the result of economic laws. Here his devotion to Spencer breaks out: "Oh, these grand, immutable, all-wise laws of natural forces, how perfectly they work if human legislators would only let them alone!" (48). And from time to time Carnegie reverts to favorite themes of his more casual writing, with quotations from classical and popular authors, references to chance (after recounting the story of the man who could have bought the site of Chicago for a pair of boots, but didn't have them: "How many chances in life do we miss just for the want of the boots. Moral: Get the boots" [53]), and love of the fatherland. He contrasts the slow early growth of Boston—by implication because it was then under monarchy—with newer areas, and concludes with a peroration he might have delivered from the platform: "When the people reign in the old home as they do in the new, the two nations will be one people, and the bonds which unite them the world combined shall not break asunder. They clasp hands. Democracy cries to democracy, 'We stand for the rights of man, the day of kings and peers is past.' ... No peal so grand as that, save one, that which proclaims the substitution of peaceful arbitration for war the world over.... Democracy goes marching on" (72–73).

Carnegie devotes thirty-five pages to describing the improvement of living conditions, especially in small towns and rural areas during the previous half century, contrasting them with such areas in Britain, and closely relating the benefits to universal suffrage. The chapter on "Occupations" is a survey to provide background, with little definite relation to the thesis of the book.

Despite his faith in education, Carnegie's chapter on that subject is largely a matter of statistics, with few comparisons to education in Britain until almost the end. At that point he lists leading colleges and universities endowed by private funds, noting the rarity of such a practice in Britain. His enthusiasm breaks out in a Biblical paraphrase: "The moral to be drawn from America by every nation is this: "Seek ye first the education of the people and all other political blessings will be added unto you" (150).

The chapter on "Religion" is largely devoted to statistics showing that America, with complete freedom, had a larger proportion of active church members than Britain, and that most denominations were better supported than even the established churches of the mother country. Concluding with "Pauperism and Crime," Carnegie notes that much of this was among those newly arrived in America and unable to find and hold jobs, and cites the close relationship between crime and poverty: "America exhibits not only the least poverty, but the best system of alleviating it" (171). Things are improving. "The next generation, or the next beyond, will probably read with horror of our inflicting the death punishment upon human beings." Many states have already abolished the death penalty. "Judged by this standard, the Democracy stands the test well" (178–79).

IV *Material Progress*

Five chapters, totaling about 130 pages, describe the progress of America in agriculture, manufacturing, mining (including oil), commerce, and transportation. Except for an undercurrent of boasting that in most of these achievements the new land far outstrips the Old World, the chapters contain little of note except statistics, interestingly presented. "Trade and Commerce" includes several pages of discussion on tariff, upholding the position of the United States, particularly against those of France and England. Each chapter ends with a few inspirational words on peace or democracy, the paragraph closing the chapter on "Railways and Waterways" serving as peroration for the whole section: "We have not travelled far yet, with all our progress upon the upward path, but we still go marching on. That which is is better than that which has been. It is the mission of Democracy to lead in this triumphant march and improve step by step the conditions under which the masses live; to ring out the Old, and to ring in the New; and in this great work the Republic rightly leads the van" (315).

V *Cultural Factors*

Carnegie's two chapters on cultural progress are notable not so much for their content as that they were included in the work at all. Few leading industrialists of the day in America would have even thought of the subject. For the most part, industrial heads were still men who had grown up in the mills, and whose schooling in many instances was not significantly better than the few years which the "white haired Scotch devil" had been able to secure before going to work at thirteen.

The fact that Carnegie gives less than fifty pages to the two chapters might indicate that he knew his side was still badly out-matched in such a contest. America was just beginning to come into its own in almost every field of cultural endeavor. What showing could he make in a contest on poetry with only Poe, Bryant, Bret Harte, and a few others to match against such giants as Browning, Scott, Byron, Wordsworth, Keats, Shelley, and a host of others Britain had produced in the same period?

Ever the canny fighter, Carnegie completely ignores such an unequal contest, and launches a flank attack. First he quotes some writers who had denigrated American art and music of the period, particularly singling out "a writer of about the same time, in the London *Quarterly Review*" who had stated that "a high and refined genius for art is indigenous to monarchies, and under such a form of government alone can it flourish, either vigorously or securely. The United States of North America can never expect to possess a fine school of art, so long as they retain their present system" (318–19).

"Art indigenous to monarchies!" Carnegie thunders. "Did anyone ever hear such an absurdity? The great law is that each shall produce fruit after its kind, but this genius makes a monarchy produce the greatest of all republics, the republic of art." He supports the argument:

Who knows or cares who Michelangelo's father was; or what was Beethoven's birth, or whether Raphael was an aristocrat, or Wagner the son of a poor actuary of police? Just imagine monarchy in art— a hereditary painter, for instance, or a sculptor who only was his father's son, or a musician, because born in the profession! . . .

This curious writer, who would have monarchy allied with art, built his theory upon the exploded idea that only monarchs and the aristocracy, which flutters around courts, could or would patronize the beautiful. That theory is unfortunate, in view of the fact that the best patrons of art are the Americans. (318–21)

He goes on to cite the growing number of art groups and museums in the United States, and the constant flow of works of art to this country. The American "is recognized now as one of the shrewdest, as well as one of the most liberal buyers." At an auction "he is no mean competitor, for he carries a pocket full of dollars, and is not afraid to spend them where he is sure of getting his money's worth" (325).

As to architecture and music, he offers similar arguments, particularly citing the great numbers of theaters and opera houses in the United States, even in small towns. From a trade guide he picks out, among others, "Centralia: ... Population one thousand five hundred," with a theater and an opera house, and Oshkosh, "away out in Wisconsin, two hundred miles from Chicago, with a population of twenty-two thousand," which showed three theaters, one the "New Opera House. Stage, forty-two by seventy feet; seats one thousand one hundred" (335).

Turning to literature, Carnegie stresses the astonishing growth of newspapers, splendid progress in the magazine field—citing *Harper's Magazine* and *Century* in particular, and in the juvenile field *St. Nicholas* and *Harper's Young People*—and the heavy sale of encyclopedias, as well as the large and increasing number of public libraries. "Triumphant Democracy is triumphant in nothing more than in this, [he concludes] that her members are readers and buyers of books and reading matter beyond the members of any government of a class, but in this particular each system is only seen to be true to its nature. The monarchist boasts more bayonets, the republican more books" (363).

VI *Government*

Finally offering an explanation of the triumphant democracy about which he has been boasting, Carnegie opens "The Federal Constellation" with the key quotation from the *Declaration of Independence*: "We hold these truths to be self-evident that all men are created free and (*sic*) equal, and are endowed by their Creator with certain inalienable rights, among which (*sic*) are life, liberty, and the pursuit of happiness" (365).[12] "Round this doctrine of the Declaration of Independence as its central sun," he continues, "the constellation of states revolves. The equality of the citizen is decreed by the fundamental law. All acts, all institutions, are based upon this idea. There is not one shred of privilege, hence no classes.... The President has not a shred of privilege which is not the birthright of every other citizen. The people are not levelled down, but levelled up to the full dignity of equal citizenship beyond which no man

can go." This, he avers, instead of creating a "dead level of
uniformity," gives leeway for the operation of natural laws.
Primogeniture and entail are unknown, and "There are but three
generations in America from shirt sleeves to shirt sleeves"
(364–66).[13]

He goes on to characterize the rights of states as "home rule,"
and describes principal features of the federal government,
beginning with the Supreme Court, and continuing with the
Legislative Department (*sic*), principally the Senate. After
listing among the powers of the latter those which relate to
making war and concluding treaties, he warns:

My American readers may not be aware of the fact that, while in
Britain an act of Parliament is necessary before works for a supply
of water or a mile of railway can be constructed, six or seven men can
plunge the nation into war, or, what is perhaps equally disastrous,
commit it to entangling alliances wthout consulting Parliament at
all. This is the most pernicious effect flowing from the monarchical
theory, for these men do this in "the king's name," who is in theory
still a real monarch, although in reality only a convenient puppet,
to be used by the cabinet at pleasure to suit their own ends.
(380–81)

He concludes the chapter with a discussion of the presidency
and praise of the Constitution.

"The Government's Non-political Work," except for its first
and final nine pages, is not written by Carnegie. He had planned,
he explains, to visit Washington and write an account of the
various bureaus, but could not spare the time. "The happy
thought occurred to me to send my secretary, Mr. [James H.]
Bridge, to perform the task, with a request to write up the
subject and see what he could make of it. He has done so well
that I cannot do better than incorporate his account" (414–15).

There follows a twenty-two-page account in a style evidently
attempting to approximate Carnegie's, but completely lacking
his frequent use of "I." Following this Carnegie continues with
some state and municipal functions, a description of the Sani-
tary Commission's work during the Civil War, and a page or
so on the Centennial of 1876. "We can confidently claim for
the Democracy that it produces a people self-reliant beyond

all others. ... The monarchical form lacks the vigor and elasticity to cope with the republican in any department of government whatever" (445).

The chapter on "The National Balance Sheet" offers a comparison of American and British budgets, deprecating the high cost of monarchy and hereditary privilege.

In the final chapter, "General Reflections," Carnegie points out that American stability is due to two factors: no one desires any change in fundamental laws, and regular elections provide a convenient way to get rid of undesirable office holders. To a considerable extent this chapter parallels the article "As Others See Us," closing with a hope for a reunion when Britain becomes a democracy.

VII *Characteristics*

In style, language, and opinions, *Triumphant Democracy* is close to the spirit and attitudes of Carnegie's articles, speeches, and letters of the same period—one of transition from Social Darwinism to a somewhat paternal capitalism. It is much more involved with statistics than any of his other works, as would appear necessary from the approach he adopted. For much the same reason it introduces fewer personal experiences, and literary references are less frequent, although appearing, when used, in his typically careless style, given from memory, often incorrect and almost never identified.

His use of humor is about the same as in his other works, not frequent, but sometimes with good effect. A fair example is "When the fair young American asked the latest lordling who did her country the honor to visit it, how the aristocratic leisure classes spent their time, he replied: 'Oh, they go from one house to another, don't you know, and enjoy themselves, you know. They never do any work, you know.' 'Oh,' she replied, 'we have such people too—tramps'" (130).

Although Carnegie occasionally admits some of the difficulties and darker spots of America, the work does in general warrant the criticism given it at the time of publication that it was all sweetness and light. Nor can this all be justified by his reply to George William Curtis's question, "Where are the shadows?" His answer was: "The book was written at high noon, when the sun casts no shadows."[14]

Triumphant Democracy is gaudy, from phraseology to Carnegie's personally designed red cover with quotations from Gladstone and Salisbury, an erect gold pyramid for "Republic," and a reversed one for "Monarchy," and on the spine a crown, upside down. But he wrote in a gaudy era. In his Preface, after telling of the labor of gathering facts, he adds that the question came to him: "Shall these dry bones live?" So he "tried to coat the wholesome medicine of facts in the sweetest and purest sugar of fancy at my command." And, as Wall rightly observes: "Curiously enough, the 'dry bones' do live, for the book is filled with anecdotes, editorial comment, and, above all, Carnegie's own ebullient personality"[15]

VIII *Who Wrote* Triumphant Democracy?

Some years ago the authenticity of Carnegie's writing—particularly *Triumphant Democracy*—came under attack, principally by Fritz Redlich in a review of R. G. McCloskey's *American Conservatism in the Age of Enterprise: A Study of William Graham Sumner, Stephen J. Field and Andrew Carnegie.*[16] His criticism reads, in part:

The question is easily formulated: can the researcher working on businessmen's minds legitimately base his investigations on their published writings? The question must be answered in the negative unless the student concerned can prove the businessman in question really wrote what was published under his name. Or to put it differently, when a wealthy and powerful American business leader "writes" a book or pamphlet, the assumption is that he hired a ghost writer.

In the case of Carnegie we know (at least for the period in which *Triumphant Democracy* was written) who the ghost writer was ... James Howard Bridge had been Herbert Spencer's secretary from 1879 to 1884. In the latter year he came to the United States, where he became Carnegie's "literary assistant,"[17] resigning from that position in 1889. How much he contributed to the *Forum* essays of 1886 and to *Triumphant Democracy*, and in turn how great was Carnegie's share therein is not known to the reviewer, nor does he know who Carnegie's later "literary assistants," were. . . . In this case the line runs from Spencer to Bridge, then to Bridge plus Carnegie. . . .

Bridge . . . may have had a more than fifty per cent share in the book, and a considerable influence on Carnegie's thinking, the latter thereby absorbing Spencerism.

This statement ignores important factors that completely vitiate Redlich's argument. 1. It was not customary in 1886 for leading industrialists to write books, or to employ ghost writers. Carnegie may well be called the first important writing industrialist. 2. Bridge was Carnegie's secretary, not an editorial assistant. Internal evidence would indicate that the "almost indispensable aid" for which Carnegie credits his "clever secretary, Mr. Bridge" (vii) was largely in gathering and arranging material. 3. The only part of *Triumphant Democracy* which Bridge wrote—and for this he received full credit—was part of one chapter, Chapter XVIII. 4. The line certainly did not run "from Spencer to Bridge, then to Bridge plus Carnegie." From 1870 (when Bridge was but fourteen years of age) Carnegie was an announced Spencerian, and by 1882 he and Spencer were close friends. 5. From his youth Carnegie had been an avid writer. His writing habits (on a pad with a stub pencil) are well known from outside evidence, as are his statements of the pleasure that he enjoyed in such work. And in the winter of 1881–82 he had written in a book for his friends and later retained when it was revised for general circulation: "If any man wants *bona fide* substantial power and influence in this world, he must handle the pen. That's flat. Truly it is a nobler weapon than the sword, and a much nobler one than the tongue, both of which have nearly had their day."[18] 6. It is utterly unbelievable that a man so personally opinionated as Carnegie, and so jealous of reputation, would have permitted his name to be used with something written by an employee, with which he might not agree. 7. The style and opinions of *Triumphant Democracy* are quite in agreement with those of his other books, articles, speeches, and letters.[19] The style of Bridge's part chapter is readily distinguishable, despite an obvious effort to agree with Carnegie's own. In fact, the Carnegian touches in this section may well have been inserted by the book's author, rather than the other way around.

IX *The Revised Edition*

Triumphant Democracy was a highly successful book in the United States, where it went through four printings and sold more than 30,000 copies. In Britain it did almost equally well, besides having a cheap edition in paperback which sold an additional 40,000. It was translated into several languages, and sold all over Europe.

To say that it created a sensation is an understatement. But not all the reaction was approving. The *Saturday Review*, for instance, denounced both the writing and the logic, and called Carnegie a snob for saying that being inferior to peers and monarch was degrading to a man.[20] And a grand jury at Wolverhampton found that a reference to the Royal family constituted a "scandalous, abominable and treasonous libel," and recommended prosecuting Carnegie and taking his book off the shelves of the Free Library.[21]

Delighted rather than discouraged by the furor, Carnegie completely revised the book following the 1890 census, with the new subtitle "Sixty Years' March of the Republic."[22] He opened the Preface with an account of Curtis's criticism[23] and his reply, adding, "Of course, everything in the Republic is not perfect. But neither is everything perfect in any land, or even in the sun. We are continually reminded that even that glorious luminare has its spots. . . . This book is not intended either to describe or dilate upon the spots upon our national sun. . . . The scope of this book is to show what we have to be thankful for, and not what we have to lament, as compared with other aggregations of the human race elsewhere."

Significantly, he omits the credit to Bridge, with whom he had quarreled in 1889, and to J. D. Champlin, Jr., of Scribner's. He expressed appreciation to four others, including a Columbia University professor and the superintendent and assistant superintendent of the census. Besides updating census data to the 1890 figures, Carnegie adapted or rewrote every chapter, including Bridge's work, which he used without giving credit.

Perhaps he was being crass and bitter over their quarrel. But Carnegie was seldom a man to hold any grudge long, and he may have withheld the name because of the task of explaining the changes. Considerable parts were omitted, including all Carnegie's original ten pages of the chapter; and he added a little more than enough to keep the length about equal to Bridge's original contribution.

The plan of the book was largely retained, but two of the sections were expanded, and the order of the second and third was reversed, apparently to give cultural progress precedence over material change. Carnegie also switched the order of the first two governmental chapters, bringing foreign affairs more to the forefront. And in place of the final observations he added two chapters—"The Record of the Decade—1880–1890," and "A Look Ahead."

Although every chapter is updated, trimmed, recast, and in most cases improved, few of the changes are radical. In the first section Carnegie has added as Chapter VI, "Wages and the Cost of Living," closely related to an article he contributed about the same time to *Contemporary Review* (September 1894) on "The Cost of Living in Britain Compared with the United States," and reprinted later in his *The Empire of Business*. The two chapters on cultural matters are expanded and made into four by separating literature, painting and sculpture, music and architecture, but without essentially altering the approach or arguments. In some cases Carnegie's more startling attacks on the British monarchy have been somewhat toned down. Except perhaps in this respect, the changes do not appear to reflect any variations in the author's thinking.

"The Record of the Decade" is a rather lively account of American progress between 1880 and 1890—and in part between 1860 and 1890, frequently compared with similar figures for Britain. In "A Look Ahead," which was published in *North American Review* a short time after the new edition appeared in book form, Carnegie reviews relations between the two countries from the time of their separation, showing a gradual and increasingly fast improvement. He closes with a stirring, yet touching vision of reunion within a fairly short time, concluding, "Let men say what they will, therefore, I say that as surely as the sun in the heavens once shone upon Britain and America united, so is it one morning to rise, shine upon, and greet again 'The Reunited States,' 'The British-American Union.' "

Although perhaps a better work than the first, the second *Triumphant Democracy* is heavier reading than its predecessor, and had a far lower sale.

X   *Other Political Writings*

The next most important collection of political writings by Andrew Carnegie appears in the latter half of *The Gospel of Wealth*,[24] a collection of twelve previously published articles and one address (which had been printed in the *Scottish Leader*). The first six chapters (he joined the second and third articles), were on economics, the latter six political. These included four on foreign affairs, and two on the mechanics of government.

Earliest of these in time, but third in order is "Democracy in England," which had originally appeared in *North American Review* for January 1886, just before the publication of *Triumphant Democracy*. Celebrating the passage of Gladstone's Reform Bills of 1884 and 1885, which provided almost universal suffrage and near-equality of representation in Britain, it comments on the results of the first election, discusses problems which may be expected to arise, and predicts that the country will fully adopt the democratic system within two decades. Immediately following is the speech, *Home Rule in America*, delivered in Glasgow in September 1887, and printed the same month. In it Carnegie discusses the American system of government in much the same way and with some of the materials used in *Triumphant Democracy*. The latter half is given to the Irish problem, to which he felt the proper answer was home rule—that is, a system like the powers of the states in the Union.

"Imperial Federation," published in *Nineteenth Century* for September 1891, is representative of a number of similar articles Carnegie wrote about this period in regard to tariffs. It was inspired by the sentiment for special trade concessions within the empire, partly in retaliation against the McKinley Tariff Act of 1890, which was sharply protective. Originally favoring free trade, especially when his mills could produce iron and steel more cheaply than those of the mother country, Carnegie had gradually come around to a position favoring free trade for England, which produced many manufactures, but little food, while for America, with its overproduction in agriculture, there should be a tariff on manufactured articles.

The two most important political articles in the collection are "Distant Possessions" and "Americanism Versus Imperialism," published in *North American Review* immediately following the Spanish-American War. Carnegie had relaxed his peace principles at the time of the conflict, on the ground that it was justified in order to give independence to Cuba. But he was horrified when the United States took over the Philippines and other island possessions at its conclusion. He became a crusader against imperialism, and especially in these articles pointed out the dangers of dependencies, citing England's problems with India and other possessions. Particularly in the earlier one, Carnegie exhibits some of his best and most persuasive writing:

If it be a noble aspiration for the . . . Cuban, as it was for the citizen of the United States himself, and for the various South American republics once under Spain, to have a country to live and, if necessary, to die for, why is not the revolt noble which the man of the Philippines has been making against Spain? Is it possible that the Republic is to be placed in the position of the suppressor of the Philippine struggle for independence? Surely, that is impossible. With what faces shall we hang in the school-houses of the Philippines our own Declaration of Independence, and yet deny independence to them? What response will the heart of the Philippine Islander make as he reads of Lincoln's Emancipation Proclamation? Are we to practice independence and preach subordination, to teach rebellion in our books, yet to stamp it out with our swords, to sow seed of revolt and expect the harvest of loyalty?[25]

CHAPTER 6

Free-Wheeling Economist: 1883-1916

BY far the most famous work by Andrew Carnegie was not really a book, but a collection of his magazine articles, along with one speech. "Wealth" had appeared in two parts in the *North American Review* in 1889, and had been reprinted in several languages in magazines and pamphlet form, for almost eleven years before *The Gospel of Wealth* was published. But as a book it made an immense impression, and went into a second edition the following year.

Although the second half of the volume was concerned largely with political questions, it was the first half dealing with economics, which proved sensational. And the important part of this dealt with unequal distribution of wealth. Carnegie had long been concerned with this vexing problem, and offered an unusual and eye-catching twist to a partial solution which he had previously advocated, and which had been suggested by others, going back as far as Tom Paine. More than a century earlier, Paine had suggested breaking up huge fortunes by heavy inheritance taxes. Carnegie agreed, but held up an ideal—that the right-minded plutocrat should not wait until death to do good with his money; to be sure it went for the desired ends, he should give it away during his lifetime.

Even this was far from new in Carnegie's thought. Two decades before writing the key article which gave the book its name, he had put down a memo embodying the plan for even a fairly modest income.[1] And two years before the publication, he had told Gladstone he "should consider it a disgrace to die a rich man."[2]

Although his father had left the Presbyterian Church, Carnegie had been surrounded in his boyhood by strong Calvinist influences, that "Protestant ethic" with its emphasis on work,

84

the duty of thrift, and the right to private property because it could be used in the divine service. His Chartist upbringing may have caused him to lean towards socialism, but only in theory, not practice. Nor did his Darwinian Socialism, to which he seems never to have paid much more than lip service, cause him to oppose some legal intervention, as well as private charity, to relieve human need. Well versed in the Bible, he adopted many Judeo-Christian principles, though rejecting the theological content.

The basis of his economic system was *The Wealth of Nations* by Adam Smith, whom he greatly admired as a thinker and fellow-Scot.[3] He was influenced on the tariff by Henry C. Carey, and later quoted favorably from John Stuart Mill's *Principles of Political Economy* when he shocked the nation by coming out in 1908 for free trade.[4] Nine months earlier his address, *The Worst Banking System in the World*, before the Economics Club of New York, had been published in *The Outlook*.[5] It was highly praised, and Carnegie had more than seventy thousand copies printed and sent to every legislator and banker in the country. The address is credited with helping stimulate the discussion which eventually resulted in the Federal Reserve System.[6] Despite his close friendship with John Bright, Carnegie blasted the doctrines of the Manchester School in an article in *Nineteenth Century*.[7]

I *The Opening Gun*

But nothing else in all Carnegie's economic writing ever created such a sensation as a pair of articles titled simply "Wealth" and "The Best Fields for Philanthropy" in the June and December issues of *North American Review* in 1889. Whatever was lacking was supplied by W. T. Stead, who titled the first article "The Gospel of Wealth" when given permission to reprint it in his *Pall Mall Gazette*. From there it was widely reprinted in several languages.

In the second article Carnegie relates the dramatic circumstances connected with the publication of the first: "The manuscript reached [*North American* editor Allen Thorndike Rice] in the morning, and late in the evening of the same day he called to say that it pleased him so much that he had determined to publish it in the forthcoming number, instead of holding it for the succeeding issue, as had been intended.... Sitting in my library, Mr. Rice expressed a wish to hear the author read his manuscript. I read and he listened from beginning to end, making but one interruption."[8]

By 1889 Carnegie had passed the stage in his writing where he felt called upon to use a fiery opening, in the newspaper style of his times. He begins "Wealth" quietly, but at the very crux of the problem to which it is addressed: "The problem of our age is the proper administration of wealth, that the ties of brotherhood may still bind together the rich and poor in harmonious relationship." Without breaking the paragraph, he approaches the problem first from a historical standpoint. True to his early training he equated capitalism with civilization, and followed Darwin on what is best for the race. In former days, as with Indians of his day, there was little difference between the living standard of the chief and his poorest brave. "The contrast between the palace of the millionaire and the cottage of the laborer with us today measures the change which has come with civilization. This change, however, is not to be deplored, but welcomed as highly beneficial. It is well, nay, essential, for the progress of the race that the houses of some should be homes for all that is highest and best in literature and the arts, and for all the refinements of civilization, rather than that none should be so. Much better this great irregularity than universal squalor. Without wealth there can be no Maecenas."[9]

A "relapse" to old conditions "would sweep away civilization with it." Carnegie continues, "But, whether the change be for good or ill, it is upon us, beyond our power to alter, and therefore, to be accepted and made the best of. It is a waste of time to criticize the inevitable" (2).

Carnegie then proceeds to trace the rise of manufacturing, and praises the result that "The poor enjoy what the rich could not before afford. What were luxuries have become the necessaries." He does not discount the social cost involved, but considers it—as well he might—both worthwhile and unavoidable. Here is the very basis of his creed as a capitalist, as well as of his "Gospel of Wealth":

The price we pay for this salutary change is, no doubt, great. We assemble thousands of operatives in the factory, and in the mine, of whom the employer can know little or nothing, and to whom he is little better than a myth. All intercourse between them is at an end. Rigid castes are formed, and, as usual, mutual ignorance breeds mutual distrust. Each caste is without sympathy with the other, and ready to credit anything disparaging in regard to it. Under the law of competition, the employer of thousands is forced into the strictest economies, among which the rates paid to labor figure prominently, and often there is friction between the employer and the employed, between capital and labor, between rich and poor. Human society loses homogeneity.

The price which society pays for the law of competition, like the price it pays for cheap comforts and luxuries, is also great; but the advantages of this law are also greater than its cost—for it is to this law that we owe our wonderful material development, which brings improved conditions in its train. But whether the law be benign or not, we must say of it, as we say of the change in the conditions of men, to which we have referred: It is here; we cannot evade it; no substitutes for it have been found; and while the law may be sometimes hard for the individual, it is best for the race, because it insures the survival of the fittest in every department. (3–4)

Soon Carnegie launches into an attack on any other systems. They are not in order "because the condition of the race is better than it has been with any other which has been tried." The effect of new substitutes is uncertain. Socialism and anarchy are attacking the foundations of civilization, because it began when "the capable, industrious workman said to his incompetent and lazy fellow, 'If thou dost not sow, thou shalt not reap'" (5–6).

At this point Carnegie really arrives at the main question: what is the proper mode of administering wealth? He feels and says that he has the only true solution: "There are but three modes in which surplus wealth can be disposed of. It can be left to the families of the decedents; or it can be bequeathed for public purposes; or, finally, it can be administered by its possessors during their lives" (8). With a severity possibly born of his own early privation and amply supported from the antics of wealthy scions of Pittsburgh and New York families, Carnegie warns of the personal and national dangers of large inheritances.

The second mode "is only a means for the disposal of wealth, provided a man is content to wait until he is dead before he becomes of much good in the world." There are many ways, however, in which the intention of a man who leaves money for public purposes may be thwarted, or the money wasted. Carnegie roguishly concludes: "Besides this, it may fairly be said that no man is to be extolled for doing what he cannot help doing, nor is he to be thanked by the community to which he only leaves wealth at death. Men who leave vast sums in this way may fairly be thought men who would not have left it at all had they been able to take it with them" (10–11).

As an alternative, Carnegie points to inheritance taxes, and urges a graduated tax, which he claims will do more good than many "trifling amounts" scattered among individuals. He cites the Cooper Institute, and hopes "[William T. Tilden's] $5 million bequest for a free library in New York may do good. But it would have been better for him to use the money while still alive." He asserts: "This, then, is held to be the duty of the man of wealth: To set an example of modest, unostentatious living, shunning display or extravagance; to provide moderately for the legitimate wants of those dependent upon him; and, after doing so, to consider all surplus revenues which come to him simply as trust funds, which he is called upon to administer, and strictly bound as a matter of duty to administer in a manner which, in his judgment, is best calculated to produce the most beneficial results for the community" (15).

Reviews of *The Gospel of Wealth* commented that the author found it "as impossible to name exact amounts or actions as it is to define good manners, good taste, or the rules of propriety."[10] Few if any suggested that it might have been ridiculous for him to name an amount that would sustain his own lavish living, or a limit of ostentation that would give approval to his castle in Scotland, with its retinue of servants and a bagpiper to wake the household for breakfast.

In the original draft of his manuscript, Carnegie had warned against giving so as to encourage "the slothful, the drunken, the unworthy," adding: "Of every thousand dollars spent in so-called charity today, it is probable that nine hundred dollars is un-wisely spent—so spent, indeed, as to produce the very evils which it hopes to mitigate or cure." When he was reading the article to Rice, the editor interrupted at this point. "Make it $950," he suggested. And that is the way the passage appeared.[11]

"Thus is the problem of rich and poor to be solved," Carnegie concludes. "The laws of accumulation will be left free, the laws of distribution free. Individualism will continue, but the mil-lionaire will be but a trustee for the poor ... administering [wealth] for the community far better than it could or would have done for itself.... The day is not far distant when the man who dies leaving behind him millions of available wealth, which was free for him to administer during life, will pass away 'unwept, unhonored and unsung,' no matter to what uses he leaves the dross which he cannot take with him. Of such as these the public verdict will then be: 'The man who dies thus rich, dies disgraced.' Such, in my opinion, is the true gospel con-cerning wealth..." (15–19).

It is a typical late-Victorian essay, complete with all the trappings: the masses do not deserve money because they are unable to use it wisely; those who have made money know how to use it and what is best for the poor; property is equated with civilization, success with goodness. What is new is the interesting presentation, along with the earth-shaking idea that a rich man ought to give away his fortune while he is alive, day by day, as he gets more than he needs—something Carnegie himself was still practicing rather gingerly. And it is buttressed by two eye-catching and glittering phrases—"The gospel con-cerning wealth," and "The man who dies rich dies disgraced" (19).

II   *Teaching Philanthropy*

In December (1889) Carnegie, happy in the international stir the article had created, had contributed a follow-up, also to the *North American Review*, to explain to his fellow mil-lionaires how to perform the duty he had urged on them. In the book he annexes it (with minor changes) to the first. Following the same thinking as before in regard to the masses, he now urges that the man who makes more millions knows better what to do with the surplus than the men who make fewer.

The article was titled "The Best Fields for Philanthropy," and opens with a defense against an attack on the earlier one. The *Pall Mall Gazette*, in a typical British comment, had said, "Great fortunes, says Mr. Carnegie, are great blessings to a community, because such and such things may be done with them. Well, but they are also a great curse, for such and such things are done with them. . . . The 'Gospel of Wealth' is killed by the acts."[12] Carnegie's very sensible reply is that the gospel of Christianity was also killed by the acts. It is no argument against a gospel that it is not lived up to, or against a law that it is broken.

Then turning to his subject, Carnegie observes that, negatively, the first requisite for a really good use of wealth is to take care that a gift shall not have a degrading, pauperizing tendency upon its recipients, and preaches further on the point. Then, positively, he lists what he considers the best uses:

1. For those who can afford very large sums, to found universities.
2. Free libraries and museums.
3. The founding or extension of hospitals, medical colleges, laboratories and similar institutions.
4. Public parks.
5. Meeting and concert halls.
6. Swimming baths.
7. Churches. (26–41)

He suggests, in addition, a wide range of unstated smaller benefactions, suited for those who cannot give large sums. He concludes, "The gospel of wealth but echoes Christ's words. It calls upon the millionaire to sell all that he hath and give it in the highest and best form to the poor. . . . So doing, he will approach his end no longer the ignoble hoarder of useless millions; poor, very poor indeed, in money, but rich, very rich, twenty times a millionaire still, in the affection . . . of his fellow-men. . . . This much is sure: against such riches as these no bars will be found at the gates of Paradise" (43–44).

III *The "Gospel" in Book Form*

Only a few months before selling out his business and retiring, Carnegie gathered thirteen of his published works, seven on economics[13] and the rest on political science, under the title *The Gospel of Wealth*, making only minor changes except for the omission of the introduction to the third article and merging it in the title chapter. As an introduction to the book he used "How I Served My Apprenticeship," an adaptation of the early portion of his autobiography which had been published in *Youth's Companion*, April 23, 1896, including the discredited story about his meeting with Woodruff, still unchanged.[14]

Several of the articles on economics are significant to Carnegie's basic position. "The Advantages of Poverty," from *Nineteenth Century*, March 1891, is an answer to criticisms in that magazine of the original "Wealth": one by Gladstone, who had praised the essay, except for its strictures against inherited wealth,[15] and Hugh Price Hughes, a Methodist divine, who had bitterly attacked it a month later in a symposium which included Archbishop Henry L. Manning and Hermann Adler, Britain's chief rabbi.[16] Hughes, asserting that "as a representative of a particular class of millionaires ... [Carnegie] is an anti-Christian phenomenon, a social monstrosity, and a grave political peril," had criticized the essay because it dealt with the administration of wealth, but not the real problem, its distribution. Carnegie's reply, not very forcible or convincing, repeats much of what was in the original two essays on wealth, and introduces his pet themes—the influence of the saintly, hardworking mother (61, 63–64); how poor boys tend to be achievers (64); and—at great length—the dangers of promoting idleness and laziness by impetuous giving (68–71).

"Popular Illusions about Trusts"[17] does not openly defend these monopolies, but holds up to ridicule any effort for laws against them. Large concentrations of capital are necessary for modern industry, and are good, because they increase efficiency. Because of the law of competition, "every attempt to monopolize the manufacture of any staple article carries within its bosom the seeds of failure. Long before we could legislate with

much effect against trusts there would be no necessity for legislation.... There should be nothing but encouragement for these
vast aggregations of capital" (101). Though the price may go
up briefly, the article affected will soon become cheaper than
before. "Generally speaking, the advance in prices would have
taken place even if no trusts existed, being caused by increased
demand" (99–100). Carnegie is still relying on his Spencerian
principles: "The great natural laws, being the outgrowth of
human nature and human needs, keep on their irresistible
course" (102).

"An Employer's View of the Labor Question" and "Results
of the Labor Struggle" were originally printed in the April and
August (1886) issues of the *Forum*; the former written just
before, the latter just after, the labor disturbances of 1886.

The former is mostly sweetness and light. Historically viewed,
things are quite hopeful, although the new employer-employee
relation system has not yet evolved. Profit-sharing is a hopeful possibility, as is arbitration of differences, but profit-sharing
is still far off: "The right of the working-man to combine and
to form trades-unions is no less sacred than the right of the
manufacturer to enter into associations and conferences with his
fellows, and it must sooner or later be conceded" (114). Carnegie finally sets up four principles:

First. That compensation be paid the men based upon a sliding
scale in proportion to the prices received for the product.

Second. A proper organization of the men of every works to be
made, by which the natural leaders, the best men, will eventually
come to the front and confer freely with the employers.

Third. Peaceful arbitration to be in all cases resorted to for the
settlement of differences which the owners and the mill committee
cannot themselves adjust.

Fourth. No interruption ever to occur in the operations.... (122)

Before the first article was in print, a rash of difficulties
broke out, including scattered violence. The second article,
"Results of the Labor Struggle," is written to provide reasons
and explain results. In it, Carnegie differentiates between the
two major outbursts. At St. Louis a leader of the Knights of
Labor had been dismissed, perhaps for union activity. In New

York employees of an elevated railway struck for shorter hours
and better pay. Carnegie appears to take sides with the workers
in both, but notes that in each the workers had lost much of
their public support by resorting to violence. As usual for that
day, he blames this on "a handful of foreign anarchists" (131–34,
142).

In summing up the situation, Carnegie makes a statement
which has been widely quoted, pro and con:

> While public sentiment has rightly and unmistakably condemned
> violence, even in the form for which there is the most excuse, I would
> have the public give due consideration to the terrible temptation
> to which the working-man on strike is sometimes subjected. To expect
> that one dependent on his daily wage for the necessities of life will
> stand by and see a new man employed in his stead, is to expect
> much. . . . In all but a very few departments of labor it is unneces-
> sary, and, I think, improper, to subject men to such an ordeal. . . .
> The employer of labor will find it much more to his interest, wherever
> possible, to allow his works to remain idle and await the result of a
> dispute. (144)

Seven years later, following the bloody Homestead strike at
his own mill, Carnegie was accused of inconsistency. Hard-
hearted he was, but not inconsistent. The paragraph sounds
very sweet and gentle. But many have overlooked another observation in the same article: "*First.* The "dead line" has been
definitely fixed between the forces of disorder and anarchy and
those of order. Bomb-throwing means swift death to the thrower.
Rioters assembling in numbers and marching to pillage will be
remorselessly shot down" (143). As usual, Carnegie was ambivalent. He had sympathy for the suffering worker, but when the decision had to be made between the striker and the steelman's
property, he was capitalist first, humanitarian afterward.

A note under the table of contents states that: "The various
articles in this volume are reprinted by permission of the publishers of the periodicals in which they originally appeared,"
and lists these as *Youth's Companion, Century Magazine, North
American Review, Forum, Contemporary Review, Fortnightly
Review, Nineteenth Century,* and *Scottish Leader.* There is
nothing from *Fortnightly Review* included, however, nor would

Carnegie's only contribution to that periodical have been appropriate. The error seems to represent one more instance of Carnegie's inexact memory, and perhaps wishful thinking, for *Fortnightly Review* was a most prestigious journal.

The book attracted even more attention on both sides of the Atlantic than had the original article, and generally drew more favorable reviews, probably because Carnegie was now far more widely known and influential than he had been eleven years previously. As the *Outlook* stated in a two-page review, "Inasmuch as the Author's income far exceeds that of the King of England, and his private fortune exceeds that of the whole people in many principalities, it is doubtful if any moralist since Marcus Aurelius has wielded greater material power; and the fact that in this case the preacher talks, not of the duties which he would lay upon others, but of those which he lays upon himself, lends to his preaching a unique interest."[18]

Commercially the book was a success, being reprinted many times in America and abroad. In 1901 Carnegie added to the volume one more essay, "British Pessimism," which had appeared in *Nineteenth Century and After* in June of that year.[19] It discusses and analyzes the severe depression in Britain at the turn of the century, rightly attributing much of the difficulty to high taxes, imperialism, and the large military and naval budget.

IV *The Later "Gospel"*

After writing the earlier articles and subsequent to their publication in book form, Carnegie had carried out his long-planned intention of retiring from business, not with his original goal of $50,000 a year in income, but with hundreds of millions in gold bonds worth more than par. Following a brief period of depression in which he bitterly regretted retirement,[20] he had flung himself again into action, writing, speaking and working out ways to give away the fortune he had made.

Seventeen years after the magazine publication of "Wealth," and six after the appearance of the book, Carnegie contributed another article in the series, again to the *North American Review*. At his request, the first essay, with its later designation, had been republished in the September 21, 1906, issue, and on December 7, the final word appeared as "The Gospel of Wealth—II."[21] No reference was made to the original second part, "The Best Fields for Philanthropy."

The article's first words reveal Carnegie's concern with the old subject, and his recognition that legislation, not evolution, would be required, both for breaking up large fortunes, and for securing better distribution: "The problem of wealth will not down. It is obviously so unequally distributed that the attention of civilized man must be attracted to it from time to time. He will ultimately enact the laws needed to produce a more equal distribution."

The article next quotes President Theodore Roosevelt's speech of April 14, 1906, urging graduated gift and inheritance taxes as a necessity to break up large concentrations of wealth. It further cites Carnegie's 1889 article. "See," Carnegie seems to be saying, "I've been telling you this, all the time." And so he had. But with age his devotion to capitalism, if not to capital, had grown stronger. He gives several pages to examples of how fortunes can be made, to show that taking part of the money is just. In every case, he decides, it is community and population growth which have really produced the wealth. He vigorously opposes, as he had before, any income tax. And now he would even except the estate of the man who made the fortune. It is only at the death of his children that the law should operate. (Carnegie now knew he might leave a widow and a daughter, but had no certainty of grandchildren.) Fortunes could be and often were unfairly made; but Britain had passed the point where they could be easily won, and so would America. For the rest, democracy would be trusted to meet such situations as might arise.

V The Empire of Business

In April 1902—perhaps heartened by the success of his earlier book—Carnegie brought out another collection of his works as *The Empire of Business*. It contains ten of his previously published articles and seven speeches dating from 1885 to 1902. The articles had appeared in the *North American Review*, the

Forum, the *New York Evening Post,* the *New York Tribune,*
Iron Age, the *New York Journal, Youth's Companion, Contempo-*
rary Review, Nineteenth Century, and *Macmillan's Magazine,*
and deal principally with industrial and economic themes.

The most interesting and persuasive one, "The A B C of
Money," had been printed in the *North American Review* in
1891, during the "free silver" agitation. It is direct and easily
understandable, presenting the argument for a gold standard
in the best tradition of the period.

In his "Interests of Labour and Capital," Carnegie stresses
education in labor, saying, "In these days of transition and of
struggles between labour and capital, to no better purpose
can you devote a few of your spare hours than to the study of
economic questions."[22] Carnegie practiced what he preached,
for he seldom missed an opportunity to learn from leading
financiers, bankers, and stockholders about business conditions,
and was even able to predict price levels six months in the
future without a second thought. But he never explained how
a man working eighty-four hours a week at fourteen cents an
hour could find "spare hours" or materials for such study.

Although Carnegie stated in "How to Win Fortune" that
"The enormous concern of the future is to divide its profits, not
among hundreds of idle capitalists who contribute nothing to
its success, but among hundreds of its ablest employees, upon
whose abilities and exertions success greatly depends,"[23] he
nevertheless believed that the best way to bind a man to his
work was not to pay him a high salary but give him a share in
the business. Carnegie was a pioneer in this policy.

Like its predecessor, *The Empire of Business* proved popular.
First issued by Doubleday in 1902, it was taken over by Harper
& Brothers the following year, and brought out in simultaneous
British and American editions. In a later edition Carnegie added
one more chapter, "Does America Hate England?" from the
November, 1897, issue of *Contemporary Review.*

VI   Problems of Today

Except for a few later portions of the *Autobiography,* the last
book Carnegie wrote was *Problems of Today.* It appeared in

November 1908, and has been largely neglected, even by his biographers. The only one who mentions it is Hendrick,[24] who dismisses it with the statement that it is a collection of "essays, touched up for republication, that had appeared in *Nineteenth Century, Contemporary Review*, the *Century Magazine*, and the like."

Examination discloses that only the first two of its ten chapters include or quote appreciable amounts of his previous material, the final one being a reprint *in toto* of "My Experience with Railway Rates and Rebates" (twenty-two pages), which had appeared in *Century* in March of the year of the book's publication. The first chapter, "Wealth," forty-six pages in length, includes thirteen and a half pages quoted from the 1889 *North American Review* article of the same title, which had formed the basis for *The Gospel of Wealth*. The next, "Labor," includes four pages quoted from "An Employer's View of the Labor Question," published in the *Forum* in April 1886, and another two and a half quoted from that article and a follow-up, "Results of the Labor Struggle," in the same magazine the following August. Elsewhere one and one-half pages are quoted from an address at the opening of the library Carnegie had given to Homestead, Pa., in 1898. Thus a total of forty-five pages of the volume's 207 were taken from his earlier writings, or just over one-fifth.

Additional chapters of the work are concerned with "Wages," "Thrift," "The Land," "Individualism versus Socialism," "Family Relations," and "The Long March Upward." Although little of it is copied or reprinted, it does not contain much that is new, even to Carnegie's thinking. The old themes are there—the advantages of poverty, necessity of thrift, importance of accumulations of capital, and evolutionary social development—the Spencerian march onward and upward.

Carnegie, as age advanced, had obviously become more and more afraid of Socialism, not because of what it might do to him, but its possible effect on the Capitalist system he had helped to shape, and which he loved. Over and over again he argues against endangering the basis of civilization (to him, property) by changing to some new, unproved, and possibly destructive way of meeting social problems. His guides are still Adam Smith, Mill, and Spencer. Changes in the tried and true system would wreck everything—even the family, which would necessarily wind up in Karl Pearson's "complete freedom in the sex-relationship, left to the judgment and taste of ... men and women."[25] Against this Carnegie sets up his well-loved theme from Goethe: "In the happiest and holiest homes of today, it is not the man who leads the wife upward, but the infinitely purer and more angelic wife whom the husband reverently follows upon the heavenly path."[26] He no longer cries out that "The problem of Wealth will not down," although he recognizes that "our Socialistic friends ... mean well.... No class is moved by worthier impulses." But the danger of Socialist change is that of revolution, as opposed to evolution: "By the nature of its being, the one rule which the human race never can persistently violate is that which proclaims, 'Hold fast to that which has proved itself good.'... We believe that the surest and best way ... is by continued evolution, as in the past, instead of by revolutionary Socialism, which spends its time preaching such changes as are not within measurable distance of attainment, even if they were desirable in themselves."[27]

VII *Summary*

Carnegie had long wrestled with the problem of unequal distribution of wealth. As a youth he was at least a lukewarm Socialist, and by 1868 was seeking to find a compromise with his conscience over practicing what he still called the "worship of the golden calf." Almost twenty years later he felt he had found an answer: the rich man should give away his surplus and inheritance taxes should prevent the family of the unwilling rich man from "burdening" his family with wealth. But no longer was the maker of wealth to give everything away beyond what he needed. Holding a large fortune was proper as long as he gave it away before his death.

By the late 1880s Carnegie had come to recognize the high cost large industry brought in the complete separation and even loss of touch between employer and employees, rich and poor. The condition was actually good, however, because it brought "progress." What later came to be known as the

"American Dream"—that any capable youth could have a Horatio Alger career and become rich—compensated for the want and suffering of others.

Following the severe labor troubles which began in 1887, and even more after bloody Homestead, Carnegie came to feel that the lives of workmen were relatively unimportant, compared with the progress of civilization, which he equated with property. Trusts, cut-throat competition were all right—everything would work out satisfactorily. At last, under the influence of Theodore Roosevelt, he came to recognize that there must be legislation to control abuses. Once again he felt that socialism had a good idea, but it would be dangerous to try it, because it was an untried and uncertain path. It was attempting to do at once what could only be accomplished by a long period of evolution. The one certain wisdom was for mankind to continue doing what had been proved so successful, the amassing of property by the capable, who would then know best what should be done with it.

CHAPTER 7

Proliferous Speaker: 1877-1912

CARNEGIE was one of the most popular and sought-after speakers of his day. Frequently he was called upon to address organizations, and not always those which were interested in acquiring donations. Early in his life, he had formed good speaking habits which helped him in his orations. One ground of Carnegie's freshness of spirit and vehement style, which gained momentum at just the right places, can be found in his own thinking:

I remember in one of my sweet strolls "ayont the wood mill braes" with a great man, my Uncle Bailie Morrison. . . . I asked him what he thought the most thrilling thing in life. He mused awhile, as was the Bailie's wont, and I said, "I think I can tell you, Uncle." "What is it then, Andrea?" . . . "Well, Uncle, I think that when, in making a speech, one feels himself lifted, as it were, by some divine power into regions beyond himself, in which he seems to soar without effort, and swept by enthusiasm into the expression of some burning truth, which has laid brooding in his soul, throwing policy and prudence to the winds, he feels words whose eloquence surprises himself, burning hot, hissing through him like molten lava coursing the veins, he throws it forth, and panting for breath hears the quick, sharp, explosive roar of his fellow-men in thunder of assent, the precious moment which tells him that the audience is his own, but one soul in it and that his; I think this the supreme moment of life."

To which his uncle responds, "Go! Andrea, ye've hit it!"[1]

Carnegie not only felt this way about speaking, but demonstrated it as well. On the public platform he was a dramatic showman. "Frequently rising to his tiptoes and pumping his short arms vigorously, to his critics in the audience he looked like a bantam rooster ready to crow."[2]

Almost from boyhood he practiced oratory. Carnegie and five

of his companions formed a debating society which met in a cobbler shop. Later this same circle of friends became members of the foremost club in the city, about which he wrote in his *Autobiography*:

Another step which exercised a decided influence over me was joining the "Webster Literary Society."

I know of no better mode of benefiting a youth than joining such a club as this. Much of my reading became such as had a bearing on forthcoming debates and that gave clearness and fixity to my ideas. The self-possession I afterwards came to have before an audience may very safely be attributed to the experience of the "Webster Society." My two rules for speaking then (and now) were: Make yourself perfectly at home before your audience, and simply talk *to* them, not *at* them. Do not try to be somebody else; be your own self and *talk*, never "orate" until you can't help it.[3]

Among the many goals that he set for himself in 1868 Carnegie listed "pay especial attention to speaking in public."[4] In so doing he discovered a Professor Churchill from Boston, who instructed him in the art of platform oratory.[5]

His convictions concerning oratory were strengthened at the time the Freedom of Dunfermline was conferred upon him in 1877. Carnegie had spoken to his Uncle Bailie Morrison concerning his speech of acceptance, confiding that he desired to say what was in his heart. To which his uncle, an orator himself, replied, "Just say that, Andra [*sic*]; nothing like saying just what you really feel." Carnegie later wrote—

It was a lesson in public speaking which I took to heart. There is one rule I might suggest for youthful orators. When you stand up before an audience reflect that there are before you only men and women. You should speak to them as you speak to other men and women in daily intercourse. If you are not trying to be something different from yourself, there is no more occasion for embarrassment than if you were talking in your office to a party of your own people—none whatever. . . . Be your own natural self and go ahead. I once asked Colonel Ingersoll, the most effective public speaker I ever heard, to what he attributed his power, "Avoid elocutionists like snakes," he said, "and be yourself."[6]

Another factor which helped to make Carnegie an articulate speaker was hearing the *Messiah* while on the great coaching trip in 1881. He was impressed by the fact that choral groups in England were more careful with their words than choruses in America: "In public as well as in private singing the purity of pronunciation struck us as remarkable. If I ever set up for a music teacher I shall bequeath to my favorite pupil as the secret of success but one word, *'pronunciation.'* "[7]

Carnegie always paid particular attention to the audience when he or others were speaking. During Matthew Arnold's lecture tour of the United States in 1883, Carnegie suggested that the visitor obtain instruction in public speaking from one of the best elocutionists in America at that time—the same Professor Churchill under whom he had studied. His remark came after an unsuccessful speech in which the front row of the audience could barely hear Arnold. Perhaps one of Andrew's best teachers was his mother. When Arnold asked Margaret Carnegie for her opinion in regard to how his oration went, her reply was, "Too meenisterial, Mr. Arnold, too meenisterial."[8]

Carnegie was an adept humorist as well as convincing speaker. As a result, his talks were frequently interrupted by applause, cheering, and laughter. Throughout his addresses, one can detect this buoyant spirit, sense the fervor of his style, and at times almost imagine the enthusiastic response that must have accompanied his orations.

Many of the addresses are available for study. Burton J. Hendrick prints sixteen of them in his collection of Carnegie works as *Miscellaneous Writings.*[9] Carnegie used one in *The Gospel of Wealth*, and seven in *The Empire of Business.*[10] In addition, at least seventy-five were distributed in pamphlet form, some by the thousands. Scores were quoted at length in contemporary newspapers.

I *Light Speeches*

Self-trained to think on his feet, Carnegie was a master of impromptu speaking and repartee. Perhaps the best example of his off-the-cuff manner still preserved is the one entitled "Industrial Pennsylvania," which was given in 1900 before the

Pennsylvania Society of New York.[11] To his surprise two earlier speakers talked on much the same subject. In his introduction Carnegie said, "I find myself most peculiarly situated here to-night. The gentleman on my left, the Assistant Attorney-General, expresses his anxiety to take my pocket book and the gentleman on my right has stolen my speech. I have given much thought to the matter but I think that I have hit upon the right plan at last for successfully disposing of my surplus wealth—I must call in the members of his profession, who are willing to talk on all sides of a question and who stand ready to take money from either or all sides."

Then he continued, "Gentlemen, I am here in a double capacity to-night, as my invitation was first as President of St. Andrew's Society, and the second as a devoted Pennsylvanian. . . . I cannot speak to you to-night of 'Industrial Pennsylvania,' because it has been spoken of twice already, and, as Governor [James A.] Beaver says, 'Pennsylvania, like Massachusetts, needs no encomium.' There she stands. She speaks for herself." (Here Carnegie was continuing the paraphrase on Daniel Webster's famous speech on secession.) Reflecting his Scottish pride, Carnegie said, "I am gratified at the evidence given here to-night of the wish of everybody who has spoken, wherever born, or however born, to be born again, that he might have at least the right to claim some Scotch blood in his veins."

Finally Carnegie proceeded to speak on his assigned subject, emphasizing the importance of home commerce. In rationalizing Pennsylvania's population decrease in the 1880 census compared to the increase in New York, he commented tongue-in-cheek, "Now it will be said by the New Yorker that [Pennsylvanians] come here for commercial reasons. Don't, gentlemen New Yorkers, entertain for a moment that mistaken idea. It is the missionary spirit which prompts them to go over to you and which is stirring the world today."

Carnegie shifted to lofty tones for his conclusion: "Pennsylvania products . . . shall continue to be the guiding star to teach all people the true path to a higher civilization and continue to give the foremost place among the states of the Union to the Keystone State, the birthplace of the nation."

"The Scotch-American," Carnegie's speech delivered at the

annual dinner of the St. Andrew's Society in New York, is another example of his light manner of speaking.[12] He launched
into his subject with, "This is, indeed, the age of instantaneous
photography. I appear before you to-night commissioned to
kodak, develop and finish the Scotsman at home, in four minutes; in four minutes more to picture him in America, and in
two minutes more to celebrate the union of the two varieties,
and place before you the ideal character of the world, the best
flower in the garden, the first-prize chrysanthemum—the Scotch-
American."

A later statement of grandiose exultation produced much
laughter: "It is only through their union that the crowning mercy
has been bestowed upon the world, and perfection at last
attained in the new variety known as the Scotch-American, who
in himself combines, in one perfect whole, the best qualities
and all the virtues of both, and stands before the world shining
for all, the sole possessor of these united talents, traits, characteristics and virtues, rare in their several excellencies and wonderful in their combination." Continuing in the same vein, Carnegie expressed his true opinion, that racial admixture made
for the improvement of society. In ridiculing European aristocracy he said, "The result of lack of fusion between the races
is seen in the royal families of Europe, most of them are
diseased, many weak-minded, not a few imbecile, and none of
them good for much." Another statement brought more laughter
and applause: "Scotch wives for American husbands is a fusion
which I am told is hard to beat, and I have a very decided
opinion, which many of you have good reason, I know, to
endorse, that Scotch husbands for American wives is an alliance
which cannot be equalled."

Reflecting his belief in the importance of meager beginnings,
Carnegie described Scotland as having the "bracing influence
of poverty, uncursed by the evils of luxury." His Scottish blood
statement, "the land of Wallace, Knox, Scott and Burns belongs
not to itself alone, but to the world," was again met with
applause.

At times Carnegie made comments even when it was not
his turn to speak. During an occasion at which he was to present
a light address, the speaker before him said, "If I am a victim

to be thrown to the lions, as our toastmaster has said, my fears are the less because men who are thrown to the lions are the ones who in the nature of the case cannot make an after-dinner speech." At this Carnegie called out clearly, "The Prophet Daniel did."[13]

II *Peace Speeches*

For many years one of Carnegie's first principles had been peace—a total prevention of war. Naturally this became a theme of many of his addresses. His hostility to war stemmed from early youth, when he attended peace demonstrations in Dunfermline. Between the years 1901 and 1910, he was involved in very many such projects, and it is not surprising that he chose the subject "A League of Peace" for his second Rectorial Address to the students of St. Andrews.[14]

He began by presenting noble ideals, encouraging the students to "leave the world a little better than you found it." He then cited some social evils which had disappeared—polygamy, slavery, and dueling—but stressed that the most hated one still remained. Commencing with a quotation from Rousseau, who said, "War is the foulest fiend ever vomited forth from the mouth of Hell," he continued to recite the words of nearly seventy other important men who held similar opinions, ending with General Sherman's "War is Hell."

Concluding with a passage from Scripture—"When men shall beat their swords into ploughshares"—he stressed the need for a League of Peace. Carnegie appealed to the students, male and female alike, to move "onward and upward," putting all else aside in their concentration on a means to end all wars—preventing them before they happen.

Five years later, on March 10, 1910, Carnegie in his speech " 'Honor' and International Arbitration" again lashed out against war.[15] Addressing the Peace Society at Guildhall in London, he began with a lament upon the recent death of King Edward, and expressed the hope that Edward's successor would continue with efforts of peace. Later in his oration he elaborated more in detail on the former sovereign.

He spoke of the changes in war, tactics not being as savage

as they formerly had been, saying that the real problem was not war itself but the ever-present threat and danger of war, its continuous preparation, and the distrust of one nation for another. Asserting that partial disarmament is not enough, he advocated peaceful arbitration as a solution. Since before the turn of the century he had stressed in many speeches and magazine articles his conviction that civilized man must substitute arbitration for war. He praised Taft for his support of arbitration: "Let all friends of peace hail President Taft as our leader." (But at a later date, when the U.S. Senate rejected the arbitration treaty, he was quick to blame the president.)[16]

Disputing the belief that "war is the nursery of heroes," Carnegie mentioned industrial heroes and his Hero Fund. He again quoted Sherman as he had in his peace speech to the St. Andrew's students, ending with his pet Spencerian notion that man moves "ever upward and onward toward perfection."

III   Gift Speeches

As a result of his many benefactions, Carnegie was called upon to speak at numerous dedications. Two early speeches are found in *Our Coaching Trip*. In his address at Dunfermline on the occasion of the laying of the memorial stone for his gift of a free library in 1881, Carnegie greeted his audience as "workingmen and women of Dunfermline." He was proud to claim the title of workingman himself, and "one who like yourselves has toiled with my hands ... you have Shakespeare, the mightiest of all intellects and you have your own genius, Burns, the ploughman." He continued, "It is impossible that any act which I may perform in after life can give me the gratification flowing from this as you, by your free and generous use of the library, enable me to indulge the sweet thought that it has been my privilege to bestow upon Dunfermline, my native town, A Free Library, which has proved itself a foundation of good to my fellow-townsmen."[17]

When he was asked to speak following the swimming competition at the baths he had donated to the town, Carnegie said, "Great Britain will continue to rule the waves about as long as I should like to prophesy any nation would rule anything;

and I think that it is incumbent for that reason that the sons of Great Britain should learn to be at home in the waves which we expect them to continue to rule. However, there is no longer any question about this, that Dunfermline has begun to see the advantage, and she will no doubt soon recognize the duty of teaching all her sons to feel this confidence at least, that they were not born to be drowned."[18]

In 1891 Carnegie spoke on the occasion of the laying of the foundation stone for the Peterhead Free Library in Scotland.[19] He stressed how other nations had followed different paths, "Some built upon military or artistic attainments, but Scotland upon the general education of her people." Using a familiar nautical theme, he said, "There is no helm that steers so true in the voyage of life as Common Sense." Later he expounded upon the influence of the sea: "To be in daily contact with the sea—nay! more weird still, nightly contact; to brave its tempests; to draw from its bosom the means of subsistence, must necessarily strengthen what is bold and adventurous in man." And, as always in referring to education, he quoted Knox: "I will never rest until there is a school in every parish in Scotland."

Disliking the dead languages all his life, Carnegie emphasized the importance of the "all-round intelligent man," as opposed to "men from college who knew so much of the skirmishes of the tribes of Greece and Rome . . . but as we say in America, 'didn't know beans.'"

He spoke of the importance of a free library, especially for working men, and stressed good labor relations, saying, "Every Free Library in these days should contain upon its shelves all contributions bearing upon the relations of labour and capital. . . ." He also emphasized the importance of a good librarian and of circulating libraries. Pointing out his belief in distributing surplus wealth, he suggested that a good way for small amounts of money to be put to the best use was through donations of books to libraries.

The Scottish blood theme came through when he spoke of immigration: "There is no land upon the earth which would not only waive all examination, but receive with open arms him who is privileged to say in response to all inquiries, 'I am a Scotchman.' The only objection might be [quoting Abram

S. Hewitt of New York], 'We are always somewhat afraid when we get a Scotchman that ere long he may own the works.' "

Scottish prejudice is reflected in his words: "The proud position occupied by our race comes, no doubt, partly from the bracing climate of the north, and from the generations of no less bracing oppression which our mountain home has had to encounter in its resolute and successful struggle for the preservation of its nationality; but, without doubt, it flows chiefly from the fact that Scotchmen have been better and more thoroughly educated for generations than other races." The words are reminiscent of a statement in his *Autobiography*: "The intensity of a Scottish boy's patriotism ... constitutes a real force in his life to the very end. If the source of my stock of that prime article—courage—were studied, I am sure the final analysis would find it founded upon Wallace, the hero of Scotland. ... The true Scotsman will not find reason in after years to lower the estimate he has formed of his own country."[20]

In 1903 Carnegie paid tribute to eight different towns within a week's time, and other gift speeches were made on the same tour. When laying the foundation stone for the free library of Dingwall, Carnegie said, "This stone is level, this stone is plumb, this stone is truly laid and may the blessing of God rest upon the work of this day and upon this library when it is completed." Adapting from Gamaliel's speech in Acts 5, he added, "If this thing be of men, it will fail, but if it be of God it must stand."[21]

At Tain for the dedication of the town hall, Carnegie reflected his Scottish prejudice in a humorous vein: "In one case I heard of, a man's claim [to Scottish blood] was based upon the fact that his grandmother had a Scotch nurse." Later in his speech he advised the young people to become interested in music and drama, taking advantage of the chance to perform in the new hall.[22]

When laying the memorial stone of the public school in Kilmarnock, Carnegie repeated the hope of John Knox for a school in every parish in Scotland. At times Carnegie wandered from his prepared text, and openly admitted it. As always, he brought up his hatred for war, saying, "That is not in my notes. But I like to make digressions when the spirit moves me." About education he said, "It has been said you can educate a man too much. You might as well tell me you could have a man too sober."[23]

At the dedication of the free library in Govan, Carnegie quoted twice the maxim "We only hate those we do not know," and prophesied that wars shall cease when nations get to know one another. Reflecting other favorite themes, he said that the "best worship of God, is serving man," and "the finest heritage for a young man is poverty."[24]

In laying the foundation stone of the free library in Waterford, Ireland, Carnegie stated that this was his first speech in that country, although he admitted, "It is not my first speech before an Irish audience, for I have had often to speak in the city of New York, and you all know what an Irish city that is. And I have spoken in Pittsburgh, and you know it is of a Scotch-Irish character." And in words of flattery he said, "All the world likes Ireland."[25]

At Limerick for the laying of the library foundation stone, Carnegie commented on how much the Irish and Scotch had in common, and went on to say, "The community that is not willing to maintain a library is not worthy of having one." He also emphasized that free libraries are especially for the toiling masses. The rich can buy books of their own. And again he stressed the importance of a good librarian versus a poor one.[26]

During the laying of the memorial stone for the library at Cork, he predicted that if the English-speaking race would reunite, they could decree peace and enforce it—again the same familar theme, peace, or, if necessary, war for peace.[27]

An important speech on this tour was addressed to the Iron and Steel Institute at Barrow, and again included the ideal of English-speaking nations enforcing peace among all nations, with Carnegie's regret that an ocean instead of a prairie separated Europe from the New World.[28]

The continuous enthusiasm of his audiences and the sense of achievement certainly must have inspired Carnegie in his dispersal of surplus wealth. To his speech at the laying of a cornerstone at Ayr, the audience responded by giving him a standing ovation, waving hats and handkerchiefs and making loud cheers before he even uttered a word.[29] Throughout the address, Carnegie gave voice to his Scottish patriotism, saying,

"All nations and all peoples love Scotland." Quoting from
Robert Burns, he reflected his dream of peace and brotherhood,
"When man to man the world o'er, shall brithers be an' a' that."

Even during his marriage tour of Scotland in 1887, Carnegie
was requested to speak on numerous occasions. Within three
days he assisted in the ceremony of unveiling the busts of
Walter Scott, John Knox, and George Buchanan in the Wallace
Monument at Stirling; lectured in Glasgow on "Home Rule in
America"; and christened a Mexican steamer and inaugurated
a public library in Grangemouth, all with the assistance of
Mrs. Carnegie. Reflecting deep-rooted patriotism in his speeches
during this tour, he emphasized especially the brotherhood of
man and the union of English-speaking nations. In his "Home
Rule" speech, he quoted Tennyson:

And where is the hope of that great day which the poet sings
of—when the drum shall beat no longer, when the battleflags are
furled,   •
In the Parliament of Man, the Federation of the World?[30]

IV *Speeches to Students*

Carnegie on many occasions was requested to speak before
student bodies. One of his favorite themes and methods of
approach in talking to his young listeners was the figure of
embarking upon the sea of life. He made frequent use of
nautical terms and expressions such as "set sail with clear
papers," "propitious gales," "wretched brothers drifting past,"
"brave mariners," "storm-beaten," and "my young untried
sailors."

Carnegie's address to the graduating class of medical students
at Bellevue Hospital in New York in 1885 is an example of this
nautical theme.[31] In the beginning he said:

No voyager ever sailed from end to end upon summer seas in
undisturbed calm. Always there come
          The visitation of the winds,
              Who take the ruffian billows by the top,
          Curling their monstrous heads.

These angry gusts call for the boldest, steadiest seamanship, a level head, a strong arm, and a heart that is not afraid.

And closing with the same theme:

Most anxious to say to you in parting the words which may prove most beneficial upon your voyage on life's troubled seas, I leave you with this expression of my deliberate opinion, which I ask you to write upon your hearts, making it at once your talisman and compass to direct your course: By maintaining intact the noble traditions of your profession, and handing these down untarnished to your successors, you can best dignify your own lives and best perform the highest service of which you are capable to humanity.

In talking to the young men at the Curry Commercial College on June 23, 1885, Carnegie told how he started his career sweeping the office: he concluded by congratulating the students upon being hard-working poor men who started at the bottom, for "He is the probable dark horse that you had better watch."[32] During this speech Carnegie encouraged the students to aim high and "be king in your dreams," just as he later would with the Cornell students in his address in 1907.[33] From experience, he pointed out several conditions essential for success—honesty, truthfulness and fair-dealing, along with avoiding liquor, speculation, and endorsement of other men's notes. And again using his pet maxim, "Put all your eggs in one basket, and then watch the basket."[34]

Likewise, Carnegie, in his lecture delivered at Union College in Schenectady, N.Y., in January 1895, emphasized, "Thou shalt earn thy bread by the sweat of thy brow," and "The richest heritage a young man can be born to is poverty."[35]

To the students of Cornell University on January 11, 1896, Carnegie again stressed that theme: the poor boy has usually one strong guarantee of future industry and ambitious usefulness—he is not burdened with wealth; it is necessary that he make his own way in the world.[36]

Gradually, as Carnegie gained confidence and realized he had become an international figure, his college speeches became less the advice of a successful man to beginners and more often pronouncements of his beliefs on world affairs, as with the rectorial address cited above.

Perhaps Carnegie's most outspoken speech for students was one on religion, intended for students at St. Andrew's University at his installation as Lord Rector in 1902. But although this address would be appropriate today, it was not acceptable then. Carnegie was ahead of his time. Due to the impact that such a speech as he had prepared would have upon the students, Carnegie was asked to write another one using a subject other than religion. As a result, he delivered one on "The Industrial Ascendancy of the World" instead.[37] Using his familiar nautical theme, Carnegie began, "My annual voyages across the Atlantic rarely yield much time for reading. I am so fond of the deck and the bridge that my time is usually spent there, reveling in the tumbling sea; the higher the waves the greater being the exhilaration."[38]

He then elaborated upon economic and business changes in the world, and the growth of nations in wealth and population, the social condition and aptitudes of their people, natural resources, prospects, ambitions, and national policy. Summing up the laws bearing upon the material position of nations, he went on to compare Europe's position with America's. As in so many other speeches, he stressed international peace, and in so doing praised the Emperor of Russia, citing the Hague Conference which established a permanent tribunal composed of able men from various nations. He also mentioned the German Emperor, who had stimulated industrial action. As a result of this speech, Wilhelm II, after reading his address, invited Carnegie to visit him for discussion upon the subject. This led to further meetings, with other influential statesmen becoming aware of Carnegie's doctrine that American-European competition is not a case of nation against nation, but instead continent against continent.[39]

The original speech he had intended for the St. Andrew's students was published at last in 1933 among Carnegie's miscellaneous writings. His confession of religious faith is personally very revealing. From this address one can readily understand the influences which helped determine his lifelong beliefs and convictions. This relatively unknown speech reveals his depth of religion and personality more than any other he wrote, and therefore should be given fuller attention.

The original speech included the nautical theme, referring to students as the "untried sailor cadets who are about to launch upon the voyage of life."[40] More importantly, in relating his own experiences to the students, he wrote—

The clear line drawn here between theology and religion, the one changing and the other surviving, the one of man and the other in man, I trust you will always keep in mind in your ministry as teachers of men.

When my young brain began to think, these were one and inseparable. I knew no difference between them. The Shorter Catechism, Confession of Faith, and all the structure of theology, the work of men of an ignorant past, were part of the one divine revelation. Theology was religion and religion was theology. . . . Today they are distinct as the stiff, dead frame is from the living picture. . . . So far have we advanced in one generation. (295)

He mentioned how he had been influenced as a working boy by the books made available to him in Colonel Anderson's library and how at this time in his life, "all at sea" with "No creed, no system . . . all was chaos," he had discovered Herbert Spencer and Charles Darwin and at last was able to say, " 'That settles the question!' I had found at last the guides which led me to the temple of man's real knowledge upon earth. . . . I was upon firm ground, and with every year of my life since there has come less dogmatism, less theology, but greater reverence" (297).

Later he discredited miracles, quoting Matthew Arnold, who wrote, "The case against miracles is closed. They do not happen," and Thomas Fuller, who said, "Miracles are the swaddling clothes of the infant church" (298, 305).

Mentioning how his travels had influenced him, Carnegie told how he had witnessed all the religions of the world, and whether in the name of "God, Jah, Jehovah, Jove, Brahm, Baal, Buddha," among others, "all the earth worships the true God. . . . The unknown had left no nation without religion" (309–11). Ever predicting and confident of the future, Carnegie added,

"New and mighty truths are yet to be revealed at intervals, and we are to see present truths more perfectly, probably to the end of time" (318).

Finally, stressing the tasks of "improving our fellow-men" and "obeying the judge within," and noting "that the worship most acceptable to God is service to man," he ended with a favorite saying, "science ... has revealed ... the divine law of his being which leads man ever steadily upward" (319).

V Speeches on the Black Problem

For a person who felt that race was of ultimate importance in explaining the progress of Anglo-Saxon peoples, Carnegie's ideas in regard to the Negro were surprisingly flexible. Although his views on black rights and achievements would be considered lukewarm by today's standards, he was nonetheless well ahead of his time in this respect. There were few Americans in that period who even considered equal rights for this minority tolerable, much less desirable.

Carnegie had never come into close contact with any blacks except for a few outstanding leaders, and had almost no knowledge of their condition in the South or even in the ghettos of the North. But he had come to have a certain respect for the yellow and brown during his trip around the world, and this may have affected his outlook on the subject. There was a popular upsurge of interest in Southern education during the first decade of the twentieth century, and his involvement in the field manifested itself alike in words and benefactions.

Speaking as chairman of a joint educational conference on this subject in New York on February 12, 1904, Carnegie began by pointing out that Lincoln did his part in freeing the slaves; but since "he only is a freeman whom education makes free ... it remains for us, the followers of that leader of men, to continue and complete it." He upheld the right of blacks in the North to vote, since they were in a minority. But in the South, where ignorant whites and Negroes made up a majority, he urged that there continue to be an educational requirement for suffrage.[41]

Again, when asked to speak in October 1907 before the

Philosophical Institution of Edinburgh, Carnegie chose as his theme "The Negro in America."[42] In this address he took up where his earlier one had left off. He cited the early history of the race in America, and assessed very aptly—for that day—the principal factors which led up to a break between North and South on the subject of slavery, the enfranchisement of blacks, and the consequent bitterness among their former masters. Now, "after a period of fifty [*sic*] years, we are tonight to inquire whether the American Negro has proved his capacity to develop and improve." Carnegie proposed three tests. Has the black proved himself able to live in contact with civilization, and increase as a freeman, or does he slowly die out like the American Indian, Maori, or Hawaiian? Could and would he seek education and achieve it? Could and would he become a man of property? All these he answered in the affirmative, supporting his answer to each with an array of statistics provided for him largely by his close friend and beneficiary, Booker T. Washington, and the faculty of Tuskegee Institute.

Next, he cited the means of improvement, and the careers of many outstanding blacks, then (and some of them until recently) neglected by historians—poets, educators, astronomers, and successful businessmen—and their part in a rapidly expanding South. As Wall has pointed out, it was like a black *Triumphant Democracy*, all "sunshine, sunshine, sunshine," which would hardly be recognizable to the Southern sharecropper or the inhabitant of the Northern black ghetto.[43] Neither was Carnegie—nor did he allow his hearers to depart—oblivious to the fact that "all that has been done, encouraging as it undoubtedly is, yet is trifling compared to what remains to be done.—The bright spots have been brought to your notice, but these are only small points surrounded by great areas of darkness. The sun spreading light over all has not yet arisen, altho there are not wanting convincing proofs that her morning beams begin to gild the mountain tops."

Carnegie did not share the general and long-standing fear of an admixture of races: "What is to be the final result of the black and white races living together in centuries to come need not concern us. They may remain separate and apart as now, or may inter-mingle." He closed by citing the words of

Lyman Abbott: "Never in the history of man has a race made such educational and material progress in forty years as the American Negro."[44]

VI   *National Mores*

On numerous occasions Carnegie spoke about social mores. His speech delivered before the Scottish Charitable Society of Boston is a fair example of this.[45]

In this charming essay, Carnegie stated that New England rightfully should be called "New Scotland," for she has more in common with that country than England. In comparing Scotland and New England, Carnegie pointed out the fact that they had a common foe, England. For independence, he said, "the Puritan left his home to establish himself in New England," and his "fellow Presbyterians in Scotland, instead of leaving the old home, remained there and fought the same issue...."

Carnegie's first speech in New York, an impromptu one before the Nineteenth Century Club, on "The Aristocracy of the Dollar," also dealt with the mores of the time.[46] After sitting through three speeches that evening, Carnegie, when given the opportunity, disputed the third speaker, Thomas Wentworth Higginson, on three points. He objected first to Higginson's statement that "the aristocracy of the dollar has inferior manners." Also disputing that the aristocracy of Britain is finer looking than any other class, he said, "Let any visitor see the House of Lords when it is filled.... Upon such an occasion one would really think, as he watches the peers pass, or rather hobble in, that every reformatory, asylum or home for incurables, in Britain, had been asked to send up to Westminster fair specimens of its inmates.... An aristocracy of birth alone has never been able to sustain itself."

He objected next to the assertion that the "aristocracy of the dollar is selfish and dangerous." Carnegie said that the aristocracy of birth is most selfish, "that class which, during life, gives the least proportion of its revenues for the good of others.... As to the danger of the Dollar I must dissent in toto.... Great fortunes ... are built up by life-long devotion to that end; by the exercise of . . . prudence, forethought, energy . . . and self-denial."

Finally, he objected to the statement that the "aristocracy of the dollar is not self-respecting." He directed the question to Higginson, "whether the reign of the aristocracy of intellect does not exist in American Society to-day, rather than that of the dollar." Carnegie concluded that "in the final aristocracy the question will not be how he was born or what he owns, nor how he has worshipped God, but how has he served Man."

Carnegie spent little time in Pittsburgh after moving to New York, but was still in close touch thirty years later when he accepted an invitation to speak before the Pittsburgh Chamber of Commerce. He opened by complimenting the city on its Chamber: "Your members are men experienced in affairs, and, therefore, upon all business questions. The united chambers of commerce, in older lands, speak upon business questions with great authority—so must they soon with us, and always with increasing authority. . . . It will not be claimed by the most extreme exponents of democracy that the masses of the people can or do form sound judgments of themselves upon intricate public questions, but . . . the masses of our people are so intelligent as to be able to weigh what men of special knowledge lay before them."[47] Pittsburgh had become a "two class" city, and the function of democracy was to teach the masses to follow the leading businessmen and men of wealth.

VII *Summary*

Ever one to grasp any opportunity to express his ideas, Carnegie was never happier than when he appeared before an appreciative and cheering crowd.

In style and general development his addresses vary little from his earliest to his latest period. He used a quick and deft approach to catch the interest and if possible the hearts of his audience; interlarded his arguments with his favorite truisms and maxims; and lightened the heavier parts with humor and witty anecdotes. Yet it is only on the surface that his speeches are the same from his early to his later periods. His rugged individualism continued, and became even stronger than in the days of his youth. But his liberalism gradually extended only toward those in agreement with him or willing to go along with his direction. The man who had believed in no aristocracy except that of ideas gradually approached belief in an aristocracy of power. His faith in democracy was skating on very thin ice when he denied that "the masses" were able to form sound judgments by themselves. When he declared himself "the Laird of Pittencrief" he considered it a joke, but few English landlords could have spoken or acted more paternally or autocratically than he.

CHAPTER 8

The Biographer: 1889-1909

THE personality of Andrew Carnegie, his likes and dislikes, his characteristics and familiar themes are clearly reflected in everything he wrote, but most obviously in his biographical works. These are five—His *Autobiography*; *James Watt*, and three pamphlet speeches: *Ezra Cornell*, *Edwin M. Stanton*, and *William Chambers*. All show the frequent use of truisms and proverbs, intensely personal outlook, flair for the dramatic and quotations from literary classics so pervasive in everything he wrote. And all attribute to the subjects many of the author's tastes, ideas and preferences.

Although his *Autobiography* and other biographic attempts are readable and enjoyable, Carnegie's work is often inexact and prone to error.[1] He continually inserts passages from Shakespeare and the Scottish poets—oftenest Scott and Burns—usually without identifying the sources. Often these are inexactly quoted, either from reliance on memory or to adapt the material to some point he seeks to make. This neglect in researching or checking a subject became more pronounced with advancing age. He would press his convictions on the reader just as he had tried to force democracy on Britain during his newspaper publishing days, sometimes omitting or changing facts in order to support his theories. The persistent nature he evidenced in business can be seen in his writing.

Among the themes that Carnegie uses most often in the biographies are devotion to his mother, Scottish patriotism, hero worship, importance of childhood poverty, significance of trifles, brotherhood, and international peace. Carnegie, with the vision of an idealist, had an irrepressible habit of predicting the future, looking forward as well as backward, and with varying accuracy in either case. A friend wrote: "His views

119

are truly large and prophetic,"[2] but the crystal ball was often clouded.

I *The Autobiography*

The question oftenest asked about any autobiography is whether the book is really necessary. There is no doubt as to the value of the *Autobiography of Andrew Carnegie*. Many encouraged him to write it. As Gilder said: "He is well worth Boswellizing, but I am urging him to be 'his own Boswell.' "[3] John C. Van Dyke, who was chosen to revise the memoirs after Carnegie's death, wrote, "He should be allowed to tell the tale in his own way, and enthusiasm, even extravagance in recitation should be received as a part of the story. The quality of the man may underlie exuberance of spirit, as truth may be found in apparent exaggeration. Therefore, in preparing these chapters for publication the editor has done little more than arrange the material chronologically and sequentially so that the narrative might run on unbrokenly to the end."[4]

Louise Whitfield Carnegie wrote in the Preface: "After retiring from active business my husband yielded to the earnest solicitations of friends, both here and in Great Britain, and began to jot down from time to time recollections of his early days.... For a few weeks each summer we retired to our little bungalow on the moors at Aultnagar to enjoy the simple life, and it was there that Mr. Carnegie did most of his writing.... He was thus engaged in July, 1914, when the war clouds began to gather.... These memoirs ended at that time."[5]

Although these can hardly be considered unbiased views, there is evidence—even aside from the book's wide public acceptance— that Carnegie's contemporaries shared the feeling. Mrs. Carnegie must have known, however, that her husband had actually begun his autobiography in 1889. Hendrick quotes from a Carnegie letter to William E. Gladstone in January 1891:

More than one wise friend has insisted that even I should dictate the story of my career. I promised, and the summer before last, when at Cluny, I spent about an hour every morning—when the conditions were not tempting for fishing or outdoor excursions—in sitting before the fire and gazing into it, recounting to myself, as it were, the incidents of my life. A clever stenographer took down the words, and at intervals transcribed them. The result was that before I knew it I had spoken about six hundred pages, making a book equal to my *Triumphant Democracy* and *Round the World* put together. This manuscript has been laid aside for some future editor—I trust a discreet one—to condense.[6]

The material revised from time to time formed the basis for Carnegie's *Autobiography* in its present form. If Carnegie's figures are accurate here, Van Dyke must have omitted from 40 to almost 50 percent of the manuscript, depending on whether the author refers to the original length of his first travel book, or the expanded form commercially published.

Van Dyke recognized and implied in his Editor's Note, that the *Autobiography* is far from being historically accurate. Relying upon a failing memory, Carnegie forgot things or wrote of events the way he wished they had been. An outstanding instance is his account of the early partnership with T. T. Woodruff in his first sleeping-car venture. In his fine biography of the steel magnate, Wall comments on the highly imaginative account taken from Carnegie's *Triumphant Democracy*: "It is a typical Carnegie story, replete with all the stock situations of popular melodrama: the shy, unworldly inventor, the crude, hand-made model, the chance meeting with the bold young business executive, who in a flashing moment of truth recognizes genius when he sees it. There is even the mysterious green bag, in which the stranger carries his invention. It is all too pat and too familiar, but those commentators upon Carnegie's life who have dealt with this incident have accepted his story in every detail. . . . Woodruff wrote to his alleged discoverer: 'Your arrogance spurred you up to make the statements recorded in your book, which is misleading and so far from the true facts of the case and so damaging to your friend of old as to merit his rebuke. You must have known before you ever saw me that there were many sleeping cars furnished with my patent seats and couches running from a number of railways. . . .' "[7]

The account that Wall and Woodruff refer to was the

following passage in Carnegie's *Triumphant Democracy*: "A tall, spare, farmer-looking kind of man ... wished me to look at an invention he had made. With that he drew from a green bag (as it were for lawyers' briefs) a small model of a sleeping berth for railway cars. He had not spoken a minute, before, like a flash, the whole range of discovery burst upon me. 'Yes,' I said, 'that is something which this continent must have.'"[8] Yet despite Woodruff's refutation, Carnegie repeats virtually the same account in his *Autobiography*.[9]

Of all Carnegie's many themes, devotion to his mother is most vivid in *The Autobiography* and overshadows much of his writing. Although he expresses admiration for his father on various occasions, he focuses most of his attention on her, with more than thirty references in the book. Illustrative of this Carnegie writes the following: "Perhaps some day I may be able to tell the world something of this heroine, but I doubt it. I feel her to be sacred to myself and not for others to know. None could ever really know her—I alone did that. After my father's early death she was all my own. The dedication of my first [commercially published] book tells the story. It was 'To my favorite Heroine My Mother'" (6). No doubt his early dependence upon his mother and hers upon him made them become more close than the average Oedipus complex. Again he writes: "Walter Scott said of Burns that he had the most extraordinary eye he ever saw in a human being. I can say as much for my mother" (32).

On a few occasions Carnegie included his father in the exaltation of his mother: "There was nothing that heroine did not do in the struggle we were making for elbow room in the western world. Father's long factory hours tried his strength, but he, too, fought the good fight like a hero, and never failed to encourage me" (37).

Fortunately for Carnegie's wife, he transferred this same adoration to her. He wrote in reference to Louise Whitfield: "It is now twenty years since Mrs. Carnegie entered and changed my life, a few months after the passing of my mother.... My life has been made so happy by her that I cannot imagine myself living without her guardianship" (209).

Another familiar theme of Carnegie's *Autobiography* is his

Scottish heritage. The description of his departure as a child from Scotland, upon hearing the old abbey bell ring, makes us realize his keen devotion to his homeland: "Never can there come to my ears on earth, nor enter so deep into my soul, a sound that shall haunt and subdue me with its sweet, gracious, melting power as that did" (26). And upon a return trip to his native land he tells of his mother's and his reaction: "Her heart was so full she could not restrain her tears, and the more I tried to make light of it or to soothe her, the more she was overcome. For myself, I felt as if I could throw myself upon the sacred soil and kiss it" (106).

The theme of patriotism may be observed early in *The Autobiography*. Seldom has there lived a man who was more patriotic than Andrew Carnegie—true-blue Scotsman and star-spangled American through and through. He readily acknowledged: "I can truly say in the words of Burns that there was then and there created in me a vein of Scottish prejudice (or patriotism) which will cease to exist only with life" (15).

Later Carnegie transferred much of this feeling to his adopted land, writing *Triumphant Democracy* to prove its superiority to Britain (318–20). Most assuredly Carnegie would not have taken such a position on first coming to America, for his strong patriotic ties surpassed any that he could have felt for another country at that time. In his own words—

It remains for maturer years and wider knowledge to tell us that every nation has its heroes, its romance, its traditions, and its achievements; and while the true Scotsman will not find reason in after years to lower the estimate he has formed of his own country and of its position even among the larger nations of the earth, he will find ample reason to raise his opinion of other nations because they all have much to be proud of—quite enough to stimulate their sons so to act their parts as not to disgrace the land that gave them birth.

It was years before he could write this, for as Carnegie said, "My heart was in Scotland" (18).

Carnegie always relied strongly on his belief in the purity of Scottish blood. For instance, he notes that upon meeting the German emperor, he had said, "Your Majesty, I now own

King Malcolm's tower in Dunfermline—he from whom you derive your precious heritage of Scottish blood" (355). Along with this intense loyalty, one can perceive Carnegie's persistent hero-worship. In his own words, when he had to face some obstacle he would ask himself, "what Wallace would have done and what a Scotsman ought to do" (34).

But the hero-worship was not confined to his Scottish inheritance. He greatly admired many of his contemporaries as well. Of his early employer and friend he writes, "Mr. [Thomas A.] Scott was one of the most delightful superiors that anybody could have and I soon became warmly attached to him. He was my great man and all the hero worship that is inherent in youth I showered upon him" (67). He also notes that in later years he showed similar devotion for Herbert Spencer and John Morley, among others.

Another popular theme that Carnegie emphasizes intermittently in the *Autobiography* is the importance of meager beginnings. He says of his family's hard times, "This is where the children of honest poverty have the most precious of all advantages of those of wealth." And then two themes in one: "The mother, nurse, cook, governess, teacher, saint, all in one: the father, exemplar, guide, counselor and friend! Thus were my brother and I brought up. What has the child of millionaire or nobleman that counts compared to such a heritage?" (30). In another passage he comments, "Let him look out for the 'dark horse' in the boy who begins by sweeping out the office" (41).

Carnegie lays emphasis on the importance of trifles, especially in regard to business matters. Remarking about a personal experience which was the deciding point in an important bid for his Keystone Bridge Company he writes, "That visit proved how much success turns upon trifles" (118). And again, concerning his merger with George M. Pullman, "One may learn, from an incident which I had from Mr. Pullman himself, by what trifles important matters are sometimes determined" (154). Along the same line he points out the significance of a slight notice of a kind word: "I am indebted to these trifles for some of the happiest attentions and the most pleasing incidents of my life" (82). And then an entire passage on the same subject, ending with, "The young should remember that upon trifles the best gifts of the gods often hang" (35).

It is easy to see the origins of these motifs in Carnegie's early experiences. But we search in vain for the source of his devotion to peace in an era in which almost every other ambitious youngster dreamed of military prowess. Frequently in his *Autobiography*, as well as his other writings and in his actions, he evidenced his concern for international peace and his love of brotherhood. These ideals he acknowledges. "Peace, at last between English-speaking peoples, must have been early in my thoughts" (270). And again he speaks of himself as "saddened and indignant that in the nineteenth century the most civilized race, as we consider ourselves, still finds men willing to adopt as a profession—until lately the only profession for gentlemen—the study of the surest means of killing other men" (323).

A disturbing factor in his life which Carnegie expresses in the latter part of the *Autobiography* is his concern over losing former friends with advancing age: "For some years after retiring I could not force myself to visit the works. This, alas, would recall so many who had gone before. Scarcely one of my early friends would remain to give me the hand-clasp of the days of old. Only one or two of these old men would call me 'Andy'" (279).

The *Autobiography* has been long held in high esteem, reprinted numerous times, and often assigned for reading in high schools and for college freshmen. Its simple, direct style, excellent choice of words, and smoothly flowing movement are exemplary. Its subject-matter and presentation are interesting to almost any type of reader. Its only serious fault is inaccuracy, and this is often unimportant except to the historian.

While its arrangement of materials is not always in the most desirable order, the complexity of the life Carnegie records is so great that anyone would find it difficult to order it better. Certainly none of the biographies of Carnegie in print have achieved any better arrangement.

From practically every point of view, literary, biographical or ideational, the *Autobiography* is Carnegie's best work. It would be interesting to know the style and content of the relatively large part which was excised by Van Dyke in editing the manuscript.

II   *The Biography of James Watt*

When in 1904 the Edinburgh and London firm of Oliphant, Anderson and Ferrier sought someone to write a life of James Watt for its *Famous Scots* series, Carnegie was the obvious choice. An internationally known author and fellow Scot, he was also a titan of industry and well versed in the mechanical trades.

For some unspecified reason, his answer was a flat rejection, based on the multiplicity of his other interests. Then the old Carnegie curiosity made itself felt. Recalling that he knew little of Watt or of steam engine history, he felt that the necessity of writing would be his surest means of making up the gap in his knowledge. He offered to write the book, and in the preface reported these facts, happily adding: "I now know about the steam-engine, and have also had revealed to me one of the finest characters that ever graced the earth."[10]

In a sense Carnegie was indebted to Watt, "the creator of the most potent instrument of mechanical force known to man,"[11] even though his father's business was ruined by steam power. Carnegie writes: "The change from handloom to steam-loom weaving was disastrous to our family. My father did not recognize the impending revolution, and was struggling under the old system." But the same steam power that caused poverty in his early years later made possible his financial success— even to getting his first start in America running a steam-engine,[12] and later as a railroad official.

In dealing with his subject, Carnegie adopted a chronological order, commencing with the time and circumstances of the Scotsman's birth. From then on, the narrative flows smoothly with few digressions from the order of events. He did not encumber his writing with excessive, long passages dwelling on technical details concerning the numerous inventions of Watt. In contrast, James P. Muirhead, from whose work Carnegie evidently drew much of his material, emphasized the inventions of Watt and wrote in detail about them. Carnegie concentrated more on the man.

Carnegie, with his passion for lofty views, wrote: "Let the dreamers therefore dream on. The world, minus enchanting dreams, would be commonplace indeed, and let us remember this dream is only dreamable because Watt's steam-engine is a reality" (21).

Throughout the book one is apt to notice Carnegie's habit of personalizing words, such as "genius steam" (21), "pride of profession" (26), "demon steam" (34), "godlike reason" (45), "chariot of progress" and "fairy girdlist" (63). And all with a fine sense of the dramatic. For herein lay the heart of Carnegie's style. In one passage concerning the inventor's hitting upon an ingenious discovery Carnegie wrote: "Many plans were entertained, only to be finally rejected. At last the flash came into that teeming brain like a strike of lightning. Eureka! He had found it" (41).

As in his *Autobiography* Carnegie injects Shakespearean strains and other literary quotations into his biography of Watt. He writes: "We may picture him reciting in Falstaffian mood, 'Would my name were not so terrible to the enemy'" (87). Always Carnegie had a tendency to adapt quotations to situations: "For this is Coilantogle ford,/ And thou must keep thee with thy sword" (22).[13] As usual Carnegie gives no reference. Being very knowledgeable about folklore and literature he sometimes mentions an author's name, but usually seems to assume that the reader will recognize the quotation and know the source.

The familiar theme of devotion to his mother, so evident throughout Carnegie's *Autobiography*, frequently appears in his work on Watt. In writing about the inventor's mother, he often projects his own feelings into what he thought the relationship between Watt and his mother must have been, when in reality it was his own mother he obviously was writing about. According to Muirhead, Watt "received from his mother his first lessons in reading; his father taught him writing and arithmetic."[14] From this Carnegie embellishes: "She taught him to read most of what he then knew, and, we may be sure, fed him on the poetry and romance upon which she herself had

fed, and for which he became noted in after life" (12). At
first the reader is led to believe that he was tutored solely
by his mother, although Carnegie does mention later that "his
wise father not only taught him writing and arithmetic, but
also provided a set of small tools for him" (14). Carnegie
ignores the fact that Watt's father lost the family fortune in
speculation.

Further reflecting upon the maternal theme, Carnegie notes:
"For what a Scotch boy born to labour is to become, and
how, cannot be forecast until we know what his mother is,
who is to him nurse, servant, governess, teacher and saint,
all in one" (9–10). And again the same refrain: "No school but
one can instil that, where rules the one best teacher . . . the
school kept at your mother's knee. Such mothers as Watt had
are the appointed trainers of genius, and make men good and
great, if the needed spark be there to enkindle: 'Kings they
make gods, and meaner subjects Kings'" (13). Nowhere in
Muirhead's biography is there a hint of such high tribute to
Watt's mother, although she was the writer's kinswoman, Agnes
Muirhead.

Projecting his own feelings into those concerning the death
of Watt's mother, Carnegie writes: "The relations between
them had been such as are only possible between mother
and son" (17).

Repeating the proverbial theme, "East or West, home is
best," Carnegie says: "Watt never ceased to keep in close
touch with his native town of Greenock and his Glasgow friends.
His heart still warmed to the tartan, the soft, broad Scotch
accent never forsook him; nor, we may be sure, did the refrain
ever leave his heart" (85). In Carnegie's words: "The heather
was on fire within Jamie's breast" (13). This fact alone would
have kindled Carnegie's adoration for the man from the very
start.

Hero-worship woven all through the biography manifests
itself in passages such as this: "When the boy absorbs . . .
Wallace, the Bruce, and Sir John Graham, is fired by the story
of the Martyrs, has at heart page after page of his country's
ballads, and also, in more recent times, is at home with
Burns's and Scott's prose and poetry, he has little room and

less desire, and still less need, for the inferior heroes.... Self-seeking heroes passed in review without gaining admittance to the soul of Watt" (15–16). Sir Walter Scott, who also drank deep of Scottish lore, worshipped Watt as well. Carnegie quotes the words of the poet in reference to the inventor, "[Watt] was not only the most profound man of science, the most successful combiner of powers, and combiner of numbers, as adapted to practical purposes—was not only one of the most generally well informed, but one of the best and kindest of human beings" (158).

Carnegie's admiration of Watt is best expressed in the following passage: "Thou art the man; go up higher and take your seat there among the immortals, the inventor of the greatest of all inventions, a great discoverer, and one of the noblest of men!" (42).

In Watt's biography as well as his own life story Carnegie stresses the importance of frugal beginnings. About Watt Carnegie observes: "But the fates had been kind; for, burdened with neither wealth nor rank, this poor would-be skilled mechanic was to have a fair chance by beginning at the bottom among his fellows, the sternest yet finest of all schools to call forth and strengthen inherent qualities, and impel a poor young man to put forth his utmost effort when launched upon the sea of life, where he must either sink or swim, no bladders being in reserve for him" (21–22). And again: "There must be something in the soil which produces such men; something in the poverty that compels exertion" (135). Also: "Not from palace or castle, but from the cottage have come, or can come, the needed leaders of our race, under whose guidance it is to ascend" (9). And "... distinguished students, who figuratively speaking, cultivate knowledge upon a little oatmeal, earning money between terms to pay their way. It is highly probable that a greater proportion of these will be heard from in later years than of any other class" (30).

Here, too, Carnegie stresses the importance of trifles, as he did in his autobiography. He wrote: "Fortune sometimes hangs upon a glance or a nod of kindly recognition as we pass" (89).

Carnegie no doubt was projecting his own feelings into those of Watt when he wrote about satisfaction in honest work done: "It is highly probable that this first tool finished by his own hands brought to Watt more unalloyed pleasure than any of his greater triumphs of later years, just as the first week's wages of youth, money earned by service rendered, proclaiming coming manhood, brings with it a thrill and glow of proud satisfaction, compared with which all the millions of later years are as dross" (25). Carnegie had written in his autobiography the same sentiment: "I have made millions since, but none of those millions gave me such happiness as my first week's earnings."[15]

Another Carnegie theme which manifests itself in the biography of Watt is his desire for international peace and brotherhood. Borrowing from Tennyson and his Scottish hero, Burns, Carnegie writes: "We may continue, therefore, to indulge the hope of the coming 'parliament of man, the federation of the world.' . . . 'It's coming yet, for a' that, that man to man the warld o'er, shall brothers be for a' that' " (21), although it had no connection with Watt.

This Carnegie habit of pressing his own convictions upon others accounts for part of his description of his subject. Himself an enthusiastic fisherman, Carnegie infers that Watt was an avid angler: "The only 'sport' of the youth was angling, 'the most fitting practice for quiet men and lovers of peace,' the 'Brothers of the angle,' according to Izaak Walton, 'being mostly men of mild and gentle disposition'" (17).

Muirhead, on the other hand—and with close ties which gave him a better chance to know the facts—offers a far different conclusion. When a boy, he agrees, Watt often fished from a jetty at the back of his father's house on the shore at Greenock. It would be pleasant to think that he continued the sport in later life. "There would have been something cheerful in associating the name of James Watt with those of . . . Izaak Walton . . . and other eminent worthies" of the pastime.[16] But the evidence, he finds, is simply on the other side. Watt had neither the time, nor apparently the inclination to follow the sport. Carnegie must have overlooked, forgotten, or determinedly chosen to make his subject like and practice something so dear to himself.

Carnegie looked forward as well as back, in his treatment

of Watt. In reference to the scientist's discovery that water was a compound, instead of an element, he goes on: "Who shall doubt, after finding this secret source of force in water, that some future Watt is to discover other sources of power, or perchance succeed in utilizing the super-abundant power known to exist in the heat of the sun?" (39).

In the final analysis, Carnegie reflected his concern for old age and retirement as he did in his autobiography. It is interesting to note that when Carnegie wrote this, he was seventy years old. He laments: "The day had come when Watt awakened to one of the saddest of all truths, that his friends were one by one rapidly passing away, the circle ever narrowing" (143–44).

Carnegie's conclusion and summary again play up the subject as an inventor, a discoverer, and finally Watt—the man. In doing so, Carnegie, who always went to the most authoritative sources, quotes from others their opinions concerning Watt. He includes Henry P. Brougham, Humphry Davy, James Mackintosh, George Hamilton-Gordon, Francis Jeffrey, William T. Kelvin, and others (148–60).

III   *Stanton, Cornell, and Chambers Pamphlets*

Carnegie's other essays at biography were three memorial addresses on Edwin M. Stanton, Ezra Cornell, and William Chambers, later issued in pamphlet form. All are reprinted in *Miscellaneous Writings of Andrew Carnegie*, Volume I.[17]

In all three Carnegie uses some of his favorite themes: the advantages of poverty in youth, importance of family, interest in reading, fortunate chance, and influence of good women. It all sounds like a replay of Carnegie's own life.

The Stanton speech[18] is little more than a patriotic hurrah, wandering off into eulogy and a highly fantasized account of Stanton's part in the Civil War, which—Carnegie would make it appear—he won almost singlehandedly. Of course, Carnegie mentions how as a messenger boy he had come to know his subject, then a Pittsburgh attorney. If, as is probable, he had met either Cornell or Chambers, he fails to mention it in the speeches on them.

The Cornell address, delivered at Cornell University on

April 26, 1907, comes nearest to honest biography. Almost at the beginning he compares his subject to the inventor of the steam engine, asserting that he was, "like Watt . . . a decidedly mechanical genius." His own interest in women's rights—he and David McCargo were "the first to employ young women as telegraph operators in the United States upon railroads,"[19] and his uxorial devotion are reflected in his comment on the admission of young women to Cornell University: "Our country is generally credited with being in advance of others in this respect. Their presence and status here in Cornell and other universities give ground for this opinion. . . . In our day, man, and notably the American man, finds in his wife the angel leading him upward, both by precept and example, to higher and holier life, refining and elevating him, making his life purer and nobler."[20]

William Chambers[21] appealed to Carnegie as a fellow Scot, and, as with Cornell, a philanthropist and donor of libraries (210–11).

IV   Related Materials

Six other related short compositions fall more or less within the realm of biography, but none of them really warrant the designation. They all lack vital statistics, background and other information which would enable the reader to reach any effective understanding of the life, thought, development and actions of the subject. Each is principally a momentary glimpse, little more.

The chapter "Characteristics" in the *Memoirs of Anne C. L. Botta*, New York, 1894, runs for six pages. It is mostly an account of Botta's kindness in introducing Carnegie, a green young Westerner in his mid-thirties, into the most cultivated society of New York. He tells of her parties, "nearest thing in the modern era to a real 'salon,' "[22] and the sadness at her funeral.

"Stevensonia," in *Critic*, January 12, 1895, is merely one of a number of brief tributes to Robert Louis Stevenson, the lament of a Scot for a fellow countryman.

Of somewhat more substance is "The Laird of Briarcliff," in the *Outlook* for May 16, 1898. It is a tribute to Walter W. Law, a New York merchant who retired from business at sixty, developing a large dairy and rose farm and establishing a model community for his employees. The pattern—son of poor Nonconformists; father an Independent and mother "of indomitable energy and enterprise," and Walter "her ain son"; put to work at fourteen; constant reader; friendly, hard working and aiding his family; youthful migrant to America; Horatio Alger type success, early retirement and philanthropy, sound almost as if Carnegie was writing about himself.

"Queen Victoria," a magazine article on the queen's death in 1901, and "William Ewart Gladstone," a memorial address at St. Deniol's Library on October 13, 1902, are mere bits of praise.

The untitled first of a group of "Tributes to Mark Twain," by prominent writers in *North American Review*, June 1910, is the best and most touching of the lot. Carnegie writes of his relationship with Clemens, and the humorist's honesty and acumen. It is a gem, but not biography.

V *Similarities of Carnegie, Watt, and Cornell*

It is a truism that one of the requirements for writing good biography is sympathy with the subject.[23] Carnegie qualified in this respect to write about both Watt and Cornell. Perhaps one reason why he became so enthusiastic about each one was the number of shared similar experiences. The adage "Like draws to like," which he quoted from time to time, well applies in regard to his admiration for each of the two.

Watt and Carnegie were born in Scotland, within fifteen months of a century apart, and died within eight days of the same period. Cornell's life spanned the gap between the pair, being contemporary with the inventor for twelve years and with the steelman for thirty-seven. All three came of good families reduced to near poverty, the Carnegies by technological change, the Cornells by business failure, and the Watts by speculation.

None of the three had more than a meager common-school education except for what he received from his family and what he later obtained by individual study. All were lifelong students, widely known for practical or scientific attainments, and became acquainted with and respected by most of the principal men of their nation and period.

Each left his birthplace when very young (Cornell and Watt at eighteen) and made his own way in strange surroundings. Cornell moved half across New York state, Carnegie to America, and Watt to England. Carnegie and Watt both suffered from ill health and repeatedly returned to Scotland on this account. Carnegie wrote: "To the old home in Scotland our hero's face was now turned in the autumn of 1756, his twentieth year.[24] His native air, best medicine of all for the invalid exile, soon restored his health."[25] In regard to himself he had written: "The cool highland air has been to me a panacea for many years. My physician has insisted that I must avoid our hot American summers."[26]

Both Carnegie and Watt had mechanical genius assistants on whom they relied and in whom they placed complete trust. For Carnegie it was Capt. William Jones and for Watt, William Murdock. Carnegie wrote: "An American Murdock was found in Captain Jones, the best manager of works of his day.... Fortunate is the firm that discovers a William Murdock or William Jones, and gives him swing to do the work of an original in his own way."[27]

All three men became philanthropists on an unexpected scale for their times. Carnegie and Cornell both endowed universities, and each donated a library to his home town. Carnegie and Cornell both built railways, and Watt designed engines for railway use. All three married happily and all were ahead of their times in regard to religion, Carnegie noted: "It seems probable that Watt, in his theological views ... was in advance of his age." "The cry was raised that [Cornell] intended to establish a Godless University." "Those who hold today that the Sabbath in its fullest sense was made for man ... are not more advanced than were my parents forty years ago."[28]

Carnegie and Cornell planned early retirement from business, the latter attaining his aim at about fifty. Carnegie, as he noted of Watt, "gracefully glided into old age. This is the great test of success in life," he added.[29] All three died of pneumonia.

VI   *Evaluation*

Literarily, Carnegie was less successful in his essays at biography than in any of his other writing. The *Autobiography*, although a classic of simple prose, is poorly organized and contains many factual errors. The Watt biography, although internationally published and reprinted several times, shows little knowledge of the background material. It contains far too much identification of author and subject, as can be said of the Cornell speech and those on Chambers and Stanton. Most modern writers on these men simply ignore Carnegie's work. And while few biographers of Carnegie have shown any tendency to question the *Autobiography* on matters of fact or interpretation,[30] those who have followed it closely have often been led into error by the practice.

Only in the *Autobiography* does Carnegie achieve the canon that biography should be "... interpretive, whether of real people, of actual events or of ideas."[31] In his other essays in the field he completely misses the aim that in "great biography ... it is the personality of the subject rather than his achievements that has made the book great."[32] He constantly violates the rule that a good biographer must keep himself in the background.[33] The one saving grace in them all is what Elizabeth Nitchie calls the crowning virtue of style in biography—simplicity.[34]

Perhaps Carnegie's principal difficulty in the field was that he never really understood biography. Travel he knew; politics and economics he learned; letter writing came to him naturally, as did journalism, which to him was a branch of the same field; and he studied assiduously to become an interesting and forceful speaker. But biography to him was never more than praise for someone he admired—including himself. That is why his one acceptable essay in the field was the *Autobiography*, whose subject he understood, even if he confused times, circumstances and other facts. *James Watt*, the Chambers, Stanton, and Cornell pamphlets, and "The Laird of Briarcliff Manor" were largely produced by putting himself into the shoes of his subjects.

CHAPTER 9

Man of Letters: 1849-1915

AN inveterate letter writer, Carnegie seems to have had a pen in hand at all times, despite the overwhelmingly active life he led. Even after he had become internationally known as a prominent figure, he wrote constantly to his family, friends, business associates, to newspapers, and public figures, including five presidents with whom he was on friendly terms.

Unfortunately, no considerable collection of his letters has ever been published, although many of them are quoted, in whole or in part, in the Wall and Hendrick biographies, and scores appeared in newspapers to whose editors he directed them.

Throughout his letter writing, from the very beginning, Carnegie used an abbreviated style. He employed the ampersand most of the time and frequently capitalized nouns and words for emphasis according to old style. In 1903 Carnegie's staccato, telegraphic style with little punctuation became exaggerated, for it was at this time that he had endowed the Simplified Spelling Board. However, by 1915 he became discouraged due to lack of progress in his revolution of the English language. He wrote, "I think I hav been patient enuf....I hav a much better use for twenty-five thousand dollars a year."[1]

But to himself he remained true. Carnegie continued to the end in his correspondence to use this abbreviated form, and even if he slipped and forgot to simplify, he would carefully erase "have" and rewrite it as "hav," in order to be consistent.

I *Family Letters*

As a youth, soon after his arrival in America, Carnegie began a trans-Atlantic correspondence with the Lauders—usually writing to his uncle, George Lauder, and cousin, George, Jr., more affectionately called "Dod." His letters were informative, keeping the Scottish relatives up to date on the family's progress, and what was happening in the New World. After a few years he began using the correspondence to debate differences in British and American ideas and political systems.

In later years he corresponded continually with his mother and brother Tom, whenever they were apart, and with Tom's widow, Lucy Coleman Carnegie, as well as with Louise Whitfield during their courtship. Following their marriage they were seldom apart for any length of time. A typical family letter is that written from Parma, Italy, on December 14, 1865, during the walking tour of Europe.[2] The letter is a good example of its kind, part business, part news, and with a reserved expression of sentiment, although he often addressed his brother somewhat as an inferior, in the manner of a superior talking to a subordinate, or at least, an older sibling giving his "kid brother" advice.

A previous letter, mailed just before the writing of this one, had recognized that business conditions were very bad, and expressed an intention of returning home in a month with Henry Phipps, who was cutting his part of the trip short, and expected to be home to Pittsburgh by the end of January. But Carnegie had just been reassured by a letter from Tom.

He opens this letter abruptly by explaining that with a new turn of events he feels no necessity for disappointing other members of the party by walking out on them. But he is willing to return if necessary:

I just feel this way about it—twenty Italys wouldn't keep me if my brother was having too much anxiety about matters. I feel that very few persons of your years have ever had such a load to travel under. I don't know anyone, I'm sure, whom I would consider able for such a task.

Carnegie expresses a hope that this letter will overtake the previous one, to avoid causing Tom unnecessary concern.

Switching to family matters, he asks for a picture of Lucy, and suggests that her father, who is to occupy the new Carnegie home until spring, handle arrangements for putting in a furnace.

He joshes about the news that a mutual friend is getting married: "Your item about John Hampton almost took the breath from me. If he goes, who is safe? I shall henceforth esteem myself not invulnerable." The letter closes with friendly chatter about how he will tease their mother on returning home. It is a cheerful, informal missive.

II *Letters to Friends*

In writing to his friends, many of whom were business associates or public figures, Carnegie was flippant at times, calling them "pard" or "chum" throughout his letters. It is obvious that even with his friends he did not mince words. While continually giving prophetic advice or pep talks, he pressed his convictions upon others.

At times he was blunt. In corresponding with British friends he spared no feelings, and warned them vehemently on differences in national policies. He wrote long, earnest letters, lecturing and preaching, with poetic effusions (quoted). When his own feelings were hurt, he would write in the suffering tones of a martyr. But when a friend expressed such feelings, he nipped self-pity in the bud.

A good example of his style is his letter to Herbert Spencer of January 5, 1897,[3] in reply to one of December 16, from Brighton.[4] Spencer had been approached on the subject of receiving public honors, and indicated he planned to refuse the gesture because of long previous neglect. Carnegie's letter, a classic of its kind, opens with New Year's greetings and closes with brief personal matters. The intervening five paragraphs are all designed to cheer up the gloomy old friend by spanking him gently.

I hope you will reconsider the whole matter, and come to the conclusion that the greater the neglect shown by your fellow countrymen, the higher the tribute to what you laid before them. . . . When have the Prophets not been stoned, from Christ down to Wagner? . . .

Why, my dear friend, what do you mean by complaining of neglect, abuse, scorn? These are the precious rewards of the teachers of mankind. The Poets fare no better. . . .

I could wish that you had been imprisoned, tortured on the rack.

This would have been no greater reward than is your due. The Philosopher who is sensitive to contemporaneous criticism is a new type, and I do not wish you to pass into history as its founder.

Carnegie urges his friend to do away with anything he has written which shows "other than a spirit of deep gratification at the neglect, scorn and abuse which you have had to suffer." The proper attitude, he urges, should be that of "lofty pity and anxiety for their reaching the light by and by, which you have discovered and in which you rest. . . ."

Referring to their early meeting,[5] he warns the philosopher: "The "Cheddar vs Cheshire" cheese story will pass into history, and prove that you are not altogether a "Brooding God" but something also of the human. But, my belief is, that one word showing disappointment, or, may I say, resentment, of the treatment you have received from your countrymen, will detract very much from the loftiness of our Guide, Philosopher and Friend." Then he closes the letter with a bit of friendly chatter having no relation to the subject of his admonition.

III *Business Letters*

The business letters of Andrew Carnegie were ordinarily (but not invariably) brief and to the point. Sometimes he opened with staggering abruptness, as in the first letter he wrote to the trustees of the Peace Fund: "I have transferred to you as trustees ... $10,000,000 of five-per-cent first mortgage bonds, value $11,500,000, the revenue of which is to be administered by you to hasten the abolition of international war, the foulest blot upon our civilization."[6]

A typical Carnegie business letter—except for the content—and a good one considering the situation, was that written to John G. A. Leishman on December 24, 1894.[7] Leishman, a partner who had been promoted to chief executive of Carnegie Steel on the resignation of Frick, was discovered to have been speculating in pig iron, a practice Carnegie considered inexcusable. Henry Phipps had passed the word along just before the letter was written. On learning the situation, Carnegie wrote as much in sorrow for an old friend as in anger:

Dear Mr. Leishman:

You have made tomorrow a sorry Christmas for me and for Mr. Phipps from whom I have heard. You have not treated me fairly as your partner. You know I often congratulated you on your *not* speculating in pig, and upon the fact that we were clear of purchases beyond this year. You kept silence and deluded me. You deceived your partner and friend, and only kept faith with him when you could deceive him no longer.

It is not the loss of money caused by your conduct, for it is better to lose than gain by speculation; neither the fact that you have involved me in speculation, which I consider dishonorable, although this hurts as you well know, but that you should have concealed our position—deceived your partner—*that* is what shakes my confidence and renders me so unhappy. What I ever did to tempt you to other than straightforward dealing with me, I cannot imagine.

I have been deceived by one whom I trusted—by a partner and a friend; do what I will, thinking over my conduct to this friend I can find nothing to justify such treatment from him.

After signing the letter, Carnegie added a postscript: "This will not be sent until your Christmas day is over. I would make it less sad than mine." The situation was patched up for a time, but almost exactly two years later Carnegie removed Leishman from his position for a similar breach of ethics.[8] Years later, however, after Leishman had lost his fortune, Carnegie placed him upon a personal pension for life.[9]

Charming as Carnegie's letters to friends might be, he was at his best when angry. Winkler quotes from instructions to Charles M. Schwab at the time that certain rival companies backed by J. Pierpont Morgan had formed a combination to control the ferrous metals trade—the situation that eventuated in Morgan's purchase of Carnegie's holdings. In part, the letter ran: "In the case of this . . . Company as in the case of the American Wire Company, if our president steps forward at the right time and in the right way, informs these people that we do not propose to be injured, on the contrary, we expect to reap great gains from it; that we will observe an '*armed neutrality*' as long as it is made to our interest to do so, but that we require this arrangement—then specify what is advantageous for us, very advantageous, more advantageous than existed before the combination and he will get it. If they decline to give us what we want, then there must be no bluff. We must accept the situation and prove that if it is fight they want, here we are 'always ready.' Here is a historic situation for the Managers to study—Richelieu's advice: 'First, all means to conciliate; failing that, all means to crush.' . . . We should look with favor upon every combination of every kind on the part of our competitors; the bigger they grow the more vulnerable they become."

IV *Letters to Public Figures*

For most of his last thirty years, Carnegie carried on a voluminous correspondence with many of the world's most prominent men, including presidents of the United States, prime ministers and secretaries of state, and even King Edward VII of England and Kaiser Wilhelm II of Germany.

With the highest public officials and other important men, his style and attitude were much the same as usual. Although he always addressed a chief executive as "Mr. President," Carnegie said exactly what he felt. Being on intimate terms with Cleveland, McKinley, Theodore Roosevelt, Taft, and Wilson, he exercised a strong influence, and often took a hand at running the government indirectly.

Carnegie's style of writing became even more forceful in age than it had been in earlier years. His words were intense and impatient at times as he became increasingly devoted to the cause of peace. Even with those in the highest positions, Carnegie did not hesitate to scold, command and at times threaten, in order to get his way.

One of the most interesting letters of this type is that addressed to the Kaiser on January 19, 1907,[10] when it appeared likely that the Emperor was preparing to block the proposed Hague conference. Far from his characteristic style, Carnegie wrote in a manner reminiscent of his 1904 magazine article, "Britain's Appeal to the Gods."[11] The missive was in two parts, the first a brief, two paragraph introduction, the other a supposed daydream sequence. Altogether, it is a most remarkable communication, written persuasively and with great charm, well planned and organized.

The letter proper, or opening part, after the salutation, begins abruptly: "In my reveries you sometimes appear and enter my brain. I then imagine myself 'The Emperor' and soliloquize somewhat as per enclosed." This is simply followed by a polite closing, varying from usual custom only in its ending: "and with profound interest in your 'Star,' which should excel all others in brilliancy if followed boldly."

The "enclosure" also opens abruptly: "The august visitor, 'The Emperor,' communes with himself thus when I am transformed:—"

Here, speaking for the Emperor, Carnegie writes in the first person. "God has seen fit to place me in command of the greatest military power ever known." Surely this is to produce good, not evil. "Thank God, my hands *as yet,* are guiltless of human blood," he continues after reflecting on the difference between the heroes of barbarism and civilization. Perhaps God has destined him to bring peace to the world.

The reverie goes on to consider the international situation, the Kaiser's opportunity, climaxing: "Yes! This is my work! Thank God. Now I see my path and am happy. To this I consecrate my life, and surely, 'The highest Worship of God is service to man.'"

At this point Carnegie wakes and becomes himself again, "but still I keep wishing that I were indeed the Emperor and had his part to play." He again lays out the Kaiser's opportunity, in new words and phrases. In conclusion, he points out that if the President had the power to take such action, Carnegie would "have been at his side long ago, urging it." Roosevelt's part will be played well, *"and for Peace, too*—but there is only one on earth to whom has been given the power to resolve and execute—the Emperor of Germany. Pitiable will be his place in history should he falter. 'Where much is given much is required.'"

After holding the letter for four days, he dispatched it through the American ambassador, Charlemagne Tower, with an offer to visit and talk with Wilhelm if so desired. Apparently Tower showed this letter, too, to the Kaiser, for on their meeting that worthy commented on a phrase in the missive to Tower: "He and our President would make a team if they were only hitched up together for the great cause of Peace."[12] In the *Autobiography*,[13] describing the week he spent at Kiel, and his talks with the Emperor, Carnegie quotes the Kaiser: "Oh, I see! You wish to drive us together. Well, I agree if you make Roosevelt first horse I shall follow." Carnegie's reply was, "Ah, no, Your Majesty, I know horse-flesh better than to attempt to drive two such gay colts tandem.... I must yoke you both in the shafts, neck and neck, so I can hold you in."

Wilhelm was friendly and gave Carnegie a good hearing—perhaps even considered his suggestion seriously. It would be interesting to know whether, during his years as a virtual prisoner in Holland, the exiled emperor ever recalled the accuracy of Carnegie's prediction as to his place in history "if he should falter."

V *Letters to Newspapers*

Very early in life Carnegie discovered the power of the press, and was deeply impressed by the opportunity it offered for one individual in a democracy to influence a situation. Success with his first letter to a newspaper, promoting Col. Anderson's library as being free to all working boys,[14] encouraged him to continue writing to editors, sometimes bitterly, always with argument, and often letters full of advice and counsel.

Beginning with missives to his home-town papers in Pittsburgh, he enlarged his field as time passed, and his influence widened. During the final thirty years of his active career he wrote frequently and at considerable length to the principal dailies of the United States, and many of those in England and Scotland as well. Apparently most of these letters were published, even those addressed argumentatively to editors in opposition to their views. He argued for causes, disputed with other writers, answered criticisms and not infrequently predicted the course of events.

Around the turn of the century Carnegie began sending New Year's greetings and warning or encouraging forecasts to newspapers around the holiday season. One of these, less than five years before his death, is a good example of his press letters. In answer to a request, he wrote to the editor of the *Pittsburgh*

Dispatch, in which his letter was printed on January 1, 1915. Writing on "America's Opportunity," his argument is: 1. America has no enmities nor rivalries for materials; 2. this happy situation results from friendly national policies; 3. she needs no large armaments or military forces; 4. there is no reason for anything but optimism in regard to the nation's future.

This letter is worth quoting almost in full, as an example of Carnegie's crisp, sententious style and optimistic nationalism, varying little from the correspondence he had carried on with Dod more than half a century earlier:

Thanks for your kind favor in which you suggest that I say a few words upon "America's Opportunity."

First: Our beloved Republic has no enemies in the world, neither personal nor national. She covets no new territory and wishes all nations peace and prosperity, setting an example to all the world. She is the foremost of nations in longing for international peace, knowing that "Peace hath her victories much more renowned than those of war."[15]

Immigrants come to her from many nations, all certain of being classed under the laws of our own citizens. She welcomes all, and shares her privileges with them. In due time, these arrivals apply for citizenship and become Americans—one man's privilege, every citizen's right—their children educated at our schools, free of expense. She is the pioneer nation proclaiming the brotherhood of man.

She needs no increase in army or navy. The latter is today quite sufficient, better that it were not so large; and as for the army—16,000,000 militia, subject to call, and if called to repel invaders, our only difficulty would be to provide for the surplus millions who would report. Men who advocate increased armies for us can be likened only to those who are afraid to step out of their homes without a lightning rod down their backs, because men have been known to be struck by lightning. Our Republic has nothing to fear, our march is onward and upward. She leads the procession, other nations must follow.

As Wall comments: "Any one of his letters to Cousin Dod, written when he was fifteen, sixty-five years before, could have been substituted for this letter, and no one would have known the difference."[16]

CHAPTER 10

Fair-Haired Scotch Angel: 1901-1919

FOR three decades after the chosen time,[1] Andrew Carnegie was unable to carry out his plan to retire from active business and devote his life to culture, politics, and philanthropy. There is no reason to question his intention. But as years passed, business matters became more and more complex, and he felt compelled to protect the best interests of partners and employees, for all of whom—despite the fiasco of the Homestead steel strike—he retained a paternal feeling. In a letter to Samuel Storey, January 3, 1883, during his newspaper venture, he said, "I am going out of business but it takes a little more time than I had bargained for—that's all."[2]

In 1899 a group led by Henry C. Frick tried to buy the Carnegie holdings for $175 million. Carnegie was willing, but the deal fell through for lack of backing, and he pocketed over $1 million in option money.[3] When J. P. Morgan proposed a major merger two years later through Charles M. Schwab, Carnegie immediately accepted, opening the way for the formation of United States Steel Corporation. He declined to remain a partner, taking as his share about $300 million in 5 percent gold bonds. These were placed in a specially built vault in Hoboken, N.J., as Carnegie never wanted to see or touch any of them.[4]

Freed from the concerns of manufacturing, on which—despite his capable officers and long absences—he had always kept a sharp watch, Carnegie turned his mind to the things he had been anticipating for so long. His self-education had replaced the plan to study at Oxford, and his newspaper venture was far behind him. There remained his writing (which by this time had become one of his best-loved activities) and his philanthropies.

His literary output continued about as usual, both in volume

145

and subject matter, although with a slight decrease in magazine articles for the first five years. Two of the three published in 1901 appeared before the sale of his business was final,[5] and the third may have been under way, as it appeared in the June issue of *Nineteenth Century*. But in 1902 he published *The Empire of Business*—a collection of his previous articles and speeches—and had two speeches printed and two articles published. From 1903 there are at least nine pamphlet speeches, all dealing with his philanthropies, but in the following year a single magazine article, on international affairs. In 1905 came his one full-scale biography, *James Watt*, one speech, and one article. With 1906 his writing activity was again at full steam, with four articles; the next year two pamphlet speeches, three magazine articles and a book introduction.[6] His articles, except for an increased emphasis on philanthropy and peace, differed little from his previous style.

I *A Remarkable Introduction*

The year 1908 brought his last book, *Problems of Today*, four articles, and an introduction to *The Roosevelt Policy*.[7] When a well-known publishing company decided to issue a two-volume edition of Theodore Roosevelt's addresses, state papers and letters on the control of corporate wealth and the relation of capital and labor, Andrew Carnegie was a natural choice to introduce it.

That "Introduction"[8] is a remarkable piece of work. Although Carnegie was in his seventy-second year when he wrote it, the thinking is as clear and the writing as crisp as if he had been many years younger. The thirteen-page (approximately 5,000-word) essay is no puff, although it speaks highly of Roosevelt. Instead it points out that the book is devoted not to the man, but to his policies. Neither does it attempt to set up arguments against those positions with which the philanthropist disagreed.

Carnegie opens with a brief assessment of the President as a man of destiny, and the times which called for such a man. He details what he considers Roosevelt's characteristics in meeting such situations, and concludes with a question which was being widely asked by thinking men.

Roosevelt like Lincoln, Carnegie points out, is a man with peculiarities beyond the usual expectation. "He would not be original if he were like anybody else. None but himself can be his parallel.... Talent can be hammered into shape and conform itself to social conventions.... With genius this is impossible— one of the marked differences between talent and genius being that talent does what it can, genius what it must" (ix).

After a century of unparalleled industrial growth in which the nation's energies had been directed to the development of its resources, legislation to safeguard public interest had been neglected. Particularly during the thirty years following the Civil War almost free rein had been given to this type of expansion on a scale never before approached by any nation. "There came the serious task of regulating interstate commerce and restricting the powers of trusts and corporations which threatened the structure of good government itself. It was at this crisis Roosevelt appeared upon the scene and became immediately a leader in the crusade,—in his gubernatorial message to the State of New York in 1899, just nine years ago. From that day to this he has hammered away and descanted always upon the same lines" (x).

After citing a dozen of the messages and papers dealing with such matters Carnegie briefly outlines Roosevelt's positions on the other principal matter involved, the relation of capital and labor (x–xi). Seeking the source of his subject's energy, he finds it in a paper, "Conduct as the Ultimate Test of Religious Belief," although wondering how the President managed to harmonize it with "the Scottish Covenanting strain in him." But Carnegie feels that all Americans are debtors to Roosevelt's teaching when he tells Christians that "more and more people who possess either religious belief, or aspiration after it, are growing to demand conduct as the ultimate test of the worth of belief" (xii).

Carnegie points out Roosevelt's principal innovations including the creation of a new cabinet post (Secretary of Commerce and Labor), pure food laws and conservation of natural resources, on which he had just called the first Governors Conference. "He reminds one of the description given of the first Naysmith steam hammer, as cracking a nut or forging an anchor

with equal facility.... The heavier the blow required, the
better ... the busiest man in the world ... is surely the President" (xiii–xiv). Roosevelt, he continues, has accomplished
wonders in the field of commerce, both interstate and corporate.
Citing the statement of Elbert H. Gary, president of the United
States Steel Corporation, that the president's principles had
"increased my feeling of responsibility toward the stockholders
I represent, toward our competitors, toward business men, and
toward the public" (xv), Carnegie cites Roosevelt's almost
unerring choice of men to be appointed to important positions,
then looks into the future:

When the day comes, as come it will, history is to record that
just as Washington, struggling for Constitutional Rights, led the hosts
that ensured national independence, as Lincoln preserved the Union
by uprooting the sole cause of disunion, so will stand Roosevelt, who
brought order out of chaos in our Interstate Commerce, and in our
industrial system elevated and purified the conceptions of fiduciary
duty in men of affairs, investigated charges and sternly enforced
honesty in the dealings of officials, enforcing everywhere a stricter
rule of conduct and a higher standard of action than that which
before his day had unfortunately prevailed. More than this will the
true historian add if he speaks the whole truth: namely, that the
President also lived up to the high standard he set for others. (xx)

In conclusion, Carnegie quotes John Morley: "I have seen
two wonders in America, Roosevelt and Niagara." He continues:

Niagara will run its course and so will Roosevelt, but both are
fresh and overflowing.... He would be a bold man who attempted
to forecast the future of either. Roosevelt's policy is already vic-
torious.... The Presidency may be held by another ... but strip him
of all external dignities and there still remains the Man in full possession of marvelous powers, high ideals, sleepless activity and bound-
less popularity.... Is he to sigh for more worlds to conquer and,
after a rest, to reappear among the champions, eager for the fray, or
to forsake public life and rust in inaction? The Sybil is silent. (xxi)

Quite unexpectedly, the "Introduction," something seldom
read, created a sensation. One journal, at least, gave it more

space in its review than it did to all the rest of the work.[9] The reviewer expressed surprise that a book attacking the undue concentration of wealth "should be sent" into the world with an introductory benediction from one of the two richest men in America. "But Mr. Carnegie, as we all know, has always been 'on the side of the angels,' and the reformers." The praise was due to Carnegie's personal admiration for Roosevelt, and agreement with some of the policies. As he was to demonstrate later the same year in *Problems of Today*,[10] his belief in the necessity of large fortunes had not changed.

From 1909 we have one pamphlet speech, 1910 two magazine articles, and in 1911 five, with a speech which was published the following year. From this time on—he was seventy-six years old—his output dropped, but before his production ceased there were five more articles, two in 1914, and one each in 1913, 1915 and 1916. Work on the *Autobiography*, published after his death, continued until the onset of World War I in 1914.

II *The Business of Philanthropy*

During these years, however, Carnegie considered philanthropy his principal work. Having amassed the world's largest private fortune, he wanted to use it while he lived. As early as 1887 he had told William E. Gladstone he considered it a disgrace to die rich.[11] Carnegie felt that at his age the major task would be enough—to disperse wisely what he already had accumulated. He underestimated the stupendous job ahead of him, and later stated that he had not worked one-tenth as hard in acquiring his fortune as he did in divesting himself of his great wealth.[12]

In his autobiography Carnegie wrote, "as usual, Shakespeare had placed his talismanic touch upon the thought and framed the sentence—'So distribution should undo excess, and each man have enough.'"[13]

His main goal was not for charitable relief, holding that this was the responsibility of the State, and often did more harm than good, but instead for the promotion of education which would help to prevent poverty and ignorance. During his lifetime, Carnegie gave over 80 percent of his fortune for educational purposes of one kind or another. To him the mind was the greatest wealth, and he set about to stimulate all those he could reach.

By 1919, Carnegie had donated 7689 organs to churches throughout the world—his first going in 1873 to the Swedenborgian Church of his father.[14]

Libraries were early in his mind. Believing that "the chief glory of a nation is its authors,"[15] Carnegie saved the library of Lord Acton (the greatest historical student in England) by buying it himself in 1890, and putting the former owner in charge. Through his efforts it finally went to Cambridge University in 1902.[16] Carnegie considered libraries his specialty and by the time of his death had given 2811 free public library buildings to English-speaking countries.[17]

Contrary to common belief, Carnegie did not require or encourage these institutions to bear his name. In his words, "I do not wish to be remembered for what I gave, but for that which I have persuaded others to give."[18] By building libraries, he inspired communities to fill them with books, asking only that they inscribe the words "Let there be light" above the entrance.[19]

Although Carnegie had in 1885 given $50,000 to establish the first great medical research laboratory in the United States at Bellevue Hospital in New York, and had at the same time sent four children at his own expense to Paris to be treated by Louis Pasteur for rabies, he did not do much medical financing. He contended that this was John D. Rockefeller's specialty. However, he did make a $120,000 donation to the Koch Institute of Berlin for medical research and gave $50,000 to Madame Curie for her work in radiology. And due to his wife's interest in Helen Keller, he contributed to the New York and Massachusetts associations for the blind.[20]

Even these benefactions did not satisfy Carnegie's philanthropic nature, and he began to distribute his wealth with more zeal, concentrating most of his efforts within the years from 1901 to 1911. Soon his headquarters in the mansion house at No. 2 East Ninety-First and Fifth Avenue, New York, became one of the most famous of all American addresses. During these years he literally received mail in bushel baskets with requests for money from all over the world. One story is told of a gentleman asking timidly for a meager $5,000. Carnegie declined, asserting, "I am not interested in the retail business!"[21]

Starting with the men in his former mills, Carnegie established a $4 million pension and relief fund, also donating a million dollars to maintain the libraries and halls he had built for the workmen. These lines, written in 1903, from the workers at Homestead, express most feelingly the sentiment inspired by his action: "The interest which you have always shown in your workmen has won for you an appreciation which cannot be expressed by mere words.... We have personal knowledge of cares lightened and of hope and strength renewed in homes where human prospects seemed dark and discouraging."[22]

As early as 1881 Carnegie offered Pittsburgh $250,000 for a library and concert hall. The offer was refused, and he made the gift to Allegheny City—now Pittsburgh's North Side. About the time the library was dedicated in 1890, Pittsburgh asked him to renew the offer. He gave $1 million, later increased many times over, to include a museum and art gallery. He also provided for a craft school for workmen, which in 1912 became a college, and is now Carnegie-Mellon University.[23]

In 1904 his second large gift was incorporated in Washington, D.C.—to promote knowledge in all universities of the United States. This Carnegie Institution was valued then at $25 million, and was the least criticized of all of Carnegie's endowments.[24]

Steeped in a lifetime of hero worship, Carnegie's next gift—his dream of a world-wide Hero Fund—soon materialized. This brain-child and pet fund was inspired mainly by the heroic but futile rescue effort in which a former Pittsburgh mine superintendent and others lost their lives at the time of the Harwick disaster in 1904. The $5 million fund, established in America in 1904, was later extended to the Carnegie Fund Trust for Great Britain and Canada, and eventually, similar ones in France, Germany, Italy, Belgium, Holland, Norway, Sweden, Switzerland and Denmark.[25]

Since Carnegie believed that the teaching profession was underpaid, he established in 1905 what he considered his fourth most important gift—The Carnegie Foundation for the Advancement of Teaching—a $15 million pension plan which organized the Teacher's Insurance and Annuity Association of America in 1917.[26]

To his native Scotland Carnegie gave $10 million in 1901 toward education—half of which was to be used to pay the fees of deserving poor students and the other half for the improvement of universities. This endowment, known as the Carnegie Trust for the Universities of Scotland, met with much opposition—one of the most controversial of all his large benefactions. Carnegie did not anticipate the bitter criticism he would encounter as a result of his philanthropy. The public thought he was doing too much.[27]

Carnegie believed that the smaller institutions and colleges were in greater need of help and preferred to concentrate on them. To Princeton, disillusioned and thinking it would receive a good portion of the Carnegie money, he gave a lake for the promotion of a rowing crew to take the young men's minds off football, a sport to which Carnegie had an aversion. The sage was not without humor, even when it came to giving. By the time of his death he had distributed $27 million among 500 institutions of learning.[28]

III *Purchasing Peace*

International peace being foremost in his mind all his life, it was only natural that Carnegie would do great services in its behalf. Along with the Hero Fund, which was intended to point out that peace as well as war had heroes, there was his attempt to revolutionize the English language, establishing The Simplified Spelling Board in 1903 for the benefit of world-wide communication which he assumed would ultimately lead to peace. This project was made sport of by humorists and the press, who ridiculed him by employing his own reasoning, spelling his name "Androo Karnage." After donating $25,000 a year toward the fund until 1915, Carnegie finally gave up this endeavor.[29]

Thinking that money could buy anything, Carnegie built three temples of peace—the Pan American Union Building opened in 1910 in Washington, D. C.; another in Cartago, Costa Rica, dedicated the same year and named the Central American Court of Justice; and the third, which was begun first, finished last and caused the most problems (1907–1913)—The Temple of Peace, at The Hague. Together with these he endowed four trusts between 1903 and 1914 for the cause of peace totaling $25,250,000.[30] In 1914 he gave over $2 million to the Church Peace Union.[31]

Among his other important benefactions were: $10 million to the Carnegie United Kingdom Trust in 1913; $1,500,000 to the United Engineering Society; $850,000 to the International Bureau of American Republics; $2,500,000 during war years to the Red Cross, YMCA, and Knights of Columbus; and sums ranging from $100,000 to $500,000 to a score of research hospitals and educational boards.[32]

The world in general places its value and emphasis on the great benefactions already mentioned, but to Carnegie himself two others gave the most satisfaction. First was his private pension fund. There was never any publicity regarding the anonymous recipients. Only Carnegie and his intimate friends knew of the gifts that were given to others in this category. At the end of his life Carnegie was distributing $250,000 a year among almost five hundred beneficiaries.[33]

The gift which probably delighted him most of all was a historic and personal triumph—giving Pittencrieff Glen to the townspeople of Dunfermline. By this time Carnegie had already given a library to his native town (1881), $25,000 for swimming baths (1873), and the Carnegie Dunfermline Trust of $4 million (1903). The community with a population of 27,000 had the largest per capita private endowment in the world as a result of Carnegie benefactions.[34]

But the park pleased him more than anything—an air-castle that indeed came true. He wrote in his *Autobiography*, "Pittencrieff Glen is the most soul-satisfying gift I ever made, or ever can make."[35] Carnegie as a youth was barred from viewing the historic ruins of Malcolm's Tower, Margaret's Shrine, and the last remains of the palace of the Stuarts, due to a feud between his grandfather Morrison and the landowner. The eccentric James Hunt decreed that no Morrison or descendant of a Morrison should "ever step foot on Pittencrieff's sacred soil." Carnegie bought the entire property in 1902, and gave it to the town the following year, retaining only a small portion, including Malcolm's Tower—the birthplace of Scottish history. Thus Carnegie became the Laird of Pittencrieff of the Royal Berg of Dunfermline.[36] A fantastic storybook ending, Carnegie loving every morsel of drama, romance, and sense of climactic outcome.

IV *Fun in Giving*

At first Carnegie considered it a game to give his money away. Whereas most men enjoyed making money, he in contrast delighted in seeing his fortune diminish. There never lived a man who had as much fun in giving away his wealth as Carnegie. One of his teasing tricks in giving was to hold his beneficiary in suspense, using the element of surprise. While playing golf with Sir Swire Smith one day in 1899, Carnegie offered to give him a stroke a hole. Later that evening Smith mentioned Keighley Institute in their conversation. Carnegie asked if there was a library in the English town. When the answer was in the negative, he asked if $50,000 would be enough, and his partner was overcome with joy. The next morning during another round of golf Carnegie said, "I have repented of the offer I made you yesterday." Smith was crest-fallen until Carnegie added, "I don't think, after all, I can give you a stroke a hole!"[37]

In 1905 Carnegie again became involved in politics, both domestic and foreign, and this took much of his time away from philanthropy. By 1906 he was tired of the game of distribution and by 1910 completely disgusted. He had to abandon his "Gospel of Wealth" as too arduous an undertaking. No longer believing that one man was capable of disposing of such a fortune, he resigned the task to the discretion of others.

Discouraged, he realized that he could not give his money away fast enough. As Wall states, "The interest on his bonds kept gaining on his dispersal of those bonds. He had given away $180,000,000 but he still had almost the same amount left." Finally in 1911 Carnegie established and gave $125 million—the bulk of his fortune—to the Carnegie Corporation of New York for the advancement of knowledge, giving freedom to the trustees on how to spend it.[38] Carnegie, with utmost modesty, never even mentioned this largest contribution of all in his *Autobiography*—Van Dyke having to insert it in a footnote.

V *Final Bequests*

Upon Carnegie's death, when the will was opened, there remained $30 million to be dispersed—two-thirds of which went to the Carnegie Corporation of New York. The remaining $10 million went for yearly pensions to Dunfermline relatives, old friends and associates, one of whom was ex-President Taft. And attempting to shame the government into action, Carnegie had included Mrs. Theodore Roosevelt and Grover Cleveland's widow. The final million dollars went to Hampton Institute, the University of Pittsburgh, Stevens Institute, Cooper Union, the Relief Fund of the Authors Club of New York, and the St. Andrew's Society.[39]

There was no other monetary bequest—nothing more except for Carnegie's real estate. He wrote, "Having years ago made provision for my wife beyond her desires and ample enough to enable her to provide for our beloved daughter, Margaret, and being unable to judge at present what provision for our daughter will best promote her happiness, I leave to her mother the duty of providing for her as her mother deems best. A mother's love will be the best guide." At the time of his death, Louise Carnegie was sixty-two. She had known, approvingly, from the beginning that her husband planned to distribute the greatest part of his fortune to mankind. The entire Carnegie benefactions amounted to over $350 million—an overwhelming sum donated by one individual. Carnegie had given away 90 percent of his fortune within his lifetime. Believing to the end it was more blessed to give than to receive, the great benefactor had not died in disgrace.

VI *Significance*

The typical industrialist of a century ago was far from being the knowledgeable, suave, college-trained man of today's world— a public figure equally at home at an interview or on the speaker's platform, ready to give voice or pen to causes which may affect his business. Ordinarily he was a man of not more

than average education; he had gone to work—often in his
father's business or mill—in his teens and learned from older men
or by experience. Very few, even up to the end of the nineteenth century, could be accurately described as truly literate,
much less learned men. If such a man used the pen it was likely
to be in brief, brusque business letters, or as an accountant.

Andrew Carnegie quickly realized that writing was easily the
most effective manner of getting his ideas and opinions before
others, and that education was the key to influence, except for
the bald, venal purchase of politicians and office holders, to
which he would not stoop.

Many of his early business superiors gave him preferment
and increased opportunity because they were impressed by what
he wrote. But there is no reason to believe that it ever occurred
to any of them to use his abilities in this respect for the benefit
of their organizations. They looked on him as bright, ambitious,
and capable, but as a little zany.

If not absolutely the first genuine writing industrialist, Carnegie was the first to use the full power of an interesting approach, trenchant style, and broad knowledge and vocabulary
in pushing his causes and making himself a public figure.

Newspaper editors—always likely to welcome "good copy"
—were the first to recognize the value of what he had to say.
His first two books, like the "story of my life" type of thing
sometimes done by other successful men of his day, were written with the expectation of a very limited audience. But their
greater interest and better style and content caught the eye of
a publisher. And with the opportunity to get them before the
public, Carnegie realized that his best way to success was by
rewriting them. Other travel writers had used the pen to express
their opinions and preferences; but few, if any, had deliberately
used their works as a means to get their ideas and convictions
a hearing, and to increase their influence in the world beyond
their own circle. From the ashes of a defeat in his attempt to
change the British opinion through organizing a newspaper
chain, Carnegie set up a great publicity caper—his coaching
trip throughout the length of England and Scotland. And when
a chance opportunity came as a result, to get his name into
the greatest of the English reviews, the article he wrote was

eye-catching and controversial enough to attract international attention. From this time forward his pen was seldom idle, despite his busy life of making a vast fortune. Where another man's words might be listened to because he was a business leader, Carnegie was widely known as a writer before it was generally recognized that he was an industrial lion.

He was the first to set the precedent of furthering industry by influencing a wide segment of the public. The vocal and literary business leader or other public figure has appeared to a large degree because of the pattern that Carnegie set. And the ghost writer, a device which many later students have supposed him to have employed, has largely come into existence because of the challenging model he so successfully set.

One other extremely significant effect of Carnegie's writing was its strong contribution (illustrated by his career) to the public belief in the so-called "American dream"—that in a democratic, capitalist society any young man who was willing to work could achieve fame and fortune.

Notes and References

Chapter One

1. Andrew Carnegie to President Woodrow Wilson, January 23, 1915, in J. F. Wall, *Andrew Carnegie* (New York, 1970), p. 971.

2. *Autobiography of Andrew Carnegie*, ed. John C. Van Dyke (Boston, 1920), p. 13.

3. Wall, p. 83.

4. Letter to George Lauder, Sr., May 30, 1852; quoted in Burton J. Hendrick, *The Life of Andrew Carnegie* (New York, 1932), I, 45–49.

5. *Autobiography*, pp. 35–37.

6. Frank C. Harper, *Pittsburgh of Today* (New York, 1931), I, 330, 333.

7. *Autobiography*, p. 43; Hendrick, I, 58–59, gives the number as 1,800 and quotes the librarian as saying 2,000.

8. Quotations are taken from the text as given by Hendrick, I, 68–70, although even casual inspection shows that it is incomplete and probably corrupt as well. Unfortunately, no file of the *Dispatch* for this period appears to have been preserved.

9. [Margaret Barclay Wilson], "Carnegie Bibliography: List of Letters by Mr. [Andrew] Carnegie, Printed in Newspapers from March 18, 1895, to February 5, 1915, n.p. [Pittsburgh], n.d. [1930]; TS in Carnegie Library of Pittsburgh.

10. Pet name for Lauder.

11. Quoted, Hendrick, I, pp. 66, 72–77.

12. *Autobiography*, pp. 68–69.

13. *Rules for the Government of the Pennsylvania Railroad Company's Telegraph* (Harrisburg, Pa., 1863). Its location since the Penn-Central Railroad bankruptcy and auction of assets is unknown.

14. A native son would have written "supper."

15. William B. Wilson, *History of the Pennsylvania Railroad Company, and Historical Sketches, etc.* (Philadelphia, 1899), p. 117; but concerning this period Wilson was writing from hearsay, and differs widely from other biographical accounts.

16. Witness the selections in William B. Wilson, *Robert Pitcairn*,

1836–1909, in Memoriam (n.p., n.d. [1913]), 8–12, 19–22, in which Pitcairn's sentence length averages nearly fifty words.

17. *Autobiography*, p. 66.

18. The others include *Problems of Today* (New York, 1908), *The Empire of Business* (London and New York, 1903), and *James Watt* (Edinburgh and London, n.d. [1905]).

Chapter Two

1. Wall, pp. 203–204.

2. Ibid. p. 227.

3. Ibid. pp. 250–51.

4. Ibid. p. 152.

5. *Autobiography*, p. 93.

6. For example, he remarked in passing up the excursion to Switzerland: "I have seen enough mountain scenery in Scotland to satisfy me for a time." Letter to Margaret Morrison Carnegie, September 2, 1865; cited in Wall, pp. 231–32.

7. Hendrick, I, 137.

8. Bayard Taylor, *Views A-Foot*, second ed. (Philadelphia, n.d. [1848]), Preface by N. P. Willis, pp. 9–12.

9. *Autobiography*, p. 137.

10. For example, "If I had time, I would write frequently for the *Commercial* but there is none"; Paris, September 29, 1865. "I have an abundance for many letters to the *Commercial* and would like to write them, but where's the time?"; Verona, December 10, 1865. Travel letters in the possession of Margaret Carnegie Miller; TS, lent by J. F. Wall, as were the further Miller letters cited below.

11. Letter of September 25, 1865; in the possession of Margaret Carnegie Miller.

12. It is uncertain whether Carnegie drew this reference from Voltaire or Swift (by 1865 he was probably familiar with both writers) or from some intermediate source.

13. Andrew Carnegie to Margaret Morrison Carnegie and Thomas Carnegie, November 5 and 12, 1865; in the possession of Margaret Carnegie Miller.

14. Andrew Carnegie to Margaret Morrison Carnegie and Thomas Carnegie, May 25, 1865; in the possession of Margaret Carnegie Miller.

15. J. K. Winkler, *Incredible Carnegie* (New York, 1931), p. 103.

16. Winkler, p. 105, referring to these sales, says, "Some of these turned out to be worthless as autumn leaves." Unless the pronoun

refers to American bonds in general, there appears to be no truth in the statement. The principal reason Carnegie found selling easy was the confidence of bankers that his offerings were always sound.

17. Wall, p. 400.

18. Hendrick, I, illustration between pp. 146 and 147.

19. Wall, p. 364.

20. Andrew Carnegie, *Before the Nineteenth Century Club upon The Aristocracy of the Dollar* (n.p., n.d.); pamphlet in Carnegie Library of Pittsburgh.

21. Wall, p. 362. Carnegie's statement in the *Autobiography*, p. 145, that the meeting came through Anne C. L. Botta is evidently a slip in an aging man's memory. He had written in 1893 in "Characteristics," *Memoirs of Anne C. L. Botta*, p. 166, that Palmer had introduced him to her.

22. *Autobiography*, pp. 144–45.

23. Van Wyck Brooks, *The Confident Years* (New York, 1952), p. 6.

24. Despite references to the contrary in *Autobiography*, p. 327 (followed by Hendrick, I, 238), Wall, p. 362, is almost certainly correct in his conclusion that these men had little if any impact on Carnegie until after he moved to New York.

25. Andrew Carnegie, "Characteristics," in *Memoirs of Anne C. L. Botta, Written by her Friends, etc.*, ed. Vincenzo Botta (New York, 1894), p. 165.

26. Ibid.

27. Ibid., p. 166.

28. Ibid., p. 165.

29. Hendrick, I, 221–22.

30. Ibid., I, 239–40.

31. *Autobiography*, p. 322.

Chapter Three

1. Hendrick, I, 137.

2. Andrew Carnegie, *Round the World* (n.p., 1879), dedication.

3. Andrew Carnegie, *Our Coaching Trip* (New York, 1882), dedication.

4. *Autobiography*, pp. 203–204.

5. The similarity in type, makeup, and binding makes it appear that *Round the World*, like *Our Coaching Trip*, was printed as a favor by Charles Scribner's Sons.

6. Wall, p. 362.

7. *Round the World,* pp. 228–29.

8. Although he had evidently planned to be away a year, Carnegie returned to New York exactly eight months from the time he had sailed from San Francisco.

9. "Vandy" was John Vandevort, a boyhood friend and associate; "Harry" was Henry Phipps, Jr., another of the Homestead and European tour group.

10. *Autobiography,* pp. 200–201.

11. *Round the World,* p. 187.

12. Ibid., p. 22.

13. Ibid., p. 24.

14. *Round the World,* second ed. (New York, 1884), p. 4. Further references to this work will be found in the text in parentheses.

15. Winkler, p. 151.

16. *Autobiography,* p. 199.

17. *Our Coaching Trip,* title page.

18. Ibid., table following p. 275.

19. Ibid., p. 9.

20. Wall, pp. 401–403.

21. *Our Coaching Trip,* pp. 186–239. Further citations of this work are provided in parentheses in the text.

22. See above, p. 42.

23. *Autobiography,* pp. 203–204.

24. Ibid., p. 204.

25. Ibid.

26. Andrew Carnegie, *An American Four-in-Hand in Britain* (New York, 1883), pp. vii–viii.

27. Hendrick, I, 237.

28. See above, p. 45.

29. Compare *Round the World,* p. 2, and *Round the World,* second edition, p. 2.

30. Compare *Round the World,* p. 37, and *Round the World,* second edition, p. 44.

31. Carnegie's implication (*Autobiography,* p. 198) that this book was published prior to *An American Four-in-Hand in Britain* is apparently the result of a faulty memory.

32. Hendrick, I, 237.

Chapter Four

1. Quoted by Hendrick, I, 264–65.

2. Ibid., I, 147.

3. Ibid., I, 261–73; Wall, pp. 429–41.

4. Andrew Carnegie, "As Others See Us," *Fortnightly Review* NS 31 (February 1, 1882), p. 156. Carnegie here appears to be trying to leave the impression that the suggestion came from William E. Gladstone, at that time Britain's prime minister. If such were the case it is doubtful if he would have avoided using the name, even if—as is quite likely—he had himself engineered the request, as he was later to do with a compliment from King Edward VII (Hendrick, II, 174). Hendrick, in a prefatory note to this article in *Miscellaneous Writings of Andrew Carnegie,* ed. Burton J. Hendrick (New York, 1933), suggests that the person was Lord Roseberry.

5. Ibid., p. 156. Further citations from this article are added in parentheses in the text.

6. Andrew Carnegie to Samuel Storey, January 3, 1883 (quoted by Hendrick, I, 267): "I've sent an article to Escott, of the Fortnightly."

7. Hendrick, II, 389, mistakenly states that the speech was before "the legislature of Pennsylvania." A recast version was given before that body in April of the same year.

8. See below, p. 94.

9. Hendrick, I, 351.

10. *Autobiography*, p. 204.

11. Quoted in Hendrick, II, 268.

12. The first two were in British reviews, and until 1890 the numbers were equal, four each in Britain and America. From this time the number originally appearing overseas declined (although all in prestigious publications) in relation to the total. In the final decade of the century sixteen appeared in this country, and ten in England, while after 1899 there were twenty-six in American magazines, eight in British journals.

13. The other participants were Thomas Wentworth Higginson, Murat Halstead, Horace Porter, Robert Collyer, James H. Eilson, and M. W. Hazeltine.

14. See below, p. 122.

15. Single quotation marks were perhaps connected with Carnegie's effort to promote simplified spelling. See below, p. 152.

16. Andrew Carnegie, "The Next Step—A League of Nations," *Outlook* 86 (May 25, 1907), 151–52, part of a symposium including President Theodore Roosevelt, Edward Everett Hale, Baron Estournelles de Constant, and Lyman Abbott.

17. October 17, 1905. Printed in pamphlet form by the Student Representative Council, St. Andrews (n.p., n.d. [1905]), for the

International Union (Boston, 1906), and reprinted by the New York Peace Society (1911).

18. *Nineteenth Century and After* 60 (August 1906), 224–33.

19. *Century* 80 (June 1910), 307–10.

20. *Cosmopolitan Student* 5 (June 1915), 333–34.

21. Hendrick, II, 171–72.

22. Ibid., pp. 170–71.

23. *Nineteenth Century and After* 55 (April 1904), 538–42.

24. Ibid., p. 538.

25. Ibid., pp. 541–42.

26. *Independent* 70 (June 1, 1911), 1183–92.

27. *Woman's Home Companion* 43 (March 1916), 19.

Chapter Five

1. See above, p. 54.

2. Wall, pp. 389–97.

3. Andrew Carnegie, *Triumphant Democracy, or Fifty Years March of the Republic* (New York and London, 1886). John C. Van Dyke, in a footnote, *Autobiography*, p. 318, erroneously gives the dates as "London, 1886; New York, 1888."

4. Andrew Carnegie, "Wealth," *North American Review* 148 (June 1889), 653–64, reprinted in the *Pall Mall Gazette* as "The Gospel of Wealth." Under this title it was widely printed in pamphlet form and with minor changes in a book of the same name. It created a tremendous impression, and has been repeatedly reissued, as recently as 1962.

5. M. G. Mulhall, *Balance Sheet of the World for Ten Years, 1870–1880*, (London, 1881).

6. *Autobiography*, pp. 318–19.

7. With Carnegie, the term "race" had nothing to do with the races of mankind, but referred to nationality and language, or loosely as a combination of heredity and culture. Specifically, he used it constantly for English-speaking peoples, and sometimes for those of Anglo-Saxon heritage.

8. *Triumphant Democracy*, pp. 18–19. Further citations of this work are provided in parentheses in the text.

9. Carnegie, although not a believer in revealed religion, was thoroughly familiar with the King James Version of the Bible, frequently using unstated references and unidentified phrases and quotations from it in his writing and speech: e.g., "unstable as water, thou shalt not excel," *Triumphant Democracy*, p. 27, quoting Genesis 49:4.

10. Matthew 7:20.

11. James 2:18.

12. The incorrect addition of "free" is derived from the Constitution of Massachusetts; the other misquotations are merely examples of Carnegie's loose method of quotation from memory.

13. John Bartlett, *Familiar Quotations*, Centennial edition (Boston, 1955), p. 891, lists the "shirtsleeves" maxim merely as "attributed to Andrew Carnegie," although it occurs three times in his published works; the other instances are in his *Memorial Address on William Chambers*, October 19, 1909, issued in pamphlet form and reprinted in *Miscellaneous Writings*, I, 209, and in the pamphlet of his address *The Aristocracy of the Dollar*.

14. Hendrick, I, 274.

15. Wall, p. 444.

16. *American Historical Review* 57 (April 1952) 707–709.

17. I have been unable to find that Bridge ever held such a post or title, or did such work except as mentioned below, p. 79. He made no such claim.

18. *Our Coaching Trip*, p. 22; *Four-in-Hand*, p. 28.

19. Joseph F. Wall, Carnegie's definitive biographer, agrees with this statement (conversation with the author, 1970), as does Robert L. Beisner, *Twelve Against Empire* (New York, 1968), p. 281.

20. "Triumphant Democracy," rev. of Andrew Carnegie, *Triumphant Democracy*, *Saturday Review* 62 (September 18, 1886), 394.

21. Information taken from two unidentified clippings pasted in a copy of the first British edition in the present author's collection, one dated in pen, "Nov. 1887."

22. Andrew Carnegie, *Triumphant Democracy Ten Years Afterward, or Sixty Years March of the Republic* (New York, 1893).

23. See above, pp. 77–78.

24. Andrew Carnegie, *The Gospel of Wealth and Other Timely Essays* (New York, 1900); see below, Chapter Six, passim.

25. Ibid., pp. 159–60.

Chapter Six

1. See above, p. 33.

2. William E. Gladstone, *Diary*, July 13, 1887; cited by Hendrick, I, p. 318.

3. "Adam Smith, who did for the science of economics what Watt did for steam"; Andrew Carnegie, *James Watt* (Edinburgh and London, 1905) p. 29.

4. Andrew Carnegie, "My Experience with and Views upon the Tariff," *Century* 77 (December 1908), 196–205.

5. Andrew Carnegie, "The Worst Banking System in the World," *Outlook* 88 (February 29, 1908), 487–89.

6. Wall, p. 960.

7. Andrew Carnegie, "The Manchester School and Today." *Nineteenth Century*, February 1898.

8. Andrew Carnegie, "The Best Fields for Philanthropy," *North American Review* 149 (December 1889), 682.

9. Andrew Carnegie, *The Gospel of Wealth and Other Timely Essays* (New York, 1900), pp. 1–2. Further citations of this work are provided in parentheses in the text.

10. "The Gospel of Wealth," rev. of Andrew Carnegie, *The Gospel of Wealth and Other Timely Essays, Outlook* 67 (March 9, 1901), 572.

11. "The Best Fields for Philanthropy," p. 682.

12. Quoted, *Gospel of Wealth*, p. 20.

13. Counting both "Wealth" and "The Best Fields for Philanthropy," which were merged in the book to form "The Gospel of Wealth."

14. See below, pp. 121–22.

15. William E. Gladstone, "Mr. Carnegie's 'Gospel of Wealth,' a Review and a Recommendation," *Nineteenth Century* 28 (November 1890), 677–93.

16. Hugh Price Hughes, Henry L. Manning, and Hermann Adler, "Irresponsible Wealth," *Nineteenth Century* 28 (December 1890), 876–900.

17. Originally printed in the *Century* 60 (May 1900), 143–49.

18. "The Gospel of Wealth," review; see above, p. 166 n. 10.

19. Andrew Carnegie, "British Pessimism," *Nineteenth Century and After* 49 (June 1901), 901–12.

20. *Autobiography*, p. 248; Wall, p. 797.

21. Andrew Carnegie, "The Gospel of Wealth—II," *North American Review* 183 (December 7, 1906), 1096–1106.

22. Andrew Carnegie, *The Empire of Business* (New York, 1902), p. 84.

23. Ibid., p. 117.

24. Hendrick, II, 268.

25. Andrew Carnegie, *Problems of Today, Wealth—Labor—Socialism* (New York, 1908), p. 165—quoting Karl Pearson, *The Ethic of Free Thought* (London, 1888), p. 445.

26. Ibid., p. 169.
27. Ibid., pp. 176–77.

Chapter Seven

1. *Our Coaching Trip*, pp. 28–29.
2. Wall, p. 448.
3. *Autobiography*, pp. 58–59.
4. Hendrick, I, 147.
5. Ibid., I, 247.
6. *Autobiography*, pp. 202–203.
7. *Our Coaching Trip*, p. 19. In repeating the story in *Four-in-Hand*, p. 25, Carnegie substituted "enunciation" for "pronunciation."
8. Hendrick, I, 246–47.
9. Burton J. Hendrick, ed., *Miscellaneous Writings of Andrew Carnegie* (New York, 1933), I, 78–125, 126–35, 176–85, 185–99, 200–13, 214–39, 240–62, 265–305; II, 1–17, 61–78, 81–87, 88–122, 203–18, 221–71, 272–88, 291–309. (This work, published as the ninth and tenth volumes of Burton J. Hendrick, ed., *Writings of Andrew Carnegie* [New York, 1933], is separately titled, and the volumes separately numbered, both on spine and title page.) Henceforth cited as *Writings*.
10. *Gospel of Wealth*, pp. 219–48; *Empire of Business*, pp. 3–18, 71–91, 125–50, 173–86, 189–225, 285–287, 291–300.
11. Andrew Carnegie, "Industrial Pennsylvania," December 18, 1900, *New York-Pennsylvania Society Year Book*, I, 33–36.
12. "Andrew Carnegie, "The Scotch-American," November 30, 1891, *St Andrew's Society Yearbook for 1892*, pp. 119–26.
13. Quoted in "Industrial Pennsylvania," p. 18.
14. "A League of Peace," October 17, 1905, *Writings*, II, 221–71.
15. " 'Honor' and International Arbitration," March 10, 1910, *Writings*, II, 272–88.
16. Ibid., introductory note, p. 272.
17. *Our Coaching Trip*, pp. 199–201.
18. Ibid., p. 236.
19. *Peterhead Free Library Speech*, August 8, 1891 (n.p., n.d.); pamphlet in Carnegie Library of Pittsburgh.
20. *Autobiography*, pp. 17–18.
21. Speech at Dingwall, Scotland, July 16, 1903, pp. 4–5. Quoted in James M. Swank, ed., *More Busy Days* (Philadelphia, 1903), pamphlet collection of Carnegie speeches, in author's collection.
22. Tain, Scotland, August 27, 1903, *More Busy Days*, pp. 30–32.

23. Kilmarnock, Scotland, September 5, 1903, *More Busy Days*, pp. 42, 44, 47.

24. Govan, Scotland, September 11, 1903, *More Busy Days*, pp. 65–66, 69, 71.

25. Waterford, Ireland, October 21, 1903, *More Busy Days*, pp. 83–85.

26. Limerick, Ireland, October 20, 1903, *More Busy Days*, pp. 96–99.

27. Cork, Ireland, October 22, 1903, *More Busy Days*, p. 113.

28. Barrow, England, September 1, 1903, *More Busy Days*, p. 132.

29. Cornerstone at Ayr speech, October 5, 1892; pamphlet in Carnegie Library of Pittsburgh.

30. *Gospel of Wealth*, p. 248. Address at Glasgow, September 13, 1887. The quotation is from Tennyson's "Locksley Hall," and as usual is inexact.

31. Speech to graduating class, Bellevue Hospital, New York, March 9, 1885; pamphlet in author's collection.

32. "The Road to Business Success," Curry Commercial College, June 23, 1885. *Empire of Business*, pp. 3–18.

33. *Empire of Business*, p. 4; *Writings*, I, 261.

34. *Empire of Business*, p. 17.

35. "Wealth and Its Uses," *Empire of Business*, p. 125.

36. "Business," *Empire of Business*, p. 193.

37. "The Industrial Ascendancy of the World," *Writings*, I, 78–125.

38. Ibid., p. 79.

39. Ibid., pp. 120–22.

40. "A Confession of Religious Faith," *Writings*, II, 291–319. Further references to this speech are inserted in parentheses in the text.

41. "White and Black in the South," *Writings*, II, 81–87.

42. "The Negro in America," *Writings*, II, 88–122.

43. Wall, pp. 973–76. Wall erroneously states (p. 976) that Hendrick omitted this speech from his edition of Carnegie's works.

44. *Writings*, II, 120–22.

45. "Old Scotland and New England," *Writings*, I, 126–35.

46. *The Aristocracy of the Dollar*, c. 1883, Nineteenth Century Club; pamphlet in Carnegie Library of Pittsburgh.

47. "Pittsburgh and Its Future," address before the Pittsburgh Chamber of Commerce, November 10, 1898; *Chamber 1899 Yearbook*, Carnegie Library of Pittsburgh.

Chapter Eight

1. For example, in regard to the nonexistent cable from the union at Homestead, *Autobiography*, p. 223; Wall, pp. 575–76.

2. *Letters of Richard Watson Gilder* (New York, 1916), p. 374; cited in *Autobiography*, p. 280.

3. Ibid., p. 375; cited in *Autobiography*, p. 328.

4. *Autobiography*, Editor's Note, p. ix.

5. Ibid., preface, p. v.

6. Hendrick, I, 415–16. This first section appears to have ended with p. 209, line 5, of the Riverside Press edition. The next paragraph begins a section apparently written in 1907, but the subsequent text gives evidence of numerous revisions.

7. Wall, pp. 139–40.

8. *Triumphant Democracy*, p. 297.

9. *Autobiography*, p. 297. Further citations of the *Autobiography* are provided in parentheses in the text.

10. *James Watt*, p. 5.

11. Ibid., p. 12. Further citations of *James Watt* in this chapter are provided in parentheses in the text.

12. *Autobiography*, 12, 37.

13. The quotation is a paraphrase from Scott's *Lady of the Lake.*

14. James Patrick Muirhead, *Life of James Watt* (London, 1858), p. 18.

15. *Autobiography*, p. 33.

16. Muirhead, pp. 25–26.

17. "William Chambers," *Writings*, I, 200–13; "Stanton the Patriot," ibid., I, 215–39; "Ezra Cornell," ibid., I, 240–62.

18. Delivered at Kenyon College on Memorial Day 1906. Following the *Ohio Archaeological and Historical Quarterly*, in which the speech was printed (15, pp. 291–311), Hendrick titles it "Stanton the Patriot." Originally (pamphlet, Kenyon College, 1906, in Carnegie Library of Pittsburgh) it simply bore the subject's name.

19. *Autobiography*, p. 66.

20. *Writings*, I, 256–57.

21. The speech was delivered October 19, 1909, at Peebles, Scotland, at the jubilee of the Chambers Institution.

22. *Memoirs*, p. 165.

23. Elizabeth Nitchie, *The Criticism of Literature* (New York, 1928), p. 259.

24. Really nineteenth. Carnegie fails to realize that Old Style (Julian calendar) dating, by which the new year began with March,

was used in Britain until 1750. Virtually all writers on Watt have fallen into the same error, especially in regard to his age at death. Ivor B. Hart, *James Watt and the History of Steam Power* (New York, 1949), says (p. 231) that Watt died "in his eighty-fourth year," instead of eighty-third. Carnegie (*James Watt*, p. 146) speaks of him at death as "aged eighty-three." Even Muirhead (p. 511), who should have known better, in contradicting Walter Scott's statement that Watt was "in his eighty-fifth year" at the time of a visit to Scotland in 1817, makes it read "eighty-second." At his death, August 19, 1819, Watt was exactly eighty-two years, seven months of age. If there were any doubt that his birth was in January 1736 Old Style—1737 New Style—it is dispelled by the notes he sent to friends early in 1819 with busts he had made after perfecting a mechanism for copying, in which he describes them as "the work of a young artist just entering his eighty-third year" (*Encyclopaedia Britannica* [1954] 23, p. 437). Watt would not have forgotten the difference between the Gregorian and Julian dating.

25. *James Watt*, p. 27.

26. *Autobiography*, p. 105.

27. *James Watt*, p. 94.

28. Ibid., p. 137; *Writings*, I, 254; *Autobiography*, p. 51.

29. *James Watt*, p. 142. (Carnegie was almost seventy when he wrote this.)

30. The notable and praiseworthy exception is Wall.

31. Nitchie, p. 256.

32. Ibid., p. 250.

33. Ibid., p. 260.

34. Ibid., p. 264.

Chapter Nine

1. Andrew Carnegie to Henry Holt, February 25, 1915; quoted by Wall, p. 893.

2. Andrew Carnegie to Margaret Morrison and Thomas Carnegie, 1865–1866; Margaret Carnegie Miller collection.

3. Quoted in Hendrick, II, 287–88.

4. Quoted, ibid., pp. 285–87.

5. See above, p. 37.

6. Letter to Trustees, Carnegie Peace Fund, December 14, 1910. Pamphlet (n.p., n.d.), Carnegie Library of Pittsburgh.

7. Quoted in Hendrick, II, 68.

8. Wall, p. 664.

9. Hendrick, II, 69.

10. Quoted in Hendrick, II, 311–13.

11. See above, pp. 63–64.

12. Quoted in Hendrick, II, 310–11.

13. *Autobiography*, p. 357. I am totally at a loss to explain Hendrick's changes from the original in quoting the passage; Hendrick, II, 314.

14. See above, pp. 19–20.

15. Adapting Milton's "To the Lord General Cromwell."

16. Wall, p. 1024.

Chapter Ten

1. Evidently the end of 1870. See above, p. 33.

2. Hendrick, I, 268.

3. Wall, pp. 720–33.

4. Ibid., pp. 792, 789.

5. "Steel Manufacturing in the United States in the Nineteenth Century," Review of the Century number, *New York Post*, January 12, 1901, and "The Woman as Queen." Hendrick, II, 391, lists this as " 'The Woman in the Queen,' *Review of Reviews*, February, 1901," and in *Writings* (I, 186) as " 'Queen Victoria,' *Review of the Republic*, February, 1901." It is not to be found in either the British or American *Review of Reviews*, and the *Union List of Periodicals* does not include any *Review of the Republic*.

6. To Hayne Davis, *Among the World's Peacemakers* (New York, 1907).

7. Theodore Roosevelt, *The Roosevelt Policy, Speeches, Letters and State Papers, Relating to Corporate Wealth etc.* (New York, 1908).

8. Ibid., pp. ix–xxi, signature at end. Further citations from this introduction are inserted in parentheses in the text.

9. "Review of *The Roosevelt Policy*," (London) *Times Literary Supplement*, July 16, 1908, p. 228; quoted by Wall, p. 970.

10. See above, p. 98.

11. See above, p. 84.

12. Wall, p. 880.

13. *Autobiography*, p. 245.

14. Wall, pp. 830, 815.

15. Hendrick, II, 277.

16. Ibid., I, 352–64.

17. Wall, p. 828.

18. Hendrick, II, 203.

19. Wall, p. 819.
20. Ibid., p. 832; Hendrick, II, 229.
21. Hendrick, II, 260.
22. *Autobiography*, p. 247.
23. Hendrick, II, 253–54; Wall, p. 864.
24. Wall, pp. 859–63.
25. Hendrick, II. 332–34.
26. Wall, pp. 871–80.
27. Ibid., 836–37.
28. *Autobiography*, p. 266.
29. Wall, pp. 891–93.
30. Ibid., pp. 904–11.
31. Hendrick, II, 353.
32. *Autobiography*, p. 278; Hendrick, II, 353.
33. Hendrick, II, 359.
34. Wall, pp. 409, 805, 854.
35. *Autobiography*, p. 278.
36. Wall, p. 847.
37. Hendrick, II, 162–63.
38. Wall, pp. 882–83.
39. Ibid., pp. 1042–43. For a further account of the will, see the *New York Times*, August 29, 1919, p. 1.

Selected Bibliography

PRIMARY SOURCES

1. Collected Works

Andrew Carnegie's College Lectures etc. Ed. Daniel Butterfield. New York: F. T. Neely, 1896.
Three Busy Weeks etc. Comp. A. S. Cunningham. Dunfermline: W. Clark and Son, 1902. Nine Carnegie speeches.
More Busy Days etc. Ed. James M. Swank. Philadelphia: Allen, Lane and Scott, 1903. Eight Carnegie Speeches.
A Carnegie Anthology. Comp. Margaret B. Wilson. New York: privately printed, 1915.
Writings of Andrew Carnegie, 10 vols. Ed. Burton J. Hendrick. New York: Doubleday, Page and Co., 1933.
Letters of Andrew Carnegie and Louise Whitfield. Ed. Helen H. Dow. New York: no publisher, 1956.

2. Books or Parts of Books

Rules for the Government of the Pennsylvania Railroad Company's Telegraph. Harrisburg, Pa.: no publisher, 1863.
Round the World. [New York]: [Charles Scribner's Sons?], 1879.
Our Coaching Trip. [New York]: [Charles Scribner's Sons], 1882.
An American Four-in-Hand in Britain. New York: Charles Scribner's Sons, 1883.
Round the World (revised and enlarged). New York: Charles Scribner's Sons, 1884.
Triumphant Democracy, or Fifty Years' March of the Republic. New York: Charles Scribner's Sons; London: Sampson Low, Marston, Searle and Rivington, 1886.
"From Andrew Carnegie," *Courtlandt Palmer*, ed. F. W. Christen. New York: Nineteenth Century Club, 1889.
Triumphant Democracy Ten Years Afterward, or Sixty Years' March of the Republic. New York: Charles Scribner's Sons, 1893.
"Characteristics," in *Memoirs of Anne C. L. Botta*, ed. Vincenzo Botta. [New York]: no publisher, 1893.

173

"Genius Illustrated from Burns," *Liber Scriptorum*. New York: The Authors' Club, 1893.

The Gospel of Wealth and Other Timely Essays. New York: Century, 1900.

"The South African Question," in James Bryce, et als. *Briton and Boer, Both Sides of the South African Question*. New York and London: Harper and Brothers, 1900.

The Empire of Business. New York and London: Harper and Brothers, 1903.

James Watt. Famous Scots Series No. 42. Edinburgh and London: Oliphant, Anderson and Ferrier, 1905.

Problems of Today, Wealth, Labor, Socialism. New York: Doubleday, Page and Co., 1908.

Autobiography of Andrew Carnegie. Ed. John C. Van Dyke. Boston: Houghton, Mifflin Co., 1920.

A. Introductions

CHITTENDEN, HIRAM M. *War or Peace*. Chicago: A. C. McClurg and Co.; London: Sampson and Low, 1911.

DAVIS, HAYNE. *Among the World's Peacemakers*. New York: Progressive Press, 1907.

ROOSEVELT, THEODORE. *The Roosevelt Policy*. New York: Current Literature, 1908.

B. Edited

CARNEGIE, ANDREW, ed. *Business*. Boston: Hall and Locke Co.

3. Magazine Articles

"As Others See Us," *Fortnightly Review*, NS 31, (February 1, 1882) 156–65.

"The McKinley Bill," *Nineteenth Century*, 29 (June 1891), 1027–36.

"Imperial Federation, An American View" *Nineteenth Century*, 30 (September 1891), 490–508.

"A Look Ahead," *North American Review*, 156 (June 1893) 685–710.

"The Silver Problem, A Word to Wage Earners," *North American Review*, 157 (September 1893) 354–70.

"The Value of the World's Fair to the American People," *Engineering Magazine*, 6 (January, 1894) 417–22.

"How I Became a Millionaire," *Cassell's Family Magazine*, Series 4, 19 (May 1896) 450–56.

"The Ship of State Adrift," *North American Review*, 162 and 163 (June and October 1896) 641–48, 496–503.

"Mr. Bryan as a Conjuror," *North American Review*, 164 (January 1897) 106–18.

"The Presidential Election—Our Duty, Bryan or McKinley?" *North American Review*, 171 (October 1900), 495–507.

"The Opportunity of the United States," *North American Review*, 174 (May 1902), 606–12.

"Britain's Appeal to the Gods," *Nineteenth Century and After*, 55 (April 1904), 538–42.

"The Anglo-French-American Understanding," *North American Review*, 181 (October 1905), 510–17.

"The Cry of 'Wolf,'" *Nineteenth Century and After*, 60 (August 1906), 224–33.

"The Gospel of Wealth—II," *North American Review*, 183 (December 1906), 1096–1106.

"The Next Step—A League of Nations," *Outlook*, 86 (May 25, 1907), 151–52.

"The Second Chamber," *Nineteenth Century and After*, 62 (November 1907), 689–98.

"The Laird of Briarcliff Manor," *Outlook*, 89 (May 16, 1908),107–11.

"Peace Versus War—The President's Solution," *Century*, 80 (June 1910), 307–16.

"Tribute to Mark Twain," *North American Review*, 191 (June 1910), 827–28.

"Dr. Golf," *Independent*, 70 (June 1, 1911), 1183–92.

"Arbitration," *Contemporary Review*, 100 (August 1911), 169–76.

"The Industrial Problem," *North American Review*, 194 (December 1911), 914–20.

"A Silver Lining to the War Cloud," *The World Today*, 21 (February 1912), 1792a.

"Hereditary Transmission of Property," *Century*, 87 (January 1914), 441–43.

"The Decadence of Militarism," *Contemporary Student*, 5 (June 1915), 333–34.

4. Published Addresses

The Aristocracy of the Dollar, Nineteenth Century Club, 1883(?). Pamphlet, n.p., n.d.

[Untitled] *to Medical Students, Bellevue Hospital*, March 9, 1885. Pamphlet, n.p., n.d.

Some Facts About the American Republic, Dundee, Scotland, March 18, 1889. Pamphlet, n.p., n. d.

The Future of Pittsburgh, Pittsburgh Chamber of Commerce, Nov. 19, 1898. Pittsburgh: Chamber of Commerce Yearbook, 1899.

The Future of Labor. American Academy of Political and Social Science, 1909. Pamphlet, n.p., n.d.

Aberdeen University Rectorial Address, June 6, 1912. Pamphlet, n.p., n.d.

SECONDARY SOURCES

ALDERSON, BERNARD. *Andrew Carnegie*. New York: Doubleday, Page and Co., 1902. Strictly sycophant work.

Andrew Carnegie Centenary. New York: Carnegie Corporation of New York, 1935. Tributes and recollections by prominent men who knew Carnegie.

ARBUTHNOT, THOMAS S. *Heroes of Peace*. [Pittsburgh]: n.p., 1935. A charming account of the origin and spirit of the Carnegie Hero Fund.

BEISNER, R. L. *Twelve Against Empire*. New York: McGraw-Hill, 1968. Good essay on Carnegie's fight against annexing the Philippines, and similar national actions.

BRIDGE, J. H. *Inside History of the Carnegie Steel Co*. New York: Aldine, 1903. Principally concerned with the Carnegie-Frick estrangement, with strong anti-Carnegie bias.

BROOKS, VAN WYCK. *The Confident Years*. New York: Dutton, 1952. Good background study of the era of Carnegie's principal activity.

BRUCE, ROBERT V. *1877—Year of Violence*. Indianapolis: Bobbs-Merrill, 1959. A good study of industry in Carnegie's early period of activity.

CHAMPLIN, JOHN D. *Chronicle of the Coach*. New York: Charles Scribner's Sons, 1886. Background history on Carnegie's first successful book.

HACKER, LOUIS M. *The World of Andrew Carnegie*. Philadelphia: Lippincott, 1968. Fine background material on Carnegie's business years.

HARPER, FRANK C. *Pittsburgh of Today, Its Resources and People*. New York: American Historical Society, 1931. City background.

HARVEY, GEORGE. *Henry Clay Frick, the Man*. New York: Charles Scribner's Sons, 1928. Sycophantic biography of Carnegie's principal later aide, and worst enemy. Strongly biased against Carnegie.

HENDRICK, BURTON J. *The Life of Andrew Carnegie,* two vols. Garden City, N.Y.: Doubleday, 1932. Done to please Carnegie's family.

———, and HENDERSON, DANIEL. *Louise Whitfield Carnegie.* New York: Hastings House, 1950. In praise of Carnegie's wife.

HILL, NAPOLEON. *Think and Grow Rich.* Cleveland: Ralston Co., 1937. Claims to be based on Carnegie's talks with writer.

HOLBROOK, STEWART. *Iron Brew.* New York: MacMillan, 1939. Material on steel industry strikes and labor problems.

HUGHES, JONATHAN. *The Vital Few.* New York: Houghton, Mifflin, 1965. Fifty good pages on Carnegie as one of the great forces in American industrial progress.

LESTER, R. M. *Forty Years of Carnegie Giving.* New York: Charles Scribner's Sons, 1941. A summary of benefactions.

LYNCH, FREDERICK. *Personal Recollections of Andrew Carnegie.* New York: Fleming H. Revell, 1920. Good material on Carnegie's cultural attitudes.

McCLOSKEY, R. G. *American Conservatism in the Age of Enterprise.* Cambridge, Mass.: Harvard University Press, 1951. Good presentation of Carnegie's place in America of his day.

MULHALL, M. G. *Balance Sheet of the World for Ten Years, 1870–1880.* London: Edward Stanford, 1881. Gazetteer of international statistics which gave Carnegie the idea and much material for *Triumphant Democracy.*

NEVIN, R. P. *Les Trois Rois.* Pittsburgh: Eichbaum, 1888. Earliest and most worshipful biographical notice of Carnegie, by a friend. Compares his career with those of two other Pittsburghers, William Thaw and George Westinghouse.

PELLING, HENRY. *America and the British Left, from Bright to Bevan.* London: A. and C. Black, 1956. Excellent brief coverage of Carnegie's moves in British politics.

Perspectives on Peace, 1910–1960. London: Carnegie Endowment for International Peace, 1960. Summary of Peace Fund work.

SPENCER, HERBERT. *Autobiography.* New York: D. Appleton, 1904. Material on friendship with Carnegie.

STEAD, W. T. "Mr. Carnegie's Conundrum," *Review of Reviews Annual.* London: Review of Reviews Corporation, 1900. In view of Carnegie's liquidation of his industrial holdings, a fellow liberal wonders if and how the millionaire will carry out his charitable plans.

WALL, JOSEPH F. *Andrew Carnegie.* New York: Oxford University Press, 1970. At long last an honest and definitive Carnegie biography.

WILLIAMS, HARLEY. *Men of Stress*. London: Jonathan Cape, 1948. Study of the ambivalent personalities of Carnegie, Lord Leverhulme, and Woodrow Wilson.

WINKLER, J. K. *Incredible Carnegie*. Garden City, N.Y.: Garden City Publishing Co., 1931. Charmingly written hatchet job.

Index

Abbot, Lyman, 116
Acton, John E., 150
Addison, Leila, 26
Adler, Hermann, 91
Aesop, 48
Alcott, Bronson, 36
Alexandria, Egypt, 41
Alger, Horatio, 99, 133
Allegheny, Pa., 17, 23, 151
Altoona, Pa., 22-23, 25
"American Dream," 99, 157
American Wire Co., 140
Amsterdam, Holland, 29-30
Anarchy, 87, 93
Anderson, Col. James, 19, 113, 143
Arnold, Edwin, 37, 39, 48
Arnold, Matthew, *36-37*, 48, 102, 113
Astor, Mrs. J. J., 34
Aultnagar, Scotland, 120
Australia, 61
Authors Club, 155
Ayr, Scotland, 109-10

Ballad of Sir Patrick Spens, 62
Banbury, England, 46
Barrow, England, 109
Beaver, James A., 103
Beck, Theodore, 46
Beecher, Henry Ward, 36
Bellevue Hospital, 110, 150
Berlin, Germany, 29, 150
Bible, the, 48, 72, 85, 105, 108
Bigelow, John, 36
"Billings, Josh." *See* Henry W. Shaw
Black, William, 48; *Adventures of a Phaeton*, 45
Blaine, James G., 60
Book, David, 28

Boston, 72, 101, 116
Boswell, 120
Botta, Anne C. Lynch, 35-37
Botta, Vincenzo, ed., *Memoirs of Anne C. L. Botta*, 36
Bridge, James H., 76, 78-81, 165n17
Brigham, C. D., 27
Bright, John, 85
Brighton, England, 44, 45, 46, 138
Brougham, Henry P., 131
Browning, Elizabeth Barrett, 48
Browning, Robert, 74
Bryant, William Cullen, 36, 48, 74
Buchanan, George, 110
Burns, Robert, 39, 40, 48, 62, 104, 106, 110, 119, 122, 130
Byron, George Noel Gordon, 74

Cable, George W., 35
Campbell, Thomas, 48
Cambridge University, 150
Canada, 21, 61
Carey, Henry C., 85
Carisbrook Castle, 27
Carlisle, England, 46
Carlyle, Thomas, 26, 48
Carnegie, Andrew (1835-1919), attacks British political system, 53; begins working, 18; bond sales, 32; born, 17; business letters, 139-40; coaching trip, *44-49*, 156; creative efforts, 62-64; death, 155; dispatches trains, 22; education, 17, 18, 23, 25, 32, 145, 153-56; embraces Darwinian Socialism, 85; emigrates to America, 17; escapes capture at Bull Run, 23; family, 17, 66, 134, 136-37; family letters, 21, *31-32*;

179